What experts have to say about

THE GUIDE TO

Hydropower Mechanical Design

"The *Guide* will no doubt join the select group of basic reference books on hydro plant equipment design—the 'Bibles' familiar to us all."

—*Carlos S. Ospina*
International Hydroelectric Consultant
Ingetec S.A., Bogota, Colombia

"*The Guide to Hydropower Mechanical Design* stands out as a modern and definitive text in this specialized field. The product of some of the finest minds in the industry, this work broadly covers hydroelectric plant and machinery for the benefit of engineers, and operations and maintenance personnel. I highly recommend this book."

—*Edward O. Groff, P.E.*
President, 1997, American Society of Civil Engineers
Principal Engineer, Northrop, Devine & Tarbell, Inc.

"The evolution of hydroelectric power engineering is marked by milestones of achievement in recording the state of the art. The *Guide* is an important new marker of progress. It achieves particular quality from the comprehensive treatment it provides of all mechanical aspects of a hydro development."

—*J. Gavin Warnock, P.E., B.Sc.(Hons), D.I.C.(Hydro Power)*
Managing Director, Monitor (International) Limited
Edinburgh, Scotland, U.K.

"The *Guide* provides extensive and valuable technical information. It is a useful and practical tool found nowhere else, which captures the experience and knowledge gained by the authors through decades of designing hydropower projects."

—Jean-Paul Rigg, Eng., M.Sc.
General Manager, Engineering and Services
Hydro-Québec
Montréal (Québec) Canada

More expert comments on back cover.

The Guide to
Hydropower Mechanical Design

Prepared by the

American Society of

Mechanical Engineers

Hydro Power Technical Committee

Published by

HCI Publications

Printed in the United States of America
ISBN 0-9651765-0-9

NOTICE

This book was prepared as a project of the American Society of Mechanical Engineers (ASME) Hydro Power Technical Committee. Neither the ASME, individual committee members, their employers, contributing organizations, HCI Publications, nor any person acting on behalf of them (a) makes any warranty, express or implied, with respect to the use of any information, apparatus, method, or process disclosed in this book or that such use may not infringe privately owned rights; or (b) assumes any liabilities with respect to the use of, or for damages resulting from the use of, any information, apparatus, method, or process disclosed in this book.

HCI Publications is the publisher of *Hydro Review* and *HRW–Hydro Review Worldwide* magazines, *Hydrowire* and *HydroWorld Alert* newsletters, books, and other materials useful to individuals and organizations having interests in hydroelectricity. HCI also organizes hydro-related meetings and events, including the HydroVision conferences.

HCI Publications
410 Archibald Street
Kansas City, MO 64111

Phone: (816) 931-1311
Fax: (816) 931-2015
E-Mail: hci@aol.com.us

Foreword

The technical information provided in this, *The Guide to Hydropower Mechanical Design,* will be useful to any technical person involved with the design of new hydroelectric plants or with the modernization of older hydroelectric facilities.

With an installed base of nearly 700,000 megawatts of hydropower worldwide, and active construction programs underway throughout much of the world, this Guide is especially timely. Other books providing guidance relevant to hydro project mechanical design are decades old, and virtually none capture the breadth and depth of experience represented in this volume.

The publication of this guide is the culmination of more than a decade long quest on the part of the American Society of Mechanical Engineers Hydro Power Technical Committee. As authors of this guide, many committee members have worked with passion and energy on the tasks leading to the Guide's completion. Committee members, and their employers, have offered their valuable time and resources to create a useful legacy to benefit future generations of hydroelectric plant designers. This is also a legacy for society at large, which will reap the rewards of safe, economical, and clean energy produced from hydroelectric plants.

The project to publish this Guide originated in 1984. It was suggested by Bernard Eyden, who was then the committee's chairman. His vision was that a mechanical design guide was needed for use by engineers who had about five years of post-graduate working experience, but who were not experienced in hydro design. In 1984, as today, there was concern that the knowledge of many very capable and experienced hydro plant designers would be lost due to the vicissitudes of reorganizations and the relentless march of Father Time. The imperative was to take action to ensure that this knowledge and experience was not lost.

Thus the Committee committed itself to the goal of capturing—in a practical, convenient, and useable form—the most useful information for guiding succeeding generations of hydro plant mechanical designers. This effort has brought the earnest contributions of many individuals, and the stalwart support of their employers. Many meetings and many thousands of hours have been devoted to creating this Guide.

In addition to the Committee member contributors and their employers, others have provided meaningful assistance to the Guide project. In particular, the Committee gratefully acknowledges the financial support of the ASME Power Division and the Factory Mutual Insurance Company. Also, the Committee appreciates the efforts of HCI Publications, especially of managing editor Marla Barnes, first-draft editor Alia Michaels, and editor-in-chief Carl Vansant, P.E.

Each chapter in the Guide is the product of multiple contributors and multiple reviews. Initial drafts of material were prepared by one or more authors; other individuals served to review the material. The Committee as a whole reviewed each manuscript section on several occasions as the Guide evolved. While the contributors listed at the front are the principal authors and reviewers, materials were solicited from others as well and have been incorporated where appropriate.

The Committee is pleased and proud to be able to provide this Guide to our present and future colleagues.

For the American Society of Mechanical Engineers
Hydro Power Technical Committee:

KERMIT PAUL, JR.
Chairman 1993-1997

Contributors

Kermit Paul, Jr., P.E.—Consulting Mechanical/Electrical Engineer. Retired Pacific Gas and Electric Company Supervising Mechanical Engineer and Project Engineer for Helms Pumped Storage Project. Chairman, ASME Hydro Power Technical Committee and past member of ASME Hydro Turbine Performance Test Code Committee. [2, 4, 11, 12]

Donald R. Webb, P.E.—Manager, Applied Hydraulic Engineering, Voith Hydro, Inc. with 27 years experience in hydraulic application, design, and testing of turbines and pump-turbines. Has served in the ASME Fluids Engineering Division as Chairman of Fluids Machinery Committee, Honors Committee Chairman, and Chairman of the Executive Committee. Vice Chairman, ASME Hydro Power Technical Committee. Editor or co-editor of eight ASME symposium volumes. [3, 12]

Glenn W. Brewer, P.E.—Engineering Manager for Raytheon Infrastructure Services in Bellevue, Washington. Formerly Project Engineer for Idaho Power Company, responsible for overall hydro construction projects including permits, engineering, major equipment procurement, cost, schedule, construction coordination, and documentation for several new hydroelectric projects on the Snake and Payette rivers in Idaho. Author or co-author of several papers for Waterpower conferences and in *Hydro Review* magazine. Secretary, ASME Hydro Power Technical Committee. [13, 14]

B.G. Ragsdale, P.E.—Mechanical Engineer/Manager of operations and maintenance of the Eastern Hydro Region power generating plants located in Tennessee and North Carolina for the Tennessee Valley Authority. Twenty-five years design experience in conventional and pumped-storage projects and design and implementation of plant upgrades, hydro modernization, and development of aerating methods to improve releases from power generating plants. Paper Review Coordinator, ASME Hydro Power Technical Committee. [13, 14]

Notes:
*–indicates retired.
**–indicates deceased.
Numbers in brackets ([]) indicate the number(s) of chapters to which the contributor has made a primary contribution.

Roger E.A. Arndt, PhD—Professor of Civil Engineering at the University of Minnesota (since 1977); served as Director of the St. Anthony Falls Laboratory until 1993. Taught Aerospace Engineering at Penn State; has worked in industry in the areas of rocket design and high speed marine vehicles. He is the recipient of several awards for teaching and research, including the Alexander von Humboldt Senior Scientist Award, the ASME Fluids Engineering Award, and the Lorenz Straub Award. He is a Fellow of the ASME. [2]

Hassine Benjannet, P.E.—Senior Project Manager, SNC-Lavalin Inc., Hydroelectric Consulting Company. Responsible for all aspects of hydroelectric developments and is involved, at both the management and technical levels, in the design of new plants and the rehabilitation of facilities built at the turn of the century, including powerhouses, dams, substations, and transmission lines. [7]

Richard E. Bishop, P.E.—Senior Marketing Specialist, Oglethorpe Power Corporation. Previously Senior Mechanical Engineer on the Rocky Mountain Pumped Storage Project. Also, Plant Engineer at the Dayton Power & Light Stuart Station, and Senior Operations Engineer for Oglethorpe Power at the Hal B. Wansley and Robert W. Scherer fossil power stations. First chairman of the Middle Georgia Group of the ASME.

John R. Chappel, P.E.*—Mechanical Engineer, Idaho National Engineering Laboratory, Idaho Falls, Idaho. Senior Engineering Specialist involved in monitoring engineering development activities of the U.S. Department of Energy National Hydropower Program. Formerly Research Supervisor at Newport News Shipbuilding Hydraulic Laboratory involved in research on and testing of hydraulic turbines, pumps, and other fluid equipment. Intermittently involved with hydropower for 41 years.

Rick J. Christensen—Mechanical Engineer, Bureau of Reclamation. Responsible for designs, specifications, and construction support for the following types of equipment: trashrakes, conveyors, selective withdrawal structures, screening equipment, screen cleaning systems, and fish-related equipment and structures. Team member of the Reclamation's research program on the Development for Better Trash Screening and Cleaning Devices. Co-author of the technical paper "Interaction of Technical and Economic Factors in Pumped Storage Evaluation."

John B. Codrington, P.E.—Senior Staff Engineer, Acres International Limited, Niagara Falls, Canada. Thirty years experience in an engineering consultancy that specializes in hydroelectric design for projects in Canada and overseas. Special interests include computer modeling of hydraulic transients and their interaction with the electrical system. Recent experience includes design for high-head hydroelectric plants for sediment-bearing rivers in Latin America, and resolution of unusual operating problems involving penstock and water column behavior. Co-author of Canadian Electricity Association research report on draft tube surges.

William H. Colwill, PhD—Vice President, Marketing, American Hydro Corporation. Responsible for marketing and sales of American Hydro's products and all hydraulic engineering including turbine and pump-turbine application, design, and quality assurance. Dr. Colwill has developed advanced application procedures for turbines and pump-turbines. Previously, Dr. Colwill has developed three-dimensional flow analysis codes for centrifugal compressors and new aerodynamic designs for compressors and new hydraulic designs for pumps and turbines. [3]

Marcus E. Crahan III, P.E.—Founder and Director of Power Engineering. Managed repair and overhaul of large hydroelectric pumped-storage and conventional turbine-generators. Proficient in the areas of system modeling, numerical analysis, field testing, and performance data acquisition. Expert in the area of rotating machinery dynamics. Authored many papers on vibration analysis of electric power generating apparatus. Member of the ASME Hydro Power Technical Committee from 1990 to present. Registered Mechanical Engineer, Member of ASME. [13]

Percy B. Dawson, P.E.—Consulting Mechanical Engineer. Formerly Chief Engineer, Pelton Division of Baldwin Lima Hamilton Corp. Senior Mechanical Engineer, Pacific Gas and Electric Company and International Engineering Company; and Project Mechanical Engineer for Morrison Knudsen Engineers.

William H. Duncan, Jr., P.E.—Mechanical Engineer, Bureau of Reclamation Hydroelectric Research and Technical Services. Leading technical expert and team leader in the operation and maintenance of hydroelectric turbines, pumps, and related mechanical equipment. Services are provided for 56 hydroelectric power plants with more than 12,000 MW of power capacity. Author or co-author of numerous USBR publications. ASME member and Vice Chairman of PTC 29 (Governor Test Code) Committee.

Bernard Eyden—P.E. (U.S.), P.Eng. (Canada), E.Eng (U.K.), Fellow ASME, Member American Society of Civil Engineers, MIMechE. Engineering Consultant. Experienced in design, procurement of equipment, construction, and testing of all types of hydropower and pumping installations, including pumped storage, in capacities ranging from 1.5 MW to 1,136 MW in the U.S., Brazil, and Korea. Served as Chief Mechanical Engineer and in other senior positions with Brown & Root, TAMS, Stone & Webster, and Brazilian L&P. Has authored several papers and articles. [8, 13]

Donald R. Froehlich, P.E.—Partner, Black & Veatch. Project Manager assigned to the Energy Group, Power Division, responsible for design and construction of hydroelectric projects. Member and former chairman of the ASME Hydro Power Technical Committee. [5, 10]

Robert G. Grubb, P.E.—Manager, Advanced Technology, Voith Hydro, Inc. Responsible for proposal, engineering, mechanical conceptual design, and advanced mechanical/structural analysis of hydro turbines and related products at the Voith Hydro facility in York, Pennsylvania. Patent holder and author and co-author of several technical papers during his 28 years in the hydro turbine field.

Gerald A. Hiett, P.E.*—Mechanical Engineer—Former Head of Hydraulic Machinery Section of the Mechanical Branch, Electrical and Mechanical Engineering Division, Denver Office, Bureau of Reclamation. Functioned as a designer, technical director, supervisor, and manager for more than 32 years in the section responsible for all mechanical equipment and piping inside hydropower plants designed by Reclamation. Member of the ASME Hydro Power Technical Committee. [6, 7, 9]

David R. Johnson—Welding Engineer, Voith Hydro, Inc. Responsible for welding engineering and metallurgical activities supporting the design, manufacture, and installation of hydro turbines. Awarded Best of Program in 1993 Lincoln Arc Welding Foundation Professional Awards Program for "Redesign of Kaplan Turbine Shaft for Weld Fabrication."

Durl E. Jones—Engineering Manager Hydro Controls, Woodward Governor Company. Responsible for application engineering for unit and plant control solutions based on digital, analog, and mechanical control systems. Past hydro positions include sales, marketing, and product development. [4]

James D. Kerr, P.E.—Chief, Mechanical Equipment and Systems Section, U.S. Army Corps of Engineers, Hydroelectric Design Center. Responsible for powerhouse support systems such as intake gates, cranes, service piping, drainage, and fish diversion equipment. Member of the ASME Hydro Power Technical Committee. [9]

Jeffrey D. Koepnick, P.E.—Senior Engineer, Wisconsin Electric Power Company. Responsible for mechanical engineering and procurement activities for specification and installation of 12 new replacement hydro turbine runners over the past five years. Served as Fossil Plant Engineer and Construction Manager for Wisconsin Electric's 300-MW Paris Generating Station combustion turbine facility.

Takashi Kubota, Dr. of Engineering—Professor of Mechanical Engineering, Kanagawa University and independent consultant. Responsible for hydrodynamic engineering on research and development of hydro turbines (especially Pelton, Francis, Kaplan, and bulb turbines) based on 34-year career in Fuji Electric Co. until 1991. Author and co-author of numerous technical papers in International Association for Hydraulic Research, Turbo-Machinery Society in Japan, and Japanese Society of Mechanical Engineers, et al.

Reese W. May**—Mechanical Engineer—Founded and managed U.S. office of Bouvier Hydro (1981-1990) for Ateliers Bouvier; served as southeast U.S. sales representative 1990-1995. Previously worked for Bailey Meter Company for 28 years, retiring as Utility Sales Manager. Other experience includes Vice President of Sales for AERO International, manufacturer of heat exchangers. Owner of R.W. May Associates with clients in small hydro and nuclear waste management. Member and Chairman of several ASME committees.

Charles A. McKee, P.E.—Project Manager, Chelan County Public Utility District No. 1 of Chelan County, Washington. Responsible for the turbine rehabilitation of the Rocky Reach Hydroelectric Project on the Columbia River in the U.S. Author, co-author of numerous technical papers and publications. [14]

Allen A. Miller, P.E.*—Mechanical Engineering Specialist—Heating, Ventilating, and Air Conditioning Systems Design—Bureau of Reclamation, Engineering and Research Center, Denver. Involved in the systems for Glen Canyon Powerplant, Flaming Gorge Powerplant, Yellowtail Powerplant, Morrow Point Powerplant, Grand Coulee Third Powerplant, Boulder (Hoover) Dam and Powerplant modifications, and many other structures.

Steve C. Peabody, P.E.—Project Engineer, Hydro Power Division, ABB Power Generation Inc. Responsible for hydro generator component assessment and uprate evaluations in the U.S. market. Assists customers with technical issues for ongoing and new ABB hydro generator projects. Served as mechanical design engineer for Bureau of Reclamation, Hydraulic Machinery Section, Turbine Division (1982-1993).

Stanislav Pejovic, P.E., Prof., PhD—University of Belgrade Mechanical Engineering, Head of Hydraulic Energy Department. Visiting Professor National University of Singapore. Design of three test rigs and 14 hydropower plants, pumped-storage plants, and pumping systems. Hydraulic transient and vibration analysis of 28 hydraulic machines and systems. Model acceptance tests of nine turbomachines. Field tests of 12 and acceptance tests of six power plants. Chief Engineer of final field tests. Author and co-author of 15 books, 115 technical papers, and six computer programs. [12]

Bernard J. Peter, P.E.*—Mechanical Engineer—Bureau of Reclamation. Responsible for designs, drawings, and specifications for procurement and installation of large structural steel wheel- and roller-mounted gates and radial gates over a 36-year career.

John Pocher, P.E.—Mechanical Engineer, B.C. Hydro. Responsible for mechanical technical services relating to the design and maintenance of projects and programs for hydroelectric generation stations within Power Supply of B.C. Hydro with a total capacity of 9,000 MW. Authored and presented technical papers on heavy mechanical equipment.

David N. Raffel, P.E.**—Protrans Consultants. Furnished practical engineering for hydroelectric and pumping plants, from small hydro to the 10,000-MW Guri plant, built in 37 states of the U.S. and 18 other countries. Best known for planning power plants and utility systems, and all aspects of selecting, specifying, erecting, testing, and operating the principal machinery. As chief executive, he managed engineering consulting firms in Argentina and in El Paso, Texas, training many engineers. He contributed to several technical publications and worked many years on ASME, American Society of Civil Engineers, and National Fire Protection Association technical committees. He was selected as an ASME Fellow. [1, 2, 11, 12]

Michael H. Ritchie, P.E.—Supervisor, Electric Supply and Delivery, Niagara Mohawk Power Corporation. Responsible for the Operation and Maintenance of nine hydroelectric developments in the Hudson Valley region of upstate New York. Served as Mechanical Engineer for Niagara Mohawk's 73 hydroelectric developments. Co-author of a paper that was awarded Honorable Mention at the Waterpower 1991 Conference. [14]

Gerard J. Russell, P.E.—Senior Engineer, Marketing, American Hydro Corporation. Has over 15 years of experience in the hydropower field. At American Hydro, his responsibilities include marketing of American Hydro's products as well as hydraulic engineering and design. Previously, Mr. Russell was involved with standardized turbine product development as an Hydraulic Engineer and was a project manager for the supply and installation of several water-to-wire turbine, generator, and control packages. [3]

Daniel Schlichting, P.E.—Mechanical Engineer—Woodward Governor Company 1982-1994. Project engineering and sales responsibility for custom controls ranging from small mechanical governors to analog governors and digital control systems. Member of ASME Hydro Power Technical Committee from 1984 to 1994. [4]

Randy V. Seifarth, P.E.—Manager, Product Engineering, Voith Hydro, Inc. Graduate in mechanical engineering from the University of Maryland. Registered Professional Engineer in Pennsylvania. Joined Allis-Chalmers (now Voith) in 1976. Has held a number of engineering positions; in 1988 he became Manager, Special Turbine Projects; responsibilities included leading the Voith engineering function for the Yacyretá project. In his current position, he is responsible for the administration and coordination of all activities of the Product Engineering Department.

James P. Sinclair, P.E.*—Mechanical Engineer—Acres International Corporation. Areas of responsibility over a 44-year career included: chief engineer, project manager, site mechanical engineer, resident engineer, mechanical department head, and mechanical design engineer for numerous large and small hydroelectric power projects ranging from 5 MW to 1,000 MW. U.S. delegate to IEC TC-4 "Hydraulic Turbines" for 14 years. Hydropower-related work led to extensive travel for assignments in the United States, Canada, Japan, Nepal, China, and Lesotho. Experience gained in all facets of mechanical design and plant layout, project management, equipment installation, and forensic review of equipment-related problems. Member of ASME for 26 years.

Thomas J. Spicher—Consultant. Provides turbine maintenance consultation and workshops with an emphasis on cavitation control. Mechanical engineer with the Bureau of Reclamation's Central Valley Project and Grand Coulee 1962-1981. With the Alaska Power Administration from 1981-1990. Authored *Hydro Wheels, A Guide to Turbine Maintenance* and numerous articles for *Hydro Review* magazine. [14]

Douglas C. Timpe—Leader, Environmental Science Unit, Black & Veatch Power Division. Responsible for licensing, environmental, and post-licensing compliance activities on several hydroelectric projects throughout the U.S. Prior to joining Black & Veatch, was employed by the FERC's Office of Hydropower Licensing as a fishery biologist, the North Dakota Game and Fish Department as a fish and wildlife biologist, and the National Marine Fisheries Service as a fisheries technician. Member of the American Fisheries Society. Co-authored "The Anatomy of Hydropower Relicensing," presented to the American Power Conference. [10]

Paul B. Tindall, P.E.—Senior Project Engineer, Wisconsin Electric Power Company. Responsible for control system upgrades, governor replacement, and specification of replacement runners for 16 hydroelectric plants in Wisconsin and Michigan. [5, 9, 11]

Edward J. Traut, P.E.**—Mechanical Engineer—Retired Supervisory Mechanical Engineer from the Large Gates and Valves Section of the Mechanical Branch, Electrical and Mechanical Engineering Division, Denver Office, Bureau of Reclamation. Functioned as a designer, technical director, and supervisor for more than 32 years in the section. Expert in the field of trashrack, stoplog, and bulkhead gate design.

Svein Vigander, PhD**—Senior Research Engineer, Tennessee Valley Authority Engineering Laboratory. Published 40 technical papers and numerous reports in the areas of flow-induced vibration, physical flow modeling, environmental fluid mechanics, and polyphase flow. Active on technical committees of the ASME, the American Society of Civil Engineers, and the International Association for Hydraulic Research. Past secretary of the ASME Hydro Power Technical Committee.

James C. Wadge, P.E.*—Mechanical Engineer—Former head of Gates and Valves Section of the Mechanical Branch, Electrical and Mechanical Engineering Division, Denver Office, Bureau of Reclamation. Functioned as a designer and supervisor for over 42 years in the section responsible for the design of Reclamation gates, valves, trashracks, and associated mechanical equipment. Participated in Reclamation's Dam Safety Training program for foreign and domestic engineers.

Noel H. Wagner, P.E.*—Senior Project Engineer, Pennsylvania Power & Light Company. Responsible for planning design and technical coordination of additions and modifications in existing fossil and hydro generating stations. Author, co-author of numerous technical papers. [1, 8, 11]

Susan K. Wagner, P.E.—Senior Engineer, Hydraulic Engineering, Voith Hydro, Inc. Responsible for hydraulic engineering activities including project application, bid proposals, model testing, and hydraulic design of turbines. Experience includes field trouble-shooting and manufacturing as related to hydraulic standards. [3]

Warren G. Whippen, P.E.*—Former Chief Hydraulic Engineer of Hydro Turbine Division of Allis-Chalmers. A Fellow of ASME and former Chairman of Fluids Engineering Division. Chairman of the Development Committee for the Hydro Power Technical Committee. Received the Henry F. Worthington Award and the Fluids Machinery Design Award from ASME.

Contents

CHAPTER 5: Major Valves

CHAPTER 6: Trashracks and Cleaning Devices

CHAPTER 7: Gates

CHAPTER 8: Lifting Equipment

CHAPTER 9: Auxiliary Systems

CHAPTER 10: Equipment for Environmental Control

CHAPTER 11: Equipment Layout Considerations

CHAPTER 12: Hydraulic Transients

CHAPTER 13: Inspection and Testing

CHAPTER 14: Maintenance and Operation Considerations

APPENDIX: Standards Organizations

INDEX

THE GUIDE TO
Hydropower Mechanical Design

Chapter 1
Introduction

This *Guide to Hydropower Mechanical Design* distills many hundreds of person-years of hydro-related mechanical design experience into this one volume. If you are responsible for designing, manufacturing, installing, or maintaining hydroelectric generating plants, this book should be very useful to you.

The book is not a code or standard, nor is it intended to restrict application of new technologies or alternative arrangements. Instead, it is a thorough overview of existing practice for the mechanical engineering portion of hydropower development. As such, it should prove helpful to engineers and other individuals involved in hydroelectric power development and production.

The guidelines in this document emphasize new project design. However, they also can be useful when rehabilitating or making design improvements to existing installations. A note of caution: many of the specific guidelines herein may require modification after considering site-specific factors. Also, some hydropower organizations have their own guides for planning and designing hydroelectric plants and, where this is the case, these also will need to be considered in any construction or rehabilitation project. Any such guides likely will enhance and/or complement the guidelines in this document.

Most hydro plants generate electricity for delivery through transmission and distribution systems. Consequently, plant conceptualization and design must fit into the economic, financial, and operational realities of the entire electrical supply system. Designers of hydro plants must provide for delivery and control of the water flow through the hydro turbine(s), and for all of the civil, electrical, instrumentation, controls, and mechanical equipment and systems needed for safe and reliable conversion of the water's energy into electricity. While these guidelines are primarily concerned with the mechanical design, it is neither possible nor desirable to avoid mentioning the related equipment and systems. Their existence and operation, together with the mechanical aspects, form an integrated entity—the hydroelectric plant.

This guide contains features to enhance its value as a reference source for engineers having interests and involvement in hydropower. The guide provides a useful overview of each aspect of hydro mechanical design. In many cases, this overview will undoubtedly be sufficient. However, if it is necessary to dig more deeply into a topic, pertinent resources are listed in a Reference and Bibliography section at the end of each chapter.

In addition, most chapters include a checklist of essential items to consider when preparing designs and specifications. Reviewing these checklists, and answering the questions posed, will aid users in giving thorough and appropriate consideration to each topic area.

Bibliography

The following list of general references contains material applicable to mechanical design, primarily concerning turbine selection and gates. The Tennessee Valley Authority reference is recommended for detailed coverage of mechanical design for hydroelectric plants.

Brown, J. Guthrie, Editor, *Hydroelectric Engineering in Practice*, Second Edition, Volumes 1, 2, and 3, London, England, Blackie & Son Limited, 1964.

Creager, W.P., and Justin, J.D., Editors, *Hydro-Electric Handbook,* Second Edition, John Wiley & Sons, New York, New York, 1950.

Kovalev, N.N., *Hydroturbines: Design and Construction,* Translated from Russian, Israel Program for Scientific Translation, Jerusalem, 1965.

Mosonyi, Emil, *Water Power Development: High Head Power Plants,* Volumes 1 and 2, Third Edition, Akadémiai Kiadó, Budapest, Hungary, 1991.

Mosonyi, Emil, *Water Power Development: Low Head Power Plants,* Volume 1, Third Edition, Akadémiai Kiadó, Budapest, Hungary, 1987.

Nechleba, Miroslav, *Hydraulic Turbines: Their Design and Equipment,* Artia Prague, Hungary, 1957.

Raabe, Joachim, *Hydropower: The Design, Use and Function of Hydromechanical, Hydraulic and Electrical Equipment,* VDI-Verlag, 1985.

Tennessee Valley Authority, *Mechanical Design of Hydro Plants,* TVA Technical Report No. 24, Volume 3, U.S. Government Printing Office, Washington, D.C., 1960.

Zipparro, Vincent J., and Hans Hasen, Editors, *Davis' Handbook of Applied Hydraulics,* Fourth Edition, New York, McGraw-Hill, Inc., 1993.

Chapter 2
Basic Considerations

Hydropower design covers many disciplines of engineering. Mechanical aspects involve designing, manufacturing, and selecting turbines, valves, bearings, governors, and other equipment necessary for converting hydraulic energy into mechanical energy. Electrical aspects encompass designing, manufacturing, and selecting generators, control systems, switchgear, and other equipment necessary for converting mechanical energy into electrical energy. Civil aspects include a variety of tasks necessary for the optimal use of the hydropower potential at a given site such as: study of site hydrology and hydraulics, site inspection and preparation, and design and construction of dams, powerhouses, and conveyance facilities. Specialists working with engineers assess and mitigate the environmental effects of a hydro plant. Engineers, planners, economists, developers, and others perform the economic analysis. A support staff operates and maintains each plant. Finally, scheduling and planning to optimize power production is performed for operation of both individual reservoirs and of complete basin-wide systems. Although these categories are not all-encompassing and there is substantial variation and overlap between disciplines, a considerable variety of engineering activities goes into developing and operating a hydropower facility.

Each hydropower site has unique characteristics. Thus, each hydroelectric project and powerhouse design is different. Each solution must be tailored to the unique characteristics of the site, the transmission and distribution system, and economic and financial resources. This chapter provides an historical review of hydro turbine development; information about the general characteristics of hydropower; and insight into the various factors used to establish the size and configuration of a given hydropower plant.

An Historical Sketch

Hydraulic turbines form the mechanical heart of a hydroelectric plant. Water turbines have a rich and varied history.[1,2] Modern turbine designs evolved from the water wheel. Originally used for directly driving machinery, water turbines began driving electric generators in 1882. Much of the early development of hydraulic turbines occurred in France, which, unlike England, did not have the inexpensive and plentiful sources of coal that sparked the industrial revolution in the eighteenth century. Nineteenth-century France found that water was its most abundant energy resource. To this day, *houille blanche* (literally, white coal) is the French term for water power.

In 1826, the Société d'Encouragement pour l'Industrie Nationale offered a prize of 6,000 francs to anyone who "would succeed in applying at large scale, in a satisfactory manner, in mills and factories, the hydraulic turbines or wheels with curved blades of Belidor."[1] Belidor was an eighteenth-century hydraulic and military engineer who, in the period 1737 to 1753, authored a monumental four-volume work on hydraulic engineering titled *Architecture Hydraulique*. The water wheels described by Belidor departed from convention by having a vertical axis of rotation. Also, the wheels were enclosed in a long cylindrical chamber approximately 1 meter in diameter. Large quantities of water were supplied from a tapered sluice tangent to the chamber and entered the chamber with considerable rotational velocity. This pre-swirl, combined with the weight of water above the wheel, was the driving force. The original "tub wheel" had an efficiency of only 15 to 20 percent.

Water turbine development proceeded on several fronts during the period 1750 to 1850. Engineers, such as John Smeaton (1724-1792) of England, and the French engineer, J.V. Poncelet (1788-1867), improved the classical horizontal-axis water wheel. These engineers created water wheels having efficiencies in the range of 60 to 70 percent. Also, Smeaton was the first to use models to obtain engineering data and was very active in windmill development. At this same time, several men considered reaction turbines (similar to the modern lawn sprinkler). The great Swiss mathematician, Leonhard Euler (1707-1783), investigated the theory of operation of these devices. Mannoury de Ectot (1777-1822) introduced a practical application of the concept in France in 1807. The theoretical analyses of Burdin (1790-1893), a French professor of mining engineering who introduced the word "turbine" in engineering terminology, contributed much toward understanding of the principles of turbine design and performance. Burdin's work underscored the principal requirements of shock-free entry and exit with minimum absolute velocity as the basic requirements for high efficiency. A student of Burdin, Benoit Fourneyron (1802-1867), was responsible for developing high-speed, outward-flow turbines with efficiencies on the order of 80 percent. The early work of Fourneyron resulted in several practical applications and gained him the coveted Société d'Encouragement pour l'Industrie Nationale's 6,000 franc prize in 1833.

As successful as the Fourneyron turbines were, they lacked flexibility and were efficient only over a narrow range of operating conditions. Both Samuel B. Howd, an inventor from Geneva, New York, and Uriah A. Boyden, an hydraulic engineer, addressed this problem in their work during the 1830s and 1840s. Their respective work evolved into the design, by James B. Francis (1815-1892), of an inward-flow motor, the precursor to the modern Francis turbine. At the same time, European engineers addressed the idea of axial flow machines, which, today, are represented by "propeller" turbines of both fixed-pitch and automatically adjustable pitch types. The automatically adjustable pitch types generally use Viktor Kaplan's hydraulic spool valve to actuate the blade-adjusting servomotor.

Just as the vertical-axis tub wheels of Belidor evolved into modern reaction turbines of the Francis and Kaplan types, the classical horizontal-axis water wheels evolved into the modern impulse turbine. In 1826, Poncelet, with his description of the criteria for an efficient water wheel, sowed the seeds of development. A group of California engineers in the late nineteenth century cultivated these ideas. One of these engineers was Lester A. Pelton (1829-1908), whose name is given to the Pelton turbine. This turbine consists of a jet or jets of water impinging on an array of specially shaped buckets that are closely spaced around the periphery of a wheel.

Turbine configurations as we know them today vary widely from those originally developed. Current technology has resulted in highly efficient devices, with efficiencies as high as 96 percent in the larger sizes of Francis units. In terms of design concept, these turbines fall into two major categories: reaction turbines of the Francis, or propeller, design; and impulse turbines of the Pelton type.

In the United States, the first hydroelectric plant using a direct-current Edison dynamo came on line September 30, 1882, in Appleton, Wisconsin. However, in *Creativity, Conflict, and Controversy: A History of the St. Paul District Corps of Engineers*, author Raymond Merritt, currently the Dean of the College of Social Sciences at St. Cloud State University in St. Cloud, Minnesota, mentions that the Minneapolis Brush Electric Company, serving an arc-lighting (street lighting) system, began operating a hydroelectric plant some 25 days earlier.[2] Transmitting power over long distances became economical in 1901 with the installation by George Westinghouse of alternating current equipment, based on a patent by Nickola Tesla, at Niagara Falls. Today, hydropower is used primarily for producing alternating current energy with the economical energy transmission through high voltage lines.

Characteristics of Hydropower

Hydropower is a renewable resource that has many advantages over other sources of electrical power. Hydropower plants have very low operating and maintenance costs because they consume no fossil or nuclear fuel and do not involve high temperature processes. For these reasons, it is normal in the industry to have hydroelectric equipment with availability exceeding 95 percent of the time. The equipment is relatively simple to operate and maintain. This means that hydro plants require fewer operating or maintenance personnel compared to thermal and thermo-nuclear power plants. Since hydropower equipment can adjust rapidly to load changes, this considerably enhances system operation flexibility. Also, the unique characteristics of hydro plants are conducive to automation, which provides even greater flexibility and reduced operation and maintenance costs (O&M). The extended equipment life provides significant economic advantages. Some hydroelectric plants installed 100 years ago still operate reliably. Because hydroelectric plants burn no fuel, there is no air pollution, acid rain, or need for hazardous waste disposal associated with their operation. The capital intensive nature of hydropower development also can be a significant hedge on inflation.

Properly designed hydro developments can provide many other benefits such as recreation, improved water quality, flood control, improved aesthetics, and improved fish habitats. Careful planning and appropriate operations can minimize environmental problems. Some of the more recently identified environmental issues provide challenges and opportunities for innovative power plant design. For example, significant environmental changes can occur even when a hydropower plant is developed at an existing dam. In this case, water normally aerated by flowing over a spillway is diverted through the turbines where the rate of aeration is essentially zero. This can have a substantial effect on downstream oxygen levels, which can, in turn, affect the fish. (Fish need dissolved oxygen, which they extract to "breathe.") However, new technology, currently under development, uses turbine aeration or aeration by other means to prevent lack of oxygen from becoming a problem.

Hydropower, accounting for approximately one-fifth of the installed electrical generating capacity worldwide, plays a very significant role in the world's supply of energy.[3] While hydropower has a relatively high initial cost compared to other energy sources, this must be balanced against fuel costs for thermal plants. Fuel costs are highly variable, and the uncertainties in the world petroleum supply can affect the supply and the cost of fuel. The capital-intensive aspect of hydropower development is a benefit in an inflationary period because long-term financing ensures relatively stable annual costs. Fuel, the most inflationary component of O&M costs at other kinds of generating plants, is eliminated. Hydropower development also is attractive when low-interest loans are available.

In summary, hydropower has the advantage of being a renewable resource that conserves depleting fuel supplies. It is a highly efficient way to generate electricity. Hydropower gives great flexibility in operating an electric system particularly because of its rapid response time and short-term and long-range storage. A century of development has resulted in highly reliable and durable equipment. Typical equipment lifetimes far exceed the amortization period, resulting in very favorable long-term economic benefits. The absence of non-renewable fuel costs, coupled with reliability and simplicity, results in extremely low O&M costs. Hydropower plants' effects on fish, water quality, and sedimentation can be successfully managed, as hydro plants throughout the world that have operated for many years demonstrate.

These benefits have to be balanced against some disadvantages. The high initial cost may impede development—depending on the economic climate at the time that financing is sought. Large, base load plants require significant impoundments that may inundate developed land and wildlife habitats. This also causes significant changes in stream regimes. Hydropower is dependent on precipitation that can be highly variable. Some sites, far from established communities, warehouses, machining facilities, and available skilled labor, may require building all of those facilities at the site—including dormitories, food services, and recreation facilities for the workers.

Each project requires a careful appraisal of environmental, institutional, and economic factors. Large impoundments can be aesthetic and can provide excellent recreational opportunities, improved fish habitats, and significant flood control, but this must be balanced against the habitat and recreation losses from inundating land.

Sometimes, more localized economic factors can play a role in the feasibility study. For example, in regions where there is a significant drain on the economy to fund importing fuel, a hydropower development can offset this economic drain. An estimate made in Minnesota, for example, indicated that every dollar spent on out-of-state coal reduces investment in the local economy by US$3. In some countries, fuel must be purchased with difficult-to-obtain foreign currency. In contrast, much of the construction materials and labor for a hydroelectric project can be paid in local currency that disperses throughout the local economy as workers spend their pay.

Factors Affecting Hydro Plant Capacity

Evaluating Head and Flow

The generating capacity of a hydro plant is usually expressed in kilowatts (kW) or megawatts (MW) and is selected based on a careful evaluation of several important parameters. The first of these is the head: the approximate vertical distance through which the water falls. The second is the flow rate (or discharge): the volume flow rate of the water that passes through the plant. Since the power output of a hydro plant is proportional to the product of the head and flow rate, the relative frequency that various head-flow combinations occur must be considered in selecting the generating capacity to be installed at a particular site. Equation 2-1 estimates the power available.

Equation 2-1:

$$P=9.8\eta QH \quad \text{(metric units)}$$

$$P=\eta QH/11.8 \quad \text{(English units)}$$

where:

P=power, kilowatts, kW;

η=efficiency (a number less than one), dimensionless;

Q=flow in cubic meters per second, cms (cubic feet per second, cfs); and

H=head in meters (feet).

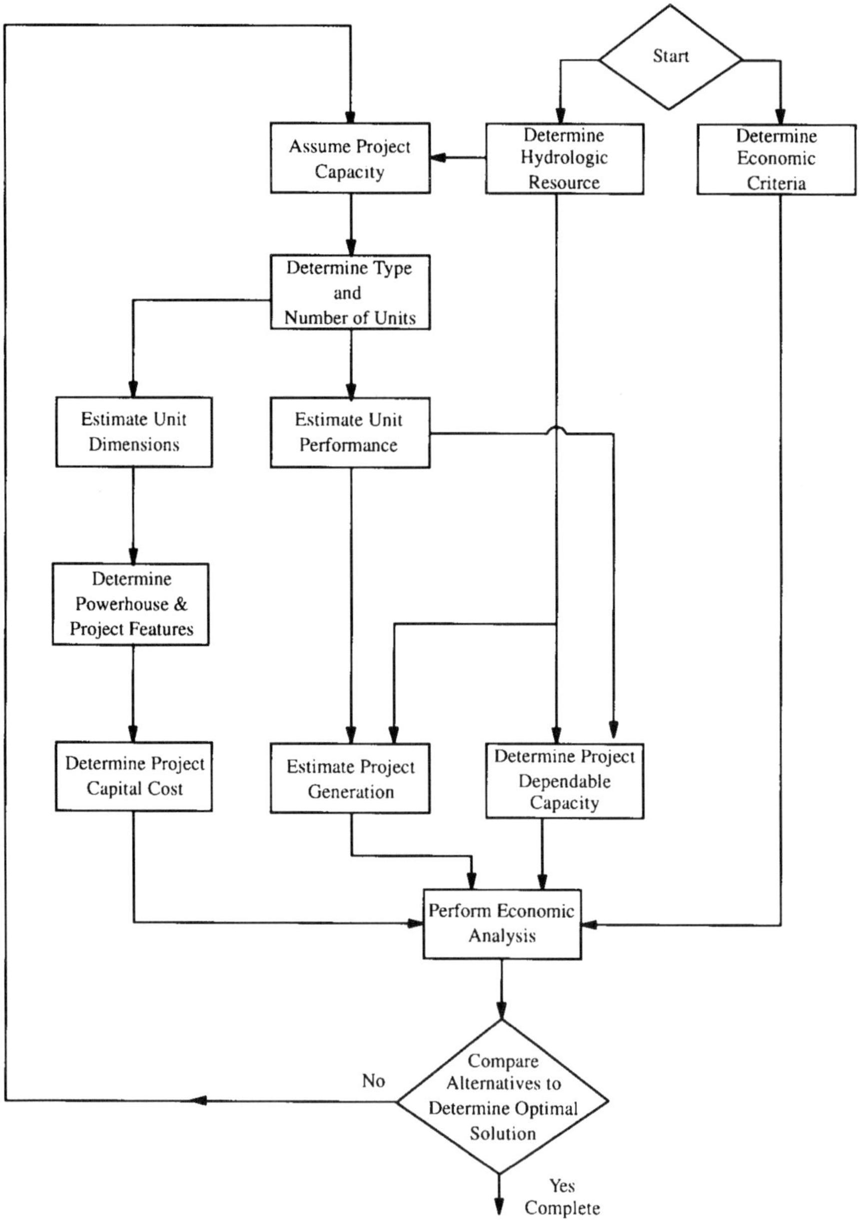

FIGURE 2-1: *Capacity Optimization Flow Chart.* *This flow chart shows the basic steps required to determine the optimal plant capacity. Using the project hydrologic data and economic criteria, project planners analyze a range of project capacities by evaluating the capital costs and economic benefits of each alternative capacity. If the project capacity appears economically feasible, the planner can evaluate other plant capacities and alternative machinery solutions in the same manner until the optimal plant capacity is determined. If the initial plant capacity is not feasible, the planner can evaluate smaller capacities in a similar manner.*

Determining Optimal Plant Capacity

The flow diagram in Figure 2-1 illustrates the basic steps in determining the optimal capacity for a hydro plant. The initial input for the process is determining: 1) the hydraulic resource or river flow, and 2) the economic criteria to evaluate the project economic feasibility, including the value of energy, value of capacity, escalation rates, discount rate, etc. The project hydrologic data (water availability versus time) is obtained by reviewing historical records or estimated by other techniques.[4] As a result of this analysis, an initial plant capacity is assumed. A trial number of turbine-generator units is also assumed at this stage. The next step is to obtain, from manufacturers or other sources, estimates of the performance characteristics of the selected turbines and generators. Such characteristics will be used in calculating the annual energy from the plant and its capacity. Outline dimensions of the turbines, generators, and other major equipment are used to estimate the powerhouse dimensions and costs. This analysis usually is done by the project developer or his architect/engineer.

Economic analyses for a range of project alternatives typically are conducted for the various capacities, annual generation, capital costs, and economic criteria. If the project appears feasible, other plant capacities and alternative machinery solutions would be evaluated in the same manner until the optimal plant capacity is determined. If the initial plant capacity is not feasible, smaller capacities would be evaluated in a similar manner. After evaluating several alternatives, a range of solutions available for consideration can provide the potential developer with a basis for selection.

Determining Annual Plant Generation

To optimize plant capacity, the revenue from annual plant generation is estimated for various plant capacities for comparison with the capital costs in the economic analysis. Estimating the annual plant generation is a complex process. The fundamentals are described in the texts *A Guide to Using Interest Factors in Economic Analysis of Water Projects* and *Hydropower Engineering Handbook*.[1,4] Computer programs have been developed and refined to aid in calculating plant generation.

The river flow available for generating electricity varies substantially owing to daily, seasonal, and annual precipitation fluctuations. These fluctuations complicate the process of estimating plant generation. Often, a flow duration curve, as shown in Figure 2-2, will be used to illustrate the fluctuation of available river flow. The shape of this curve depicts the percentage of time (on average) that a given flow is exceeded. A simple example is illustrated in Figure 2-2. Referring to the curve labeled "River Discharge," it can be seen that the river flow is greater than 28 cubic meters per second, cms (1,000 cubic feet per second, cfs) about 35 percent of the time. Twenty percent of the time the river can be counted on to deliver 45 cms (1,600 cfs). Because the headwater and tailwater elevations vary differently with a change in river flow rate, the available head also varies with flow rate. This also is depicted in Figure 2-2, showing that the available head is less

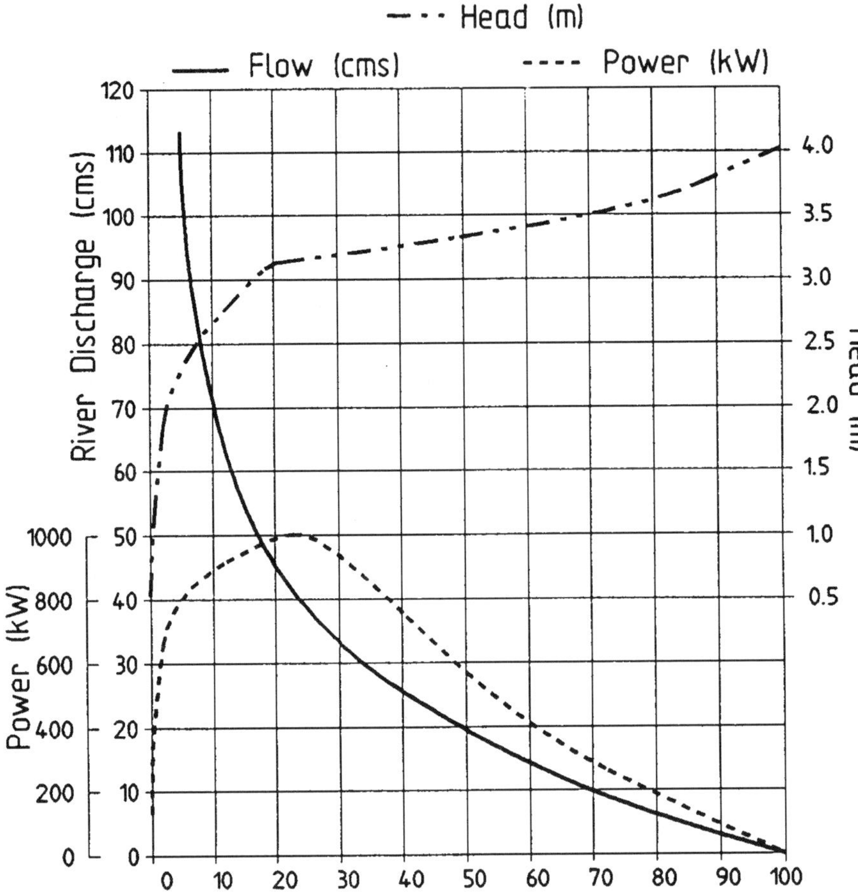

FIGURE 2-2: ***Flow Duration Curve.*** *This curve for Crookston Dam, Minnesota, shows that the river flow varies substantially owing to daily, seasonal, and annual precipitation conditions. These fluctuations complicate the process of estimating plant generation. Because the headwater and tailwater elevations vary differently with a change in river flow rate, the available head also varies with flow rate. A power duration curve can be derived from the variation in flow and head, as illustrated in the figure.*

than 3 meters (10 feet) about 20 percent of the time because of excess flow. The depicted variation in head and flow associated with the flow duration characteristics illustrated in Figure 2-2 can be used to calculate a power duration curve, as is also depicted in the graph for the curve labeled "Power (kW)."

The example cited is typical of run-of-river sites with low head (less than 15 meters (50 feet)). As shown, the head available to do work is affected significantly at high flows because the tailwater elevation increases more than the headwater elevation.

Upstream reservoirs can be used to store water during high flow periods for later release in low flow conditions. In this way, storage reservoirs can reduce the adverse effect of fluctuations and can increase the total potential annual energy. Storage reservoirs maintain a higher head and permit the available high flows, which might otherwise be wasted as spill, to be recovered as energy.

The use of the flow duration curve method is not recommended for a project with significant variations in reservoir level unless the specified reservoir level requirements are considered in the calculation. Many projects are required to maintain the reservoir level within a specified range that can vary with the time of year. The Sequential Power Analysis and Modified Flow Duration Analysis methods have been developed to carry out energy estimates for storage reservoir projects.[1,4]

In calculating the average annual energy production, it is important to recognize the differences between net head and gross head and the definition of efficiency in Equation 2-1. Gross head is the difference in elevation between the upstream pool and the tailrace downstream of the plant. The net head on the unit is the gross head less all hydraulic losses (in the turbine mode) or plus all losses (in the pumping mode) of a pumped-storage scheme. As an approximation, hydraulic losses can be expressed as proportional to the square of the rate of flow (or velocity). (See Chapter 3, Hydraulic Turbines, for a detailed discussion of net head.)

If gross head is used in Equation 2-1, the efficiency is defined as plant efficiency. Plant efficiency incorporates all the losses due to the flow passage from the reservoir into the plant inlet through the turbine and draft tube, and the energy required for discharging the water into the tailrace. In addition, station service losses and the loads from other mechanical and electrical equipment and systems needed to operate the plant must be included in the calculations. The gross power output from a plant is not necessarily revenue-earning, saleable power because transmission and distribution losses are not accounted for. The terms in the power sales agreement contract, if one exists, will influence the revenue from the plant.

Hydropower plants may be designed to meet different objectives. Turbine efficiency is only one of several factors. At certain low-head projects, a less efficient turbine that can pass more water at times of high river flows may yield more total annual energy than a more efficient turbine releasing less water when high river flows prevail. Thus, selecting the turbine or turbines with the highest efficiency does not guarantee that the optimal plant output will be obtained.

Types of Projects

There are two main types of hydropower schemes that can be categorized in terms of how the flow at a given site is controlled or modified. These are:

1. Run-of-river plants (no active storage); and
2. Plants with significant storage.

In a run-of-river project, the natural flow of the river is relatively uncontrolled.

In a storage project, the filling and emptying of the impounded storage along with the pattern of the natural streamflow controls the flow in the river downstream from the storage impoundment. Ocean tidal schemes, mixed schemes, and other hydropower variations exist, but they basically fall into one of these two categories.

Run-of-river plants can be located near or at a dam or lake outlet, or at the downstream end of a canal, open flume, or pipeline diverting the stream's flow around a dam or falls. The available flow governs the capacity of the plant. The plant has little or no ability to operate at flow rates higher than that available at the moment.

Plants with significant storage can either be conventional or pumped-storage projects.

In a conventional plant, a dam, which stores water in a reservoir or lake impoundment, controls the river flows. Water is released according to electric, irrigation, water supply, or flood control needs. Constructing a dam and storage reservoir can increase the percentage of time that a project can produce a given level of power. Base load plants—those operated at relatively constant output—may have either a small capacity relative to the river flow or may have a significant storage reservoir. Storage reservoirs can be sized for storing water during wet years or wet seasons. Alternatively, they can be sized to provide water for weekly or daily peak generation. A storage reservoir allows using available energy that might otherwise be wasted as spill.

A pumped-storage project is, in effect, a large storage battery. Water is pumped from a lower stream or reservoir to an upper reservoir during periods of low power demand when there is an excess of peak capacity from other plants. During periods of high demand, water flows in the opposite direction (upper reservoir to lower reservoir), producing power.

A typical pumped-storage plant is a net consumer of energy: it returns approximately 3 kilowatt-hours (kWh) of electricity for each 4 kWh required for pumping. However, it offers the following important benefits:

- The energy generated during peak periods has a higher monetary value than the energy required for pumping during off-peak periods;
- It permits continuous operation of the highest efficiency plants in the utility's system;
- It provides rapid and flexible response to system load changes. Typically, very large load swings can be accommodated; and
- The utility's overall fuel consumption is reduced because the pumped-storage plant's on-peak generation avoids or displaces generation at the least efficient thermal plants in the system.

Normally, the pumping and generation modes at modern pumped-storage plants use the same turbo-machinery and generator-motor equipment. There are cases, however, where separate pumps and turbines are used. Most separate pump and turbine applications are in Europe.

Components of a Typical Hydropower Plant

A large variety of hydropower configurations is possible because each site is different. There are, however, certain components that are common to most projects.

Figure 2-3 shows the details of a typical hydropower plant. The heart of the plant centers around the turbine-generators. (See Chapter 3, Hydraulic Turbines, for a discussion about turbines.) Wicket gates (on reaction turbines) or one or more needle nozzles (on impulse turbines) may control the flow through the turbine. Some turbines are built without wicket gates; consequently, the available head controls their discharge. Often, a shut-off valve is provided for shutdown and repair. With reaction turbines, the flow is discharged from the turbine runner into a draft tube. Primarily, the draft tube acts as a diffuser. The expanding water passage reduces the flow velocity and recovers the kinetic energy still remaining in the water leaving the runner. In effect, this increases the total dynamic head available to the turbine. The secondary purpose of the draft tube is to direct the flow from the runner exit to a tailrace channel, a pond, or directly into the river. (See Chapter 3, Hydraulic Turbines, for more details concerning the function and design of draft tubes.)

Wicket gates, service gates, or valves that can be located upstream or downstream from the turbine usually can shut off the flow to the unit. Gates, valves, or bulkheads or stoplogs can be used for isolating and dewatering the turbine. Trashracks prevent debris from entering the intake structure.

River flow in excess of turbine capacity must be stored or diverted. If a spillway is used, various kinds of gates can be used to control upstream water levels. (Chapter 7, Gates, describes several different types of gates.) Usually there is some provision made to draw the reservoir down below crest level using various types of bottom outlets.

Switchgear and relays protect the generators from electrical faults. Generally, plants have transformers to increase the generator voltage to match the transmission line voltage, which is usually higher.

A typical low-head project has a forebay and an intake structure containing two sets of slots: the upstream slot for bulkhead gates or stoplogs and the down stream slot for a slide gate. Placed in front of the inlet structure, trashracks prevent debris from flowing into the turbine. Depending on the configuration, the flow leaving the inlet either will enter the turbine directly or enter a penstock that passes water to the turbine.

In higher head installations, the penstock and other water conduits can be very long, and rapid changes in the rate of flow through the turbine can cause pressure increases known as waterhammer. For example, when the electrical load on the plant is disconnected, the turbine(s) will rapidly shut down to prevent over-speeding the generator(s). The water conduit system may need one or more surge tanks to minimize waterhammer pressures.

Key

1. *Generators*
2. *Hydraulic Turbines*
3. *Shut-Off Valves*
4. *Intake Gates*
5. *Intake Gate Hoists*
6. *Trashrake Cars with Hoist*
7. *Trashrakes*
8. *Crest Gates*
9. *Bypass Valves*
10. *Gate Hoists*
11. *Draft Tube Gates*
12. *Relief Valves*
13. *Exciters*
14. *Switchgear*
15. *Transformer*
16. *Circuit Breaker*

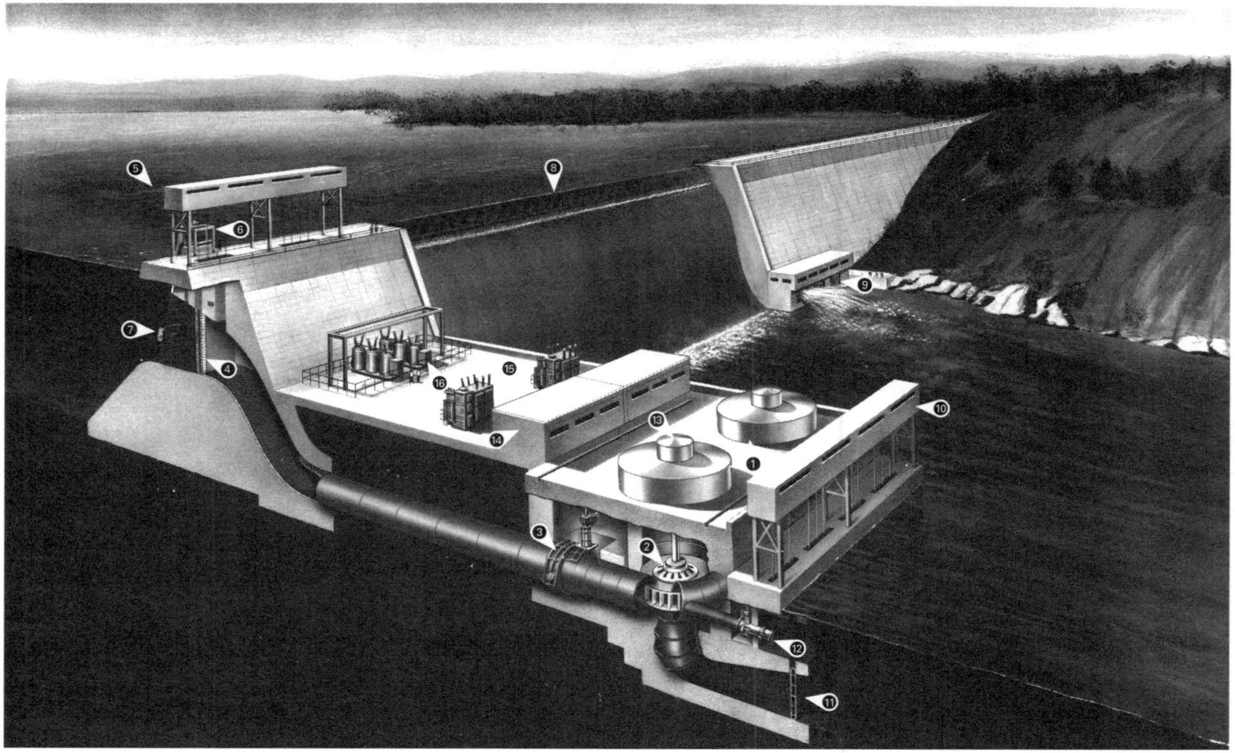

FIGURE 2-3: *Typical Hydroelectric Power Plant.* *This illustration shows the details of a typical hydroelectric plant. The heart of the plant centers around the turbine-generators. Wicket gates control the flow through the turbine. A shut-off valve is provided for shutdown and repair. The flow is discharged from the turbine runner into a draft tube. Gates, valves, or bulkheads can be used for isolating and dewatering the turbine. Trashracks prevent debris from entering the intake structure.*

(Courtesy Voith Hydro)

REFERENCES

[1]Gulliver, J.S., and R.E.A. Arndt, *Hydropower Engineering Handbook,* McGraw-Hill, Inc., New York, New York, 1991.

[2]Merritt, Raymond, *Creativity, Conflict, and Controversy: A History of the St. Paul District Corps of Engineers,* U.S. Army Corps of Engineers, Washington, D.C., 1979.

[3]*1992 Energy Statistics Yearbook,* United Nations, New York, New York, 1994.

[4]Glenn, B.P., and E. Barbour, *A Guide to Using Interest Factors in Economic Analysis of Water Projects,* U.S. Bureau of Reclamation, Denver, Colorado, December 1970.

Chapter 2

BIBLIOGRAPHY

Hydroelectric Cost Estimating Manual, U.S. Army Corps of Engineers, 1981.

Hydroelectric Power Evaluation, Federal Energy Regulatory Commission, Washington, D.C., 1979.

Mosonyi, Emil, *Water Power Development: High Head Power Plants,* Volumes 1 and 2, Third Edition, Akadémiai Kiadó, Budapest, Hungary, 1991.

Mosonyi, Emil, *Water Power Development: Low Head Power Plants,* Volumes 1 and 2, Third Edition, Akadémiai Kiadó, Budapest, Hungary, 1987.

Reynolds, Terry S., "Hydro before the Turbine: The Rise and Fall of the Water Wheel," *Hydro Review,* Volume XI, No. 7, December 1992, page 52.

Shortridge, Robert W., "Francis and His Turbine, *Hydro Review,* Volume VIII, No. 1, 1989, page 24.

Shortridge, Robert W., "Lester Pelton and His Water Wheel," *Hydro Review,* Volume VIII, No. 5, 1989, page 22.

Shortridge, Robert W., "Looking Back: Some Early History of Hydroelectric Power," *Hydro Review,* Volume VII, No. 3, 1988, page 30.

Shortridge, Robert W., "Viktor Kaplan and the Adjustable-Blade Propeller Turbine," *Hydro Review,* Volume VIII, No. 6, 1989, page 38.

Vansant, Carl, "Hydropower: Tapping the Potential, Facing the Challenges of the World's Leading Source of Sustainable Energy," *Hydro Review,* Volume 14, No. 4, July 1995, pages 58-68.

Zipparro, Vincent J., and Hans Hasen, *Davis' Handbook of Applied Hydraulics,* Fourth Edition, McGraw-Hill, Inc., 1993.

Chapter 3
Hydraulic Turbines

To evaluate the power potential of a project site, it is necessary to determine the total hydraulic energy available as characterized by the relationship of head and flow quantities and their variations versus time. Economic factors and other issues, such as intended operational modes and environmental constraints, determine the power to be developed at a given site.

Selecting the type, kind, configuration, size, and number of turbine units that best suit a project is a detailed process. This involves technical, environmental, financial, and political considerations. The most inexpensive turbine may not be the best solution to use the available head and flow. Information exchange and close cooperation between the owner and manufacturer(s) is essential to determine the hardware that will allow a site's full potential to be realized. This exchange should begin in the early stages of project development.

This chapter discusses the basic principles of hydraulic turbines and reversible pump turbines. The turbine theory and relevant fundamental equations are available elsewhere and will not be repeated here. The bibliography at the end of the chapter lists suggested hydraulic texts and handbooks. Due to space limitations, certain explanations and examples in this chapter focus on only one type of turbine.

Site Conditions

Head

Gross head. The difference in elevation between the headwater surface (upstream) and the tailwater surface (downstream) of a project is known as the gross head. To establish the equivalent still water surface elevation, each of these elevations must be corrected for the local velocity head when water passes downstream in a project, as given in Equation 3-1.

EQUATION 3-1:

$$H_v = V^2/2g$$

where:

H_v=local velocity head, meters (feet);
V=local velocity of water, meters/second (feet/second); and
g=local gravitational constant, meters/second squared (feet/second squared).

Head losses. Not all of the gross head can produce power because losses occur when delivering the water into the turbine. Also, some residual head is required to discharge the water back into the tailrace. Such losses are velocity dependent and, therefore, controllable within limits. In other words, the larger the water passage, the smaller the losses. Project costs for enlarging the water passages must be evaluated against the revenue gain for the increased energy, which is a result of lower velocities and reduced head losses.

Various texts, test codes, and handbooks that refer to loss coefficients have been published. The reference section at the end of this chapter mentions several. These references cover conduits of different shapes, various bend angles, gradual and sudden contractions and enlargements, plus the cumulative effects of these losses in series.

Net head. The effective head available to the turbine unit for power production is called the net head. The American Society of Mechanical Engineers (ASME) Test Code PTC-18 and the International Electrotechnical Commission (IEC) test codes exclude the draft tube exit loss from the definition of net head. Therefore, according to this convention, draft tube exit losses are not chargeable to the turbine, as shown in Figure 3-1. To minimize this loss, it is recommended that a maximum velocity or minimum area at the draft tube exit be specified in the procurement documents.

The net head determination made during a laboratory or field test consists of measuring the total energy at each of the applicable stations referenced previously. That is, the test must find the difference between the sums of the pressure, velocity, and elevation energies at the two station locations. See Figure 3-1 for a drawing that illustrates the definition of net head.

For the exact definition of net head used in the mentioned test codes, refer to the codes in the reference list.

IEC publications use the term "specific hydraulic energy" instead of head. Specific hydraulic energy, with the symbol E, is defined as the total energy of water available between the high pressure (at section number 1 on Figure 3-1) and the low pressure (at section number 2 on Figure 3-1) of the machine. This takes into account the influence of the compressibility.

This guide uses the term, "head," because it is far more commonly used than "specific hydraulic energy."

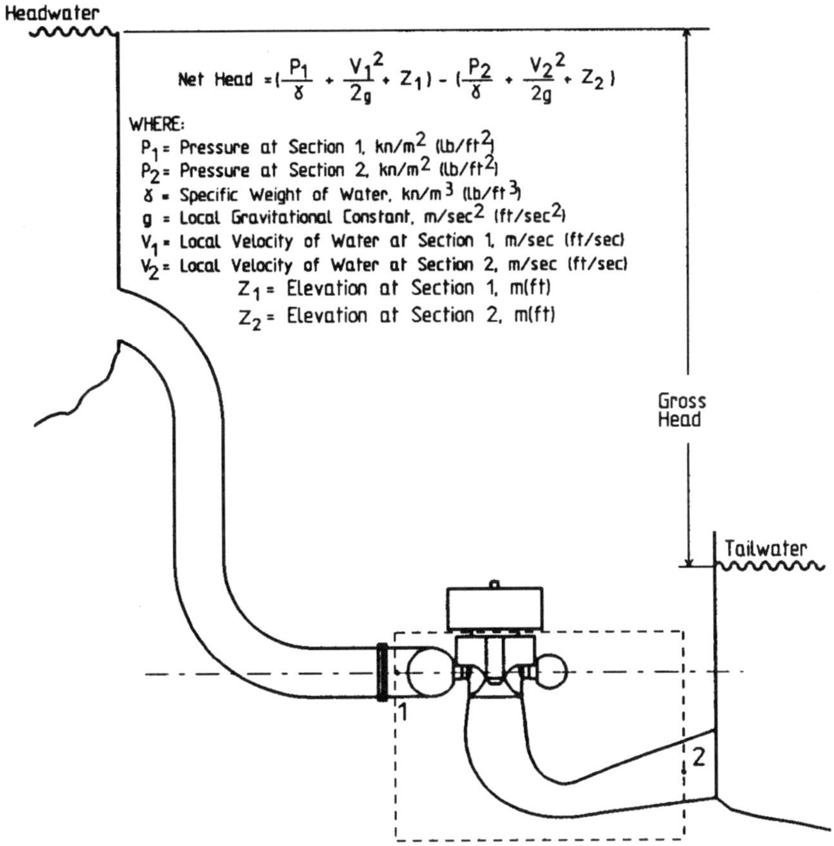

FIGURE 3-1: ***Definition of Net Head.*** *This drawing illustrates the definition of net head, the effective head available to the turbine unit for power production. A given turbine under a given gross head and flow rate generates the same power regardless of the net head definition.*

Discharge (Flow)

As discussed in Chapter 2, Basic Considerations, the flow varies with the type of project: run-of-river or storage. Also, the flow can vary substantially during the year.

TURBINE TYPES AND THEIR MAJOR COMPONENTS

The hydraulic turbine is a mechanical device that converts the potential energy contained in an elevated body of water (a river or reservoir) into rotational mechanical energy.

Turbines can be either reaction or impulse types. The turbine type indicates the manner in which the water causes the turbine runner to rotate. Reaction turbines operate with their runners fully flooded and develop torque because of the

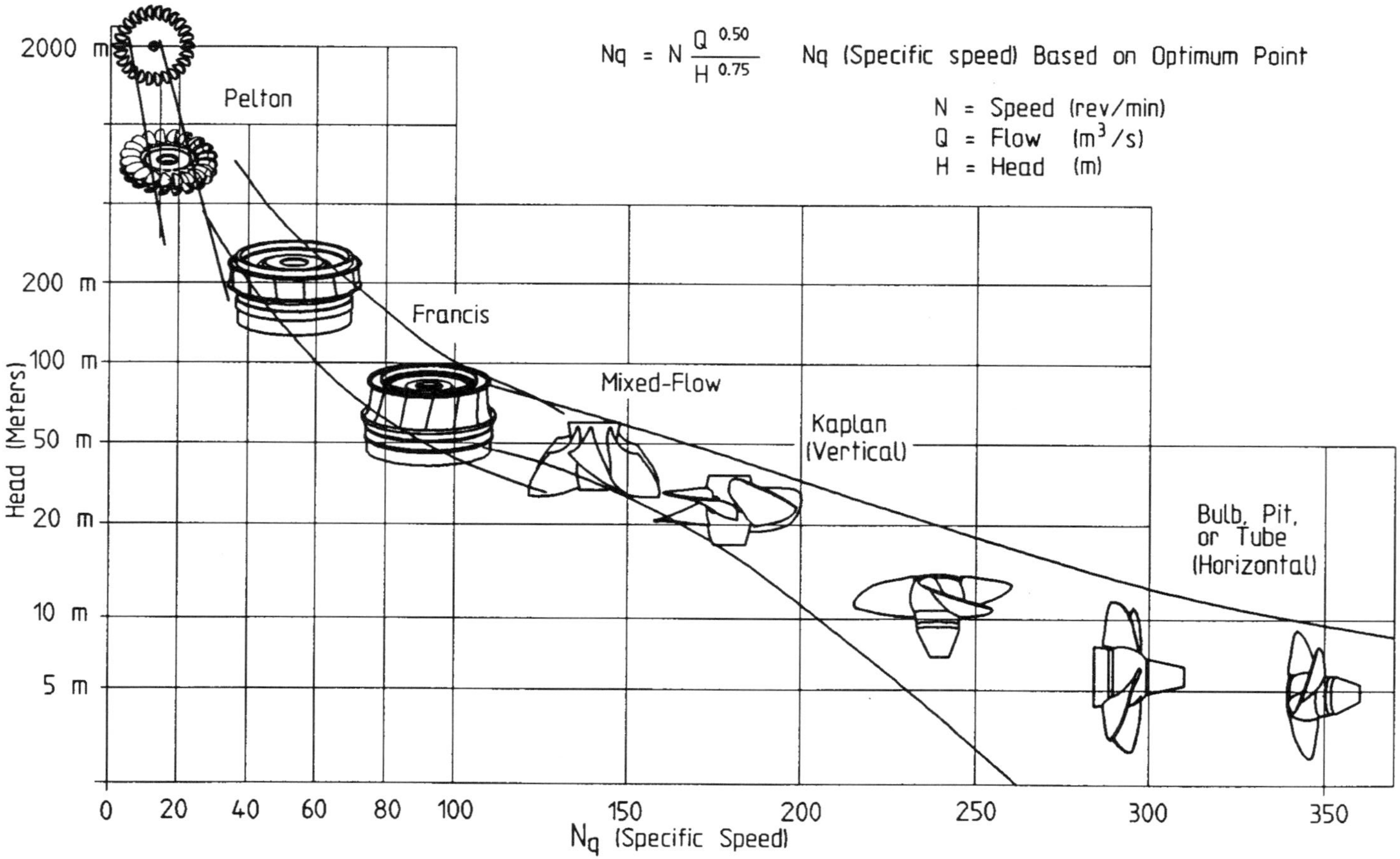

FIGURE 3-2: ***Nq versus Head.*** *This figure shows the various turbine types as a function of specific speed (Nq) and head. This figure should be used as a guideline, as there is overlap between the various turbine types with respect to their operating ranges.*

reaction of water pressure against runner blades. Impulse turbines operate with their runners in air and convert the water's pressure energy into kinetic energy of a jet that impinges onto the runner buckets to develop torque.

These two turbine classifications have various sub-types available for different applications. Reaction turbines include propeller turbines, Kaplan turbines, Deriaz turbines, turbines with a rim generator, Francis turbines, and reversible pump-turbines. Impulse turbines include Pelton, Turgo, and Banki (or crossflow) turbines.

The term specific speed is often used in classifying types of turbines and characteristics of turbines within types. As shown in Figure 3-2, the specific speed value defines the approximate head range of application for each turbine type. The exact definition is given later in the Similitude section of this chapter. Low-head units tend to have a high specific speed, and high-head units to have a low specific speed.

Reaction Turbines

The basic water passage components of a reaction turbine, listed in the order in which the water flows through them, are a spiral (or semi-spiral) case or open flume, fixed stay vanes, movable wicket gates, a runner, and a draft tube. Figures 3-3 and 3-4 show the two typical spiral case designs with an elbow draft tube (in each figure, the plan view of the spiral case is on top and elevation view is on bottom). There are a few units without stay vanes and some without wicket gates. The position of the wicket gates, which are adjusted by the governor to correspond to the desired output, control the rate at which water is supplied to the runner. As the water flows through the runner, a series of curved vanes on the runner deflects the flow. This deflection creates a dynamic force on the vanes, and this force causes the runner to rotate.

For a turbine with a full spiral case, the entrance diameter of the spiral case is chosen so that the velocity in the fully cylindrical portion immediately upstream of the first stay vane is about 20 percent of the available spouting velocity at maximum flow rate. This can be determined by Equation 3-2.

EQUATION 3-2:

$$V = 0.20\,(2gH_n)^{0.5}$$

where:

V=velocity, meters/second (feet/second);
g=local gravitational constant, meters/second squared (feet/second squared); and
H_n=net head, meters (feet).

Larger spiral cases will result in higher efficiencies and higher costs.

As water passes around the spiral case, the water flows inward toward the runner, and the cross-sectional area of the spiral case is reduced appropriately in the direction of flow. Design theories regarding velocity distribution around the spiral case differ among manufacturers.

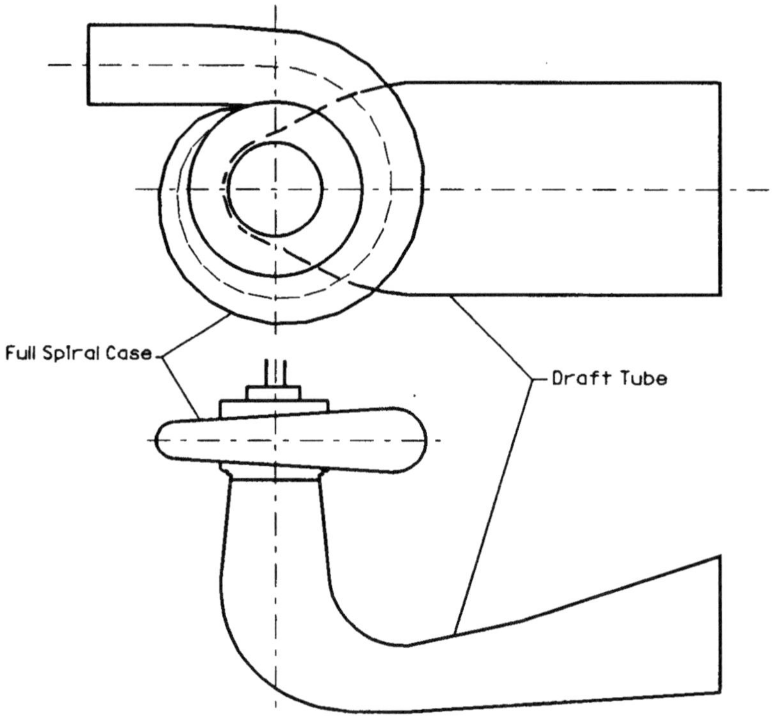

FIGURE 3-3: ***Full Spiral Case and Elbow Draft Tube.*** *This figure shows the typical style spiral case and elbow draft tube that are used on vertical shaft reaction turbines. A steel full spiral case design is used for higher head applications where a penstock connects to the spiral case.*

The shaft or axis in a reaction turbine may be either vertical, horizontal, or inclined. When the shaft is vertical (so the runner rotates in a horizontal plane), the turbine is called a vertical turbine. When the shaft is horizontal and the runner rotates in a vertical plane, the turbine is called a horizontal turbine. The inclined shaft turbine is referred to as a inclined axis unit.

The vertical unit (see Figure 3-12) is the most common configuration in the industry; therefore, the following description focuses on the vertical unit.

The primary function of the stay ring is to provide structural continuity between the upper and lower portions of the spiral case. Secondarily, it guides the flow of water, with minimal interference, into the wicket gate system.

Hydraulic design considerations determine the height of the stay ring opening at the inner diameter and the shape of the vanes. The intent of some designs is to direct the flow from its approximately tangential path into a more radial direction as it passes from the spiral case and approaches the runner through the wicket gate. Because the runner's performance depends heavily on the flow rate and direction of flow of this input energy, the delivery must be done with minimum loss.

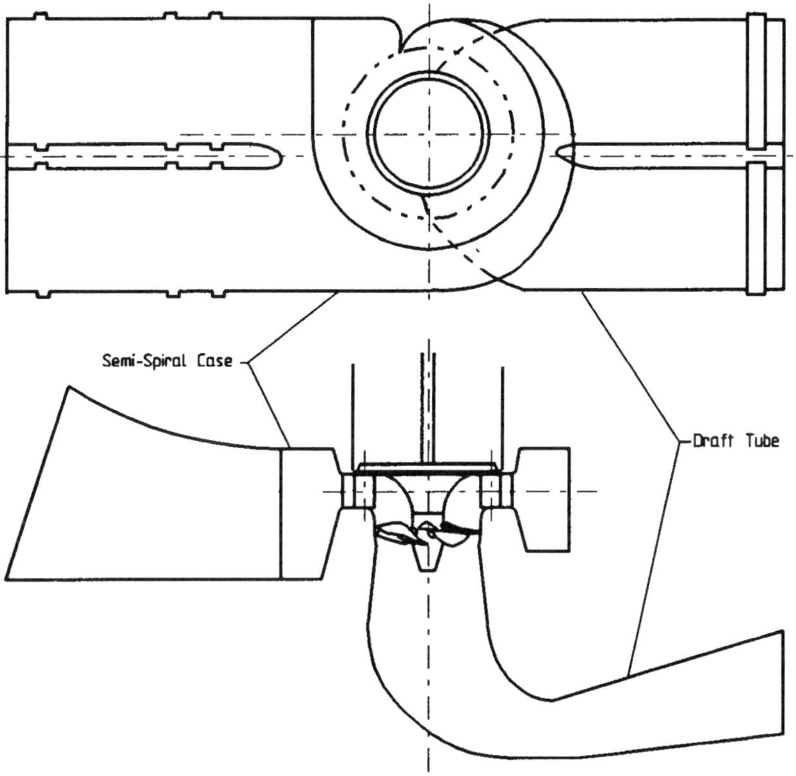

FIGURE 3-4: ***Semi-Spiral Case and Elbow Draft Tube.*** *This figure shows a semi-spiral case along with an elbow draft tube. The semi-spiral case design, which is constructed of concrete, is typically used for low-head applications.*

The wicket gates control the quantity and direction of the water entering the runner. Usually 16 to 24 in number, they are arranged circumferentially on the inside of the stay ring and are guided by bushings in the headcover and discharge (or bottom) ring. The hydraulic requirements of the project determine the water passage height and cross-sectional shape for the wicket gates. They move to control the flow upon instruction from the governor to the servomotors.

The water leaving the runner is conveyed to the tailwater through the draft tube. The draft tube serves two purposes. It acts as a diffuser to recover kinetic energy still remaining in the discharge from the runner. It also creates an unbroken streamflow to the tailrace, reduces the exit loss, and influences the pressure under the runner. For these reasons, the draft tube must be considered an integral part of the turbine.

Because the function of the draft tube is to efficiently decelerate the water velocity after it leaves the runner and to return it to the river, the shape must be designed carefully. Normally, there is: 1) an area increase before entering the elbow; 2) a constant or slightly reducing area through the elbow; and 3) gradual

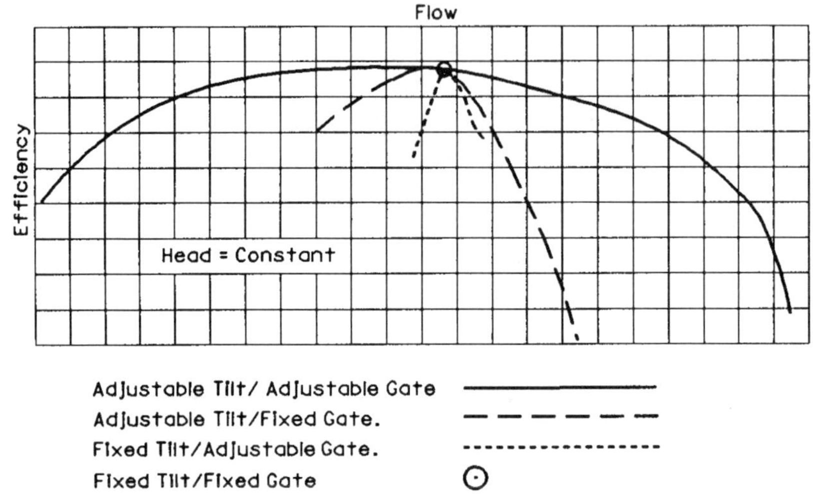

FIGURE 3-5: ***Relative Performance of Axial Flow Machines.*** *Axial flow hydraulic turbines are those in which the flow through the runner is aligned with the axis of rotation. This includes turbines with fixed or adjustable blades, fixed or adjustable gates, and with a horizontal, inclined, or vertical axis. This graph shows the comparative performance of axial flow machines with and without adjustable runner blades and wicket gates.*

expansion, again, out to the exit. These changes in shape must be examined carefully in all views to avoid flow separation at the walls and a resulting deterioration in efficiency. The roof at the exit needs to be submerged sufficiently to prevent ventilating the tube owing to turbulence in the tailrace.

Axial Flow Turbines

Axial flow hydraulic turbines are those in which the flow through the runner is aligned with the axis of rotation. This includes turbines with fixed or adjustable runner blades, fixed or adjustable wicket gates, and with a horizontal, inclined, or vertical axis. Arrangements typically used include propeller (fixed blades), Kaplan (adjustable and synchronized blades and gates), bulb units, and S-turbines. See Figure 3-5 for comparative performance of some of these options.

Figures 3-6 through 3-9 show cross-sectional drawings through typical Kaplan, propeller, bulb, and pit units, respectively.

The Kaplan principle applies to all units that have their runner blade angle and wicket gate opening coordinated for extending the range of optimal performance. Although the patent for the Kaplan turbine did not involve head, the Kaplan turbine adjustable runner blades are coordinated automatically with operating adjustable wicket gates to obtain the most efficient operation under varying head, flow, or load conditions. This principal can be used in any configuration.

A natural variation of the bulb arrangement, shown in Figure 3-8, is the pit unit shown in Figure 3-9. The generator, while still encased in the water passage,

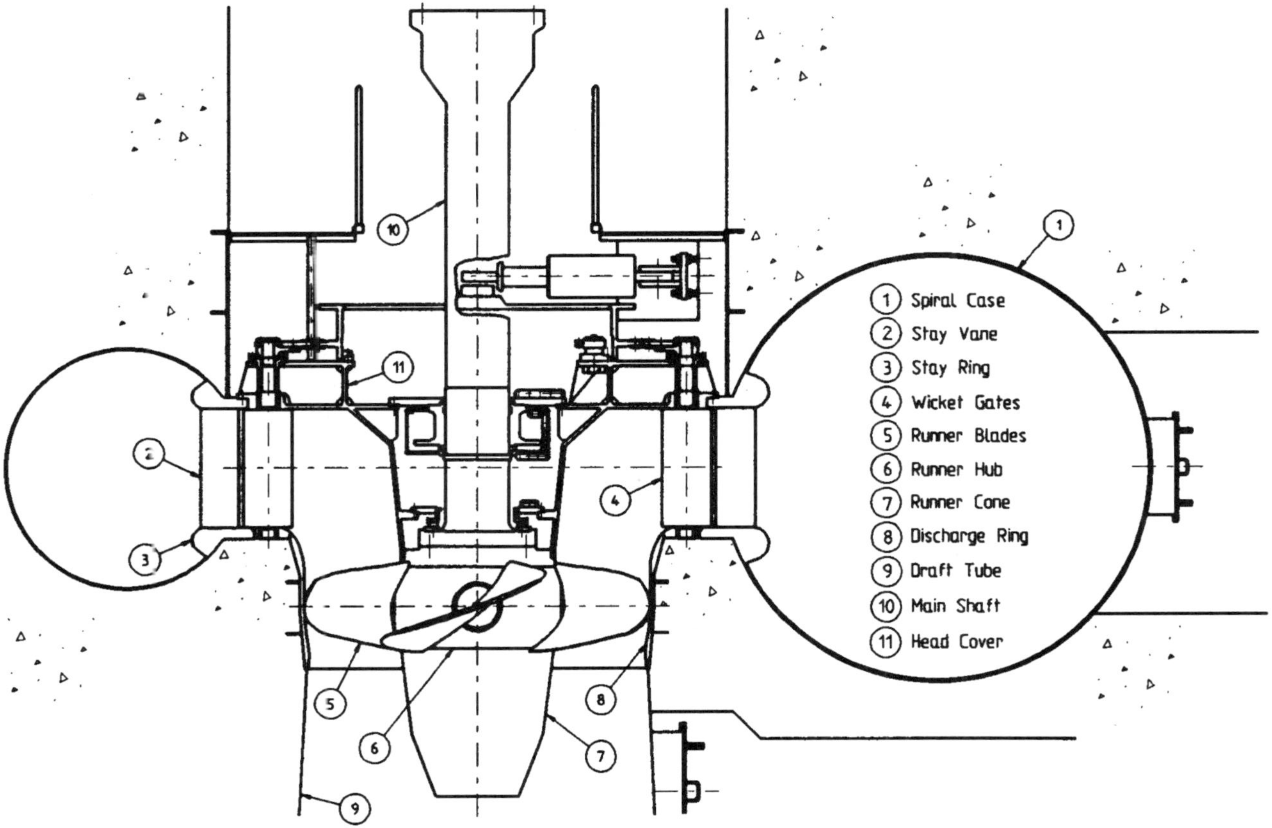

FIGURE 3-6: *Kaplan Turbine.* *This figure shows a cross section of a typical Kaplan turbine. A Kaplan unit has adjustable runner blades, which are coordinated with adjustable wicket gates to obtain the most efficient operating point under the given head, flow, or load conditions.*

Chapter 3

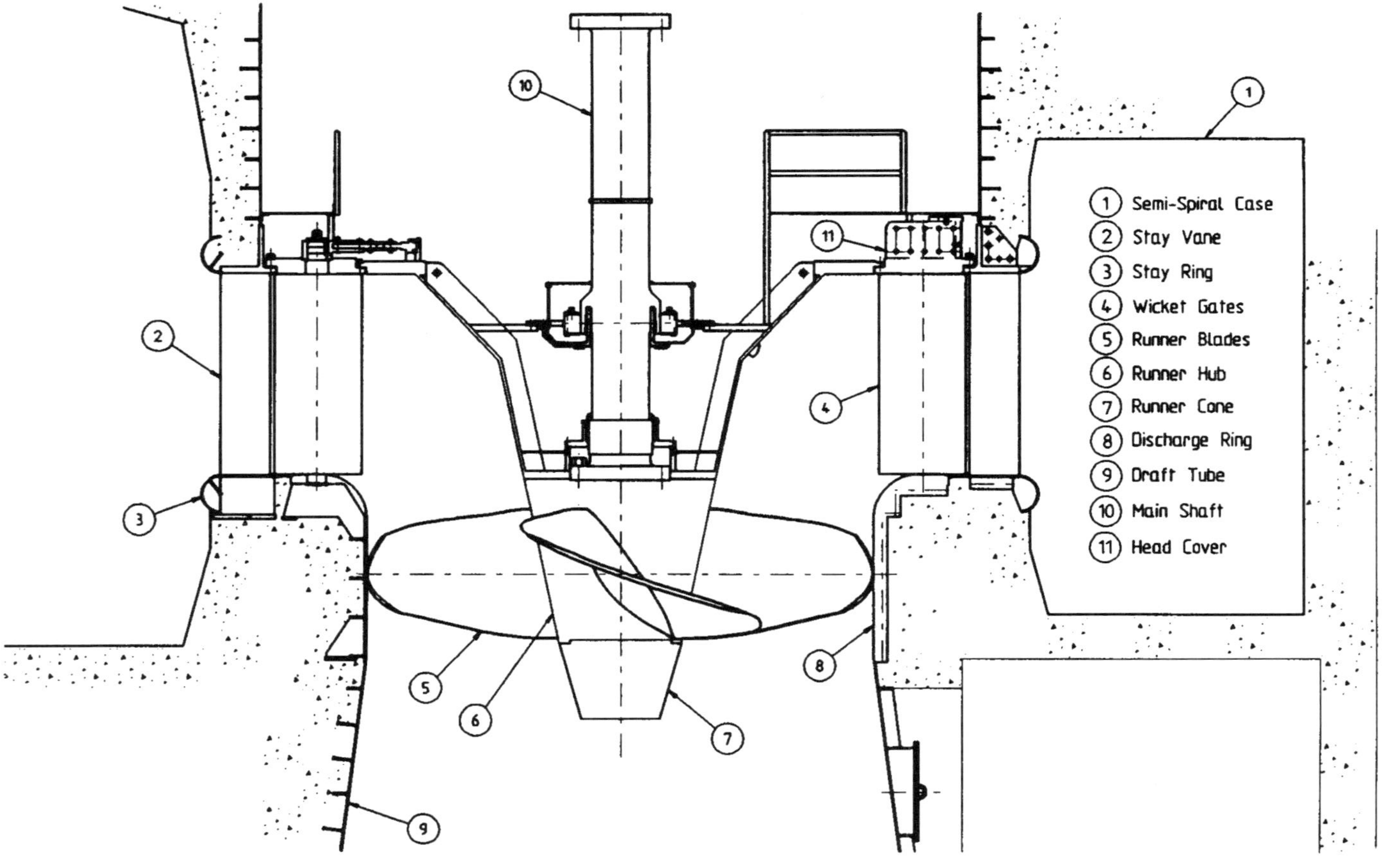

FIGURE 3-7: *Propeller Turbine.* *This figure shows a cross section of a typical propeller unit. A propeller unit has fixed runner blades. The wicket gates may be adjustable or fixed. Typical of this type of unit is a cylindrical discharge ring and no blade trunnion.*

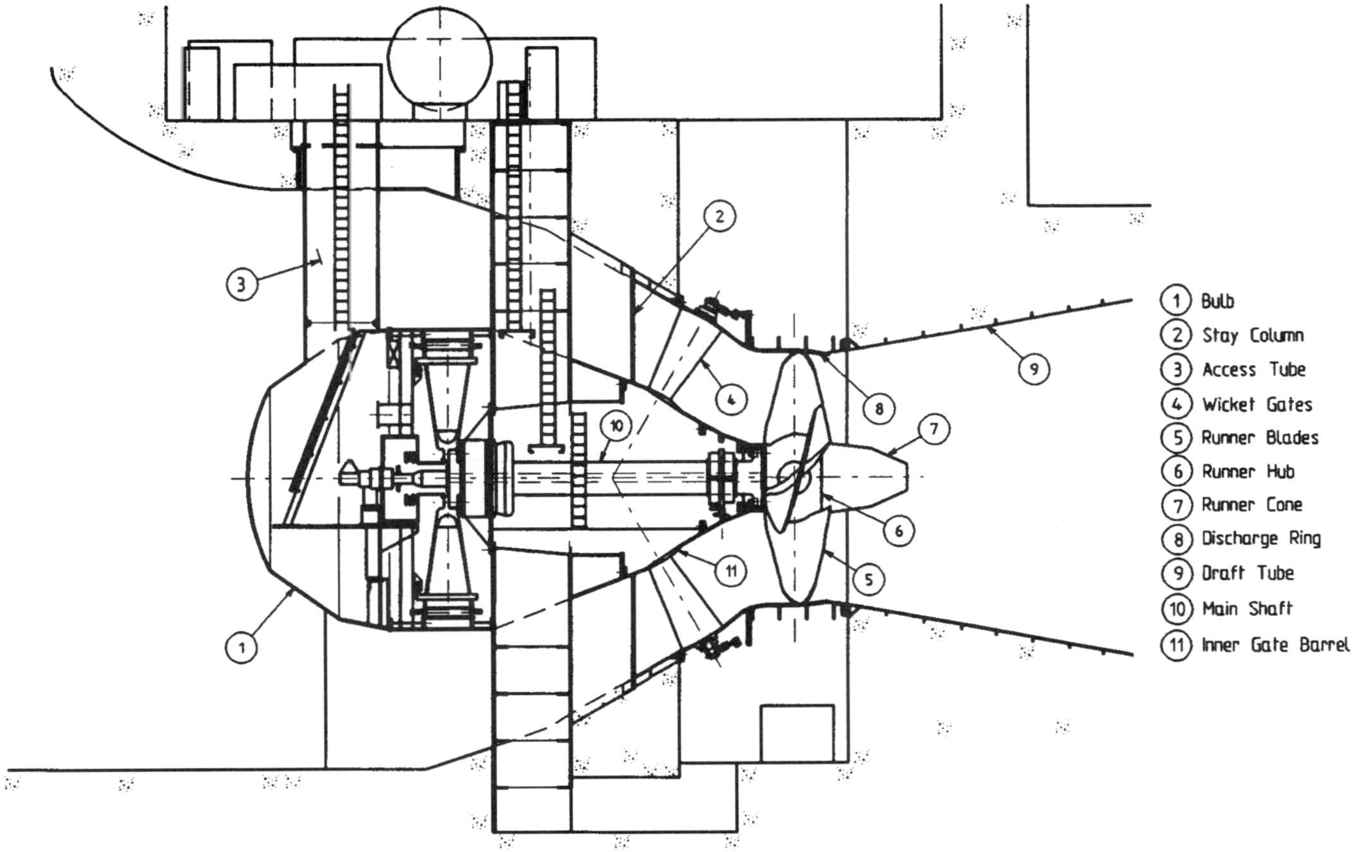

FIGURE 3-8: *Bulb Unit.* *This figure is a cross-sectional drawing of a bulb unit. A bulb unit is a horizontal unit with a conical-shaped draft tube. Used in low-head applications, it reduces excavation and other civil costs.*

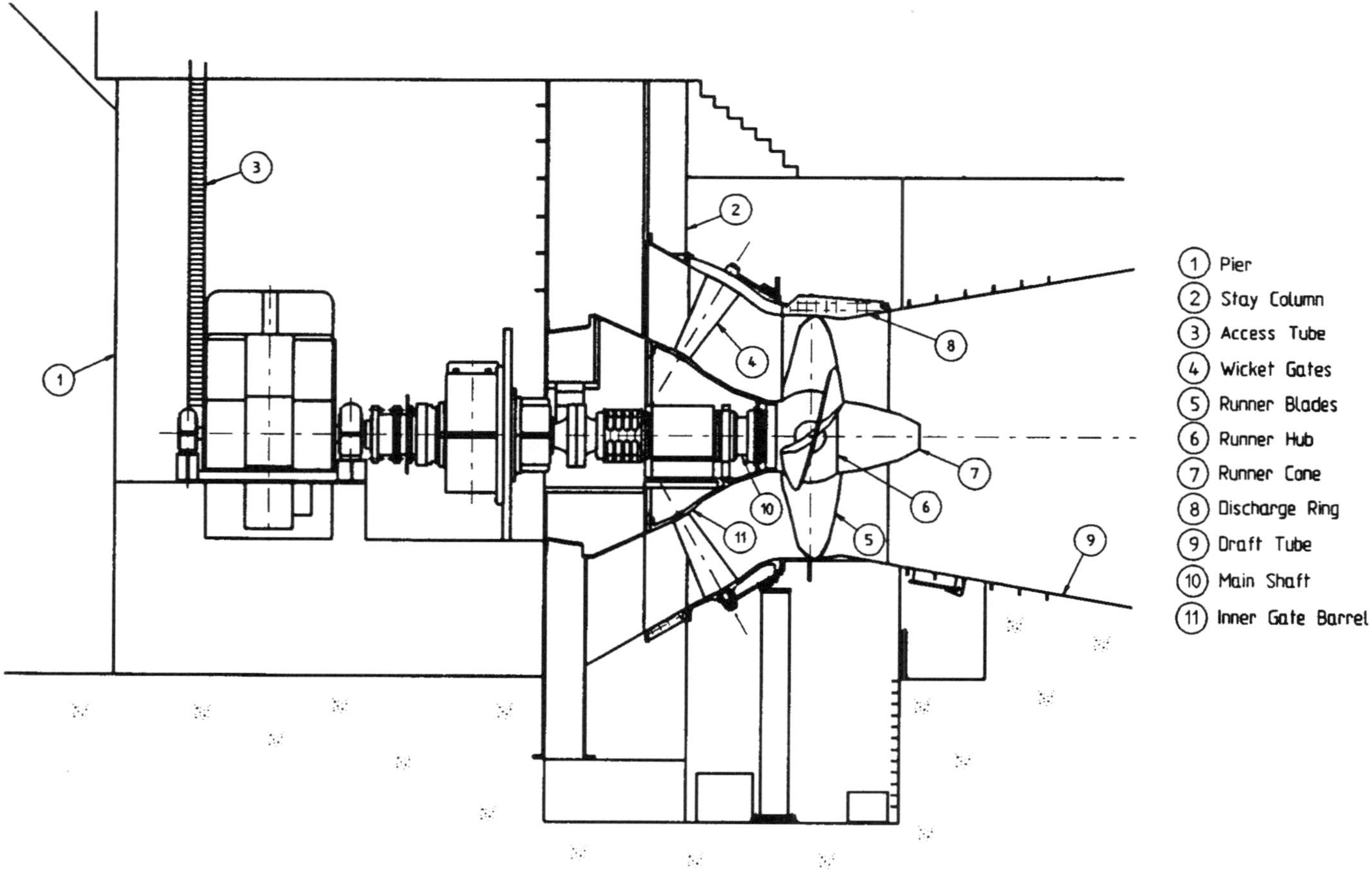

FIGURE 3-9: ***Pit Unit.*** *This figure shows the cross section of a pit unit, which is a natural variation of the bulb arrangement. The generator, while still encased in the water passage, is now housed in a streamlined island that is typically open the full width and length vertically to the upper level of the powerhouse. Most pit units have speed increasers to reduce the size of the generator and enclosing pit.*

is now housed in a streamlined island that is typically open the full width and length vertically to the upper level of the powerhouse. Most pit units have speed increasers to reduce the size of the generator and enclosing pit.

Axial flow hydraulic turbines have been used for net heads up to 75 meters (250 feet) with power output up to 200 MW (270,000 horsepower). However, they are generally used in head applications below 35 meters (115 feet). S-turbines are used below 30 meters (100 feet) head and 8 MW (10,000 horsepower) capacity. Bulb and pit units can be used to about 25 meters (85 feet) head and 100 MW (135,000 horsepower) capacity. Specific mechanical designs, civil construction, and economic factors must be given full consideration when selecting among these three axial flow turbine arrangements.

The simplest and lowest cost arrangement consists of a constant speed, fixed blade runner with fixed gates. Such a unit has one point of operation at a given head. This unit is well-suited to the multiple-unit sites with relatively small head ranges, where capacity can be adjusted to varying load conditions by controlling the number of units on line. Owing to the complex mechanisms used to control and coordinate the blade and gate movement, the Kaplan turbine is the most costly of the group.

Horizontal or inclined axial flow units such as S-turbines, bulb units, and pit units generally have higher operating speeds and greater capacities for a given water passage width. This permits the powerhouse structure to be substantially smaller. The need for deep foundations is minimized, reducing excavation and other civil construction costs. The building can be blended into the dam and its surroundings, or it may be located underground or within the dam. Unit spacing may be only two-thirds of the conventional vertical turbine for the same head and turbine output.

At higher head and flow conditions where high power output is needed, large vertical turbines are the most advantageous. This is because the structure and the mechanical support equipment are now relatively less costly than on multiple pit or bulb units. Also, the generator can be designed with sufficient diameter and mass to provide inertia consistent with system stability requirements. Cooling and access requirements for large, high capacity generators are not restricted. Neither of these last two advantages is available to pit or bulb units.

Deriaz Turbines

Paul Deriaz developed an unusual configuration of the Kaplan principle that has found limited use. As shown in Figure 3-10, the runner hub is spherical and the blades are mounted and pivoted about centerlines that are elements of a cone. Likewise, the gates (flaps) that control the incoming flow are mounted conically so that the flow approaches the runner at approximately 45 degrees from the horizontal. This configuration permits the spiral case to be smaller in overall width than a conventional machine while still retaining the same area at each cross-section.

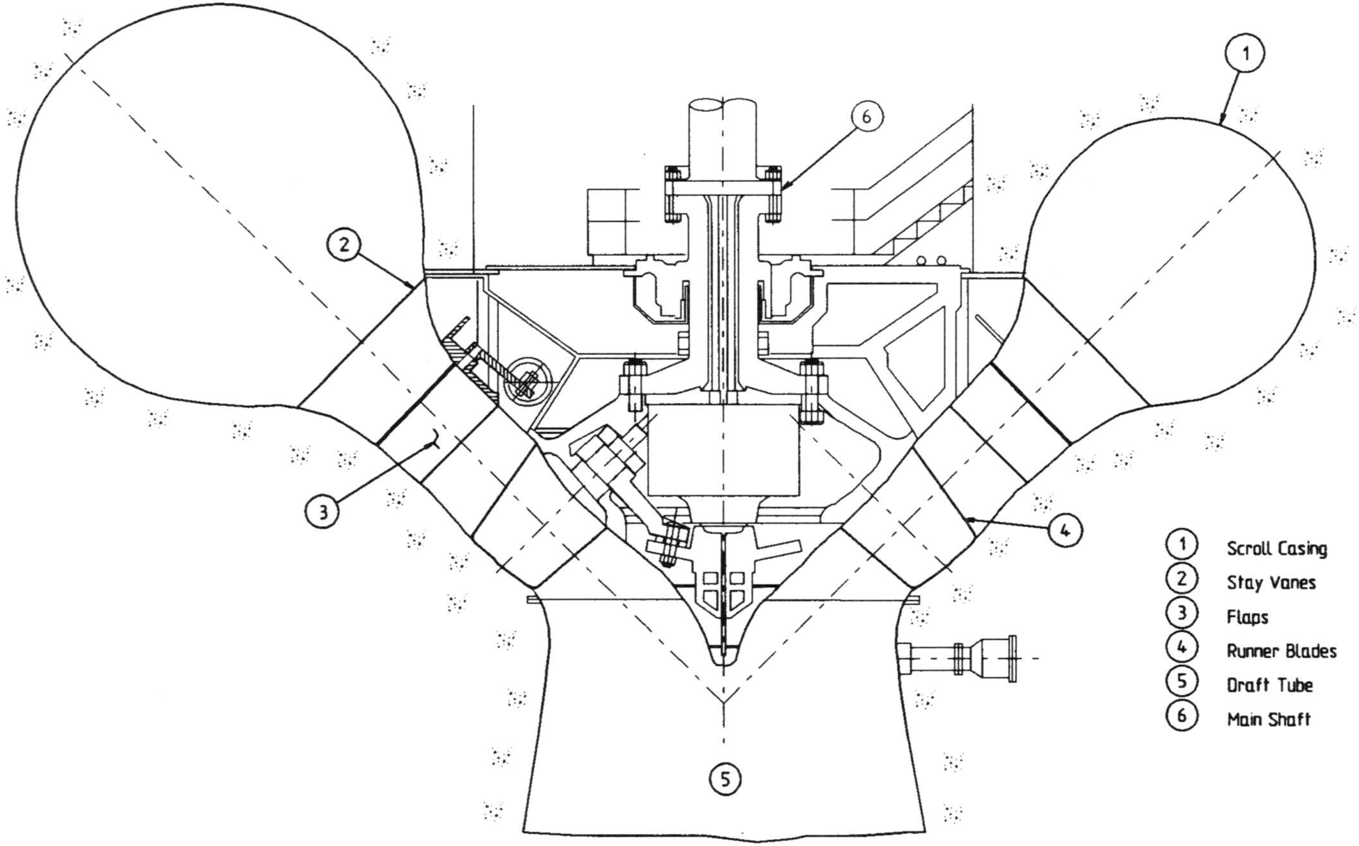

FIGURE 3-10: ***Deriaz Turbine.*** *This figure shows a cross section through a Deriaz unit. This unit is a variation of a typical Kaplan design. Its advantage is that it permits the overall spiral case dimensions to be smaller than a conventional unit.*

Turbines with Rim Generators

The rim generating unit, called the straight flow turbine, works with an infrequently used axial flow arrangement. Leroy Harza, the individual who founded the Harza Engineering Company in 1920, originally patented this turbine.

This arrangement includes a generator rotor surrounding and attached to the runner blade periphery and sealed from the water passageway on the upstream and downstream ends. The stator is located around the rotor, also concentric with the runner. Because the runner and generator are concentric, the powerhouse can be very short. Also, the draft tube can be a conventional horizontal cone.

Mixed Flow Turbines

Mixed flow turbines are those in which the flow has both radial and axial components in its passage through the runner. Although several configurations of this turbine were developed originally, the most common configuration is the Francis turbine. It exists in large numbers throughout the world. It is applied at head ranges generally from about 15 to 750 meters (50 to 2,500 feet) and in power ranges from about 0.25 to 820 MW (335 to 1,100,000 horsepower) per unit. There are also numerous small units at very low heads, for example, 1.5 meters (5 feet), and output, for example, 2 kW (3 hp), but these are not typical of the current practice.

There is a general relationship between the proportions of a Francis runner and the value of its specific speed. That is, the numerical value of specific speed characterizes the general design appearance of the runner. This relationship is indicated in Figure 3-11. While only three specific speed values are shown, it is to be understood that the profile changes are gradual and continuous with changes in specific speed.

Francis turbines have been arranged in a variety of configurations to accommodate local site conditions. Early applications generally involved small-sized runners and, therefore, required multiple runners if a high power output was needed. These were horizontal, open flume units with commonly two, four, or more runners per shaft. A few sites have used two runners in a vertical setting. When higher head sites became achievable, practicality led to replacing the open flume with a cylindrical pressure case and penstock. These directed the water from the forebay to the turbine and finally into the tailrace. Ultimately, the spiral case was developed, resulting in more efficient delivery of the water into the runner. The vertical shaft, spiral case, and elbow draft-tube style of Francis turbine is the predominant arrangement used today as seen in Figures 3-3 and 3-12.

Reversible Pump-Turbines

A popular application of turbine and pump technology is the reversible pump-turbine. The reversible unit achieves a double benefit for its owner by using low-cost, off-peak output typically from thermal power stations to drive the unit in

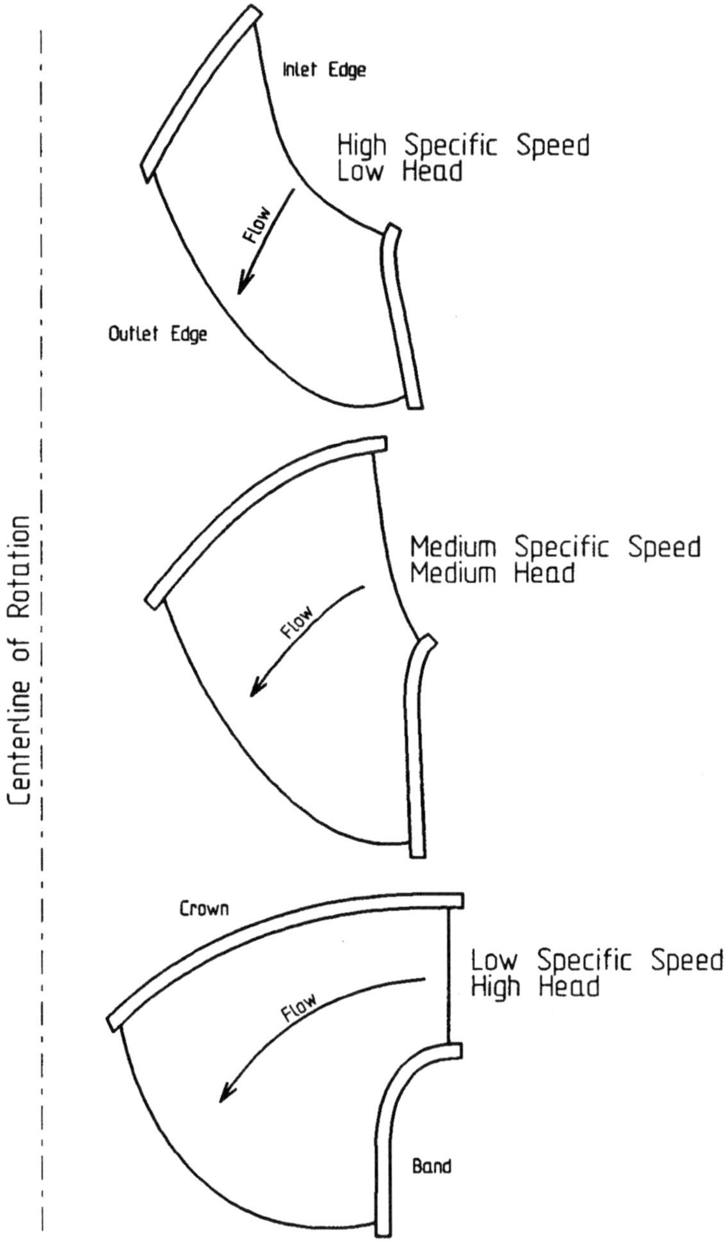

FIGURE 3-11: ***Francis Turbine Types versus Head.*** *This graph illustrates the general relationship between the runner proportions and the value of its specific speed. That is, the numerical value of specific speed characterizes the general design appearance of the runner. While only three specific speed values are shown, the profile changes are gradual and continuous with changes in specific speed.*

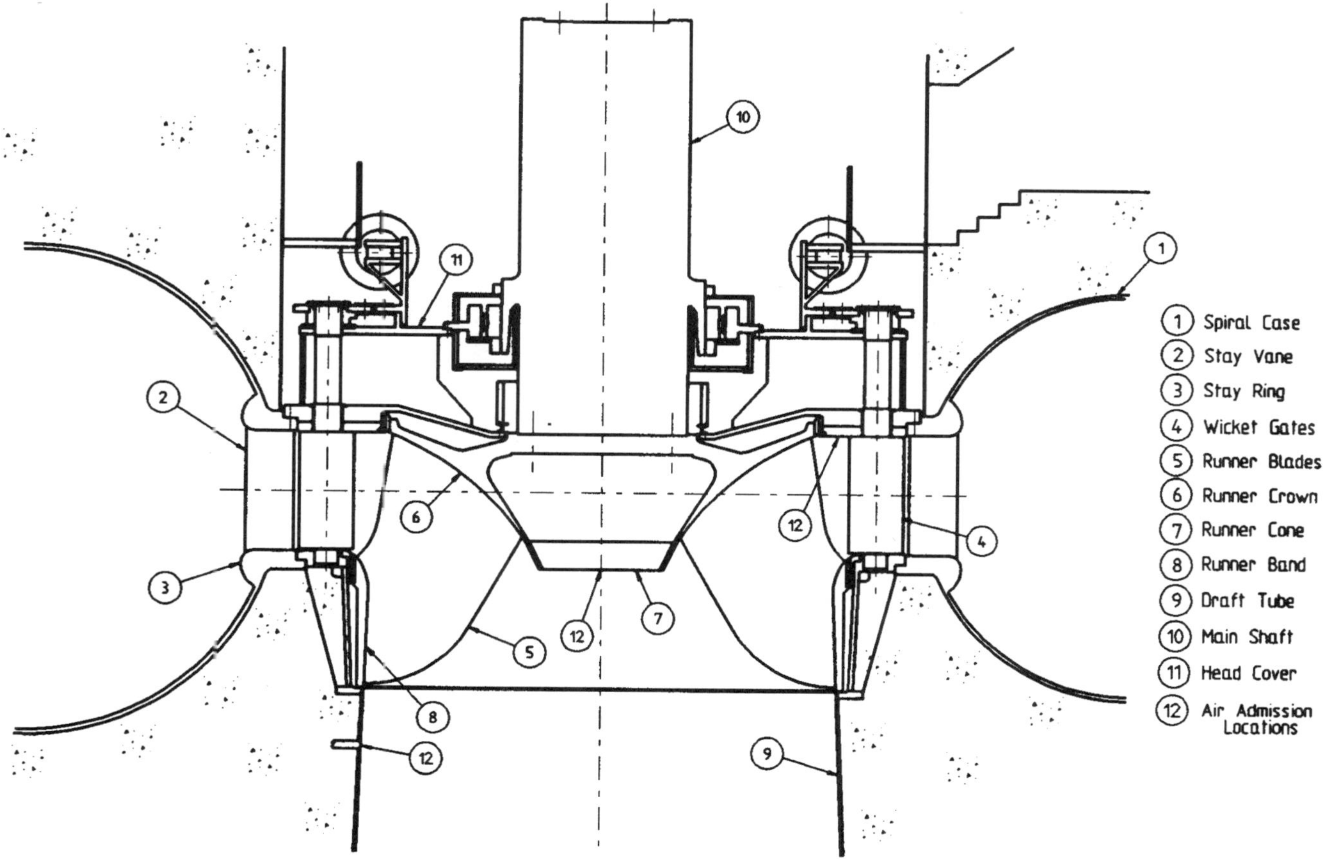

FIGURE 3-12: ***Francis Turbine.*** *This figure shows a cross section through a Francis turbine, the most common mixed flow configuration.*

This is an example of a medium specific speed Francis runner.

reverse as a pump. The quantity of water returned to the upper reservoir can be used to generate on-peak power that is marketed at a higher price. By using off-peak power to drive the pump cycle, thermal base load units can maintain higher efficiency levels at almost constant output level and, thereby, delay costly outages.

Because the pump-turbine unit must develop optimal performance for both modes of operation, its design is a compromise for both modes as seen in Figure 3-13. The stay vanes and wicket gates appear somewhat like a pump diffuser. The number of runner buckets is reduced to approximately half the number used when only generating. But, these fewer buckets will be much longer in the direction of flow. Pump-turbine runner proportions more closely resemble a pump than a turbine.

With a common rotational speed in both directions, the pump and turbine cycles inherently have their optimal efficiencies occur at somewhat different values of head. The selection of design criteria must then consider the requirements of both cycles. The project owner needs a detailed analysis of the future power market coupled with the best estimate of the project's hydrological future and the present value of equipment cost to choose between the two options. Recently, a few pump-turbine installations are being constructed with adjustable speed generator-motors that can overcome this need to operate at constant speed and thus enhance operation in both the pump and turbine cycles.

The method for starting the pump-turbine in the pump mode has a significant effect on the initial cost of the starting equipment and a lifelong effect on system operations. There are many methods available to the project planner for starting as a pump and bringing it to synchronizing speed. These methods include full voltage, reduced voltage, variable frequency, part winding, reactor, starting motor, synchronous, and turbine starting. Each of these has its advantages and disadvantages, its costs and rewards, the discussion of which is beyond the scope of this chapter. Water in the runner chamber is usually depressed with compressed air prior to a pump start to reduce motor starting power requirements. (Chapter 9, Auxiliary Systems, describes this depression procedure in more detail.)

Impulse Turbines

Pelton

The runner of a Pelton turbine normally operates in air at near atmospheric pressure with one to six jets of water impinging tangentially on the runner. The jet(s) are cut into successive slugs of water as the buckets pass through the tangential jet(s). The bucket splitter divides each slug of water so that the discharge from each bucket half is symmetrical with, virtually, no axial thrust. At any instant the water from this jet is acting on several buckets simultaneously. Figures 3-14 and 3-15 show a typical Pelton turbine.

Pelton turbines can be arranged in a number of configurations. Typically, lower specific speed horizontal shaft units have one or two runners with one or two jets per runner. Typically, the higher specific speed vertical shaft units have

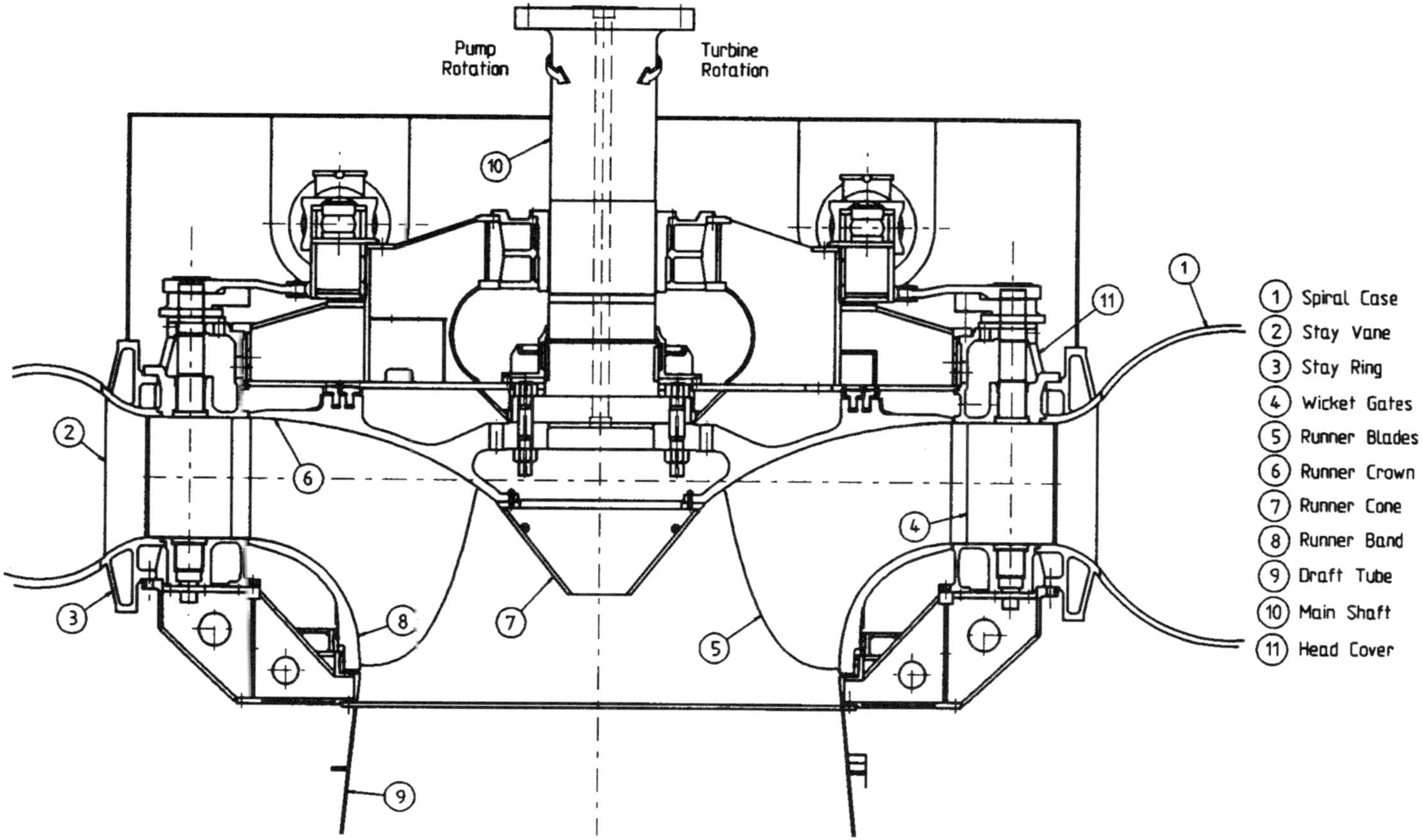

FIGURE 3-13: *Reversible Pump-Turbine.* *This figure shows a reversible pump-turbine. Its configuration is similar to a Francis unit, except the stay vanes and wicket gates appear somewhat like a pump diffuser. There are also fewer, and longer, runner buckets than on a Francis runner.*

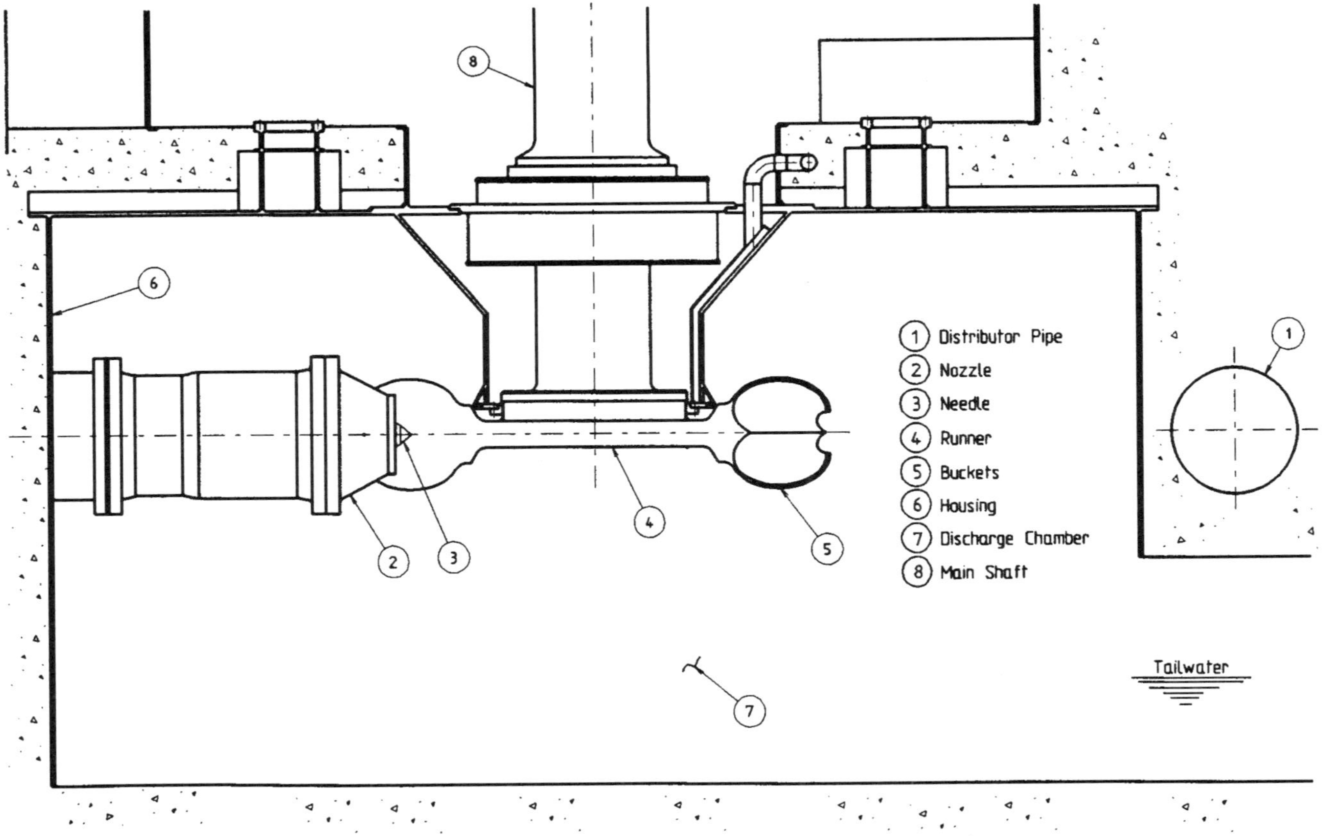

FIGURE 3-14: *Pelton Turbine—Elevation View.* *This figure shows a cross section through a vertical shaft, multi-jet Pelton unit. Unlike a reaction turbine, which is fully submerged in water, a Pelton turbine runner operates in air.*

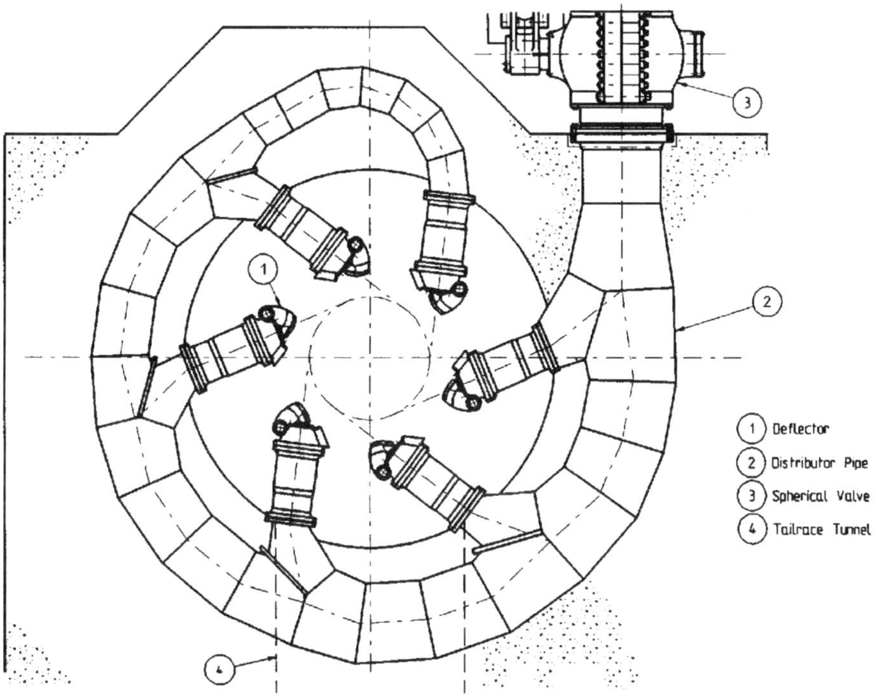

FIGURE 3-15: ***Pelton Turbine—Plan View.*** *This figure shows the plan view of a vertical shaft Pelton unit. These high specific speed vertical shaft Pelton units typically have three, four, five, or six jets.*

single runners with three, four, five, or six jets. Also, the higher the specific speed, the fewer the number of buckets, and the smaller the runner diameter.

Comparison of model tests with field tests indicates some part load increase in efficiency with size. The International Electrotechnical Commission (IEC) Test Code No. 193 allows no scaling up in efficiency for Pelton turbines.

Vertical shaft multi jet turbines are generally selected for large flow installations, whereas horizontal shaft turbines are suitable for those applications that have less water available.

Multi-jet turbines are slightly more efficient than single jet turbines; however, the more rapid accumulation of stress cycle alternations justify a more conservative runner design. Abrasive material entrained in the water will erode the buckets of a multi-jet turbine more rapidly than in the case of a single jet per runner.

For the same rated head and flow conditions, increasing the number of jets results in a smaller runner and a higher operating speed. Therefore, whether vertical or horizontal, multi-jet turbines tend to be less costly for comparable outputs because the cost of the runner represents up to 20 percent of the cost of the entire turbine.

Distributor Features. Regardless of the configuration, it is imperative that the inlet water passage be designed for minimum losses. The turbine shut-off valve often is a spherical valve, rather than a butterfly valve, to avoid the obstruction in the waterway and to economically accommodate the higher pressure. The shapes of the needle and nozzle are designed for rapid acceleration at the exit end and for assuring a uniform jet shape at all openings. The needle-nozzle assembly is placed as close to the wheel as possible to avoid dispersal of the jet due to air friction.

A deflector is normally used to cut into the jet when rapid power reductions are required such as a complete loss of connected-load. The deflector is mounted close to the runner on the nozzle assembly and typically is provided with its own servomotor.

Bucket. While the bucket's shape is approximately ellipsoidal and symmetrical about the splitter, it is not symmetrical inside of and outside of the pitch diameter. A cut-out is made at the outermost bucket edge to give the jet free passage through the incoming bucket. This is so it can enter the previous bucket as close to the pitch diameter as possible. The manufacturing tolerances on shape, finish, and waviness of the cut-out edge are very tight. Properly contoured buckets do not pit from cavitation. Abrasive material in the water can alter the bucket shape eventually, which can lead to cavitation and pitting.

Each runner bucket is subjected to both bending and torsional stresses as each jet strikes it. Additionally, there is a tensile stress due to centrifugal force. Owing to the nature of the runner shape, the critical stress zone occurs where the bucket attaches to the disk. It is of utmost importance that the entire casting, but especially this region, be free of defects. Experience has shown that runners will have fatigue failure from very low alternating stresses when a defect is present in a critical zone.

Integral cast runners are stronger than runners built with buckets bolted to the disk because the bolt holes reduce the resisting section in a bolted-on-bucket runner. The casting of a sound runner is difficult because when the metal contracts during cooling, it tends to tear the metal between buckets at the roots. A fracture analysis of the design (to establish acceptable flaw sizes) and extensive inspection procedures are recommended to ensure that a runner casting is free of dangerous flaws. A material widely used for Pelton runners is a martensitic stainless steel with major alloys being 13-4 (13 percent chrome, 4 percent nickel steel). This material has high strength, good corrosion fatigue properties, and good cavitation resistance. Also, with proper precautions, it is repair weldable. However, other materials such as aluminum bronze or 18-8 steel may be more appropriate, depending on size and application.

Frequent inspections of the runner are recommended to detect any potential failures that may have revealed themselves since the last inspection. The ease of access and repair make this preventive maintenance highly cost-effective.

Housing. The shape and dimensions of the runner housing have an important influence on performance. The water discharging from Pelton buckets has some residual velocity that carries entrained air into the water in the wheel pit. The resultant air-water mixture in the wheel pit has a lower specific gravity than the water in the tailrace from which the air bubbles have escaped. This causes a buildup of the air-water mixture and resultant higher water surface in the wheel pit. It is, therefore, essential to allow air to enter the wheel pit above the discharging water by way of the tailrace. Preferably, additional air should be admitted near the shaft at the center of the wheel pit. The runner rotating in air acts as a fan with low pressure in the center and higher pressure at the periphery.

Turgo Impulse Turbines

Another type of impulse turbine is the Turgo impulse. This turbine is higher in specific speed than the typical impulse turbine. Eric Crewdson originally patented this turbine in 1920. The difference between a Pelton unit and a Turgo is that, on a Turgo unit, the jet enters one side of the runner and exits the other side. The Turgo unit operates at a higher specific speed, which means for the same runner diameter as a Pelton runner, the rotational speed can be higher. (See Figure 3-16.) The application head range for a Turgo unit is 15 meters (50 feet) to 300 meters (1,000 feet). Turgo units have been used for applications up to 7,500 kW.

Several of the following sections of this chapter will apply to reaction turbines more than to other types.

SIMILITUDE

Laws of Similitude

Fluid flow theory alone does not provide an easy understanding of hydraulic turbine and pump behavior when changing parameters such as geometry and fluid viscosity. Experimental investigation can find definitive relationships, but the combinations and variations to be tested are nearly limitless. In the early 1900s, Buckingham (Buckingham's Pi-theorem) applied the principles of dimensional analysis to the problem, thereby reducing the number of test parameters to a manageable few.

The results of this work, known as the laws of similitude, define geometric, kinematic, and dynamic similarity. Geometric similarity requires a common ratio of all corresponding dimensions. Kinematic similarity requires a common ratio of all corresponding velocities. Dynamic similarity requires a common ratio of all corresponding forces (for example, friction, gravity, and surface tension). Those relationships most useful to the hydraulic turbine and pump industry are outlined in the following equations.

Equation 3-3, Q-N-D:

$$\frac{Q_1}{N_1 D_1^3} = \frac{Q_2}{N_2 D_2^3}$$

Equation 3-4, P-N-D:

$$\frac{P_1}{N_1^3 D_1^5 \rho_1} = \frac{P_2}{N_2^3 D_2^5 \rho_2}$$

Equation 3-5, D-N-H:

$$\frac{D_1 N_1}{(g_1 H_1)^{0.5}} = \frac{D_2 N_2}{(g_2 H_2)^{0.5}}$$

Equation 3-6, Q-H-D:

$$\frac{Q_1}{D_1^2 (g_1 H_1)^{0.5}} = \frac{Q_2}{D_2^2 (g_2 H_2)^{0.5}}$$

Equation 3-7, P-H-D:

$$\frac{P_1}{(g_1 H_1)^{1.5} D_1^2 \rho_1} = \frac{P_2}{(g_2 H_2)^{1.5} D_2^2 \rho_2}$$

where:
Q=flow rate, cubic meters/second (cubic feet/second);
N=speed of rotation, revolutions/minute, also written as rpm;
D=diameter, meters (feet);
P=power, kilowatts (hp);
ρ=water density, kilograms/cubic meters (slug/cubic feet);
g=local gravitational constant, meters/second squared (feet/second squared);
H=head, meters (feet);
subscript 1=machine 1 or condition 1; and
subscript 2=machine 2 or condition 2.

The same units must be used on both sides of equations. The expressions for specific speed used in earlier sections of this chapter are developed combinations of several of these laws. Many other useful dimensionless groups have been developed from the basic laws. They include the Weber number, Froude number, Reynolds number, Mach number, and Cauchy number.

The IEC code gives tolerances for geometric similarity between a model and prototype.

Presentation of Model Data. Most turbine model data, unless otherwise requested, are presented using the last three relationships, as shown in Equations 3-5, 3-6, and 3-7. The relationship of size (D), speed (N), and head (H) is defined by N_{11}, which is proportional to the inverse of the square root of the head. N_{11} is defined as the speed at which the machine would rotate if the runner

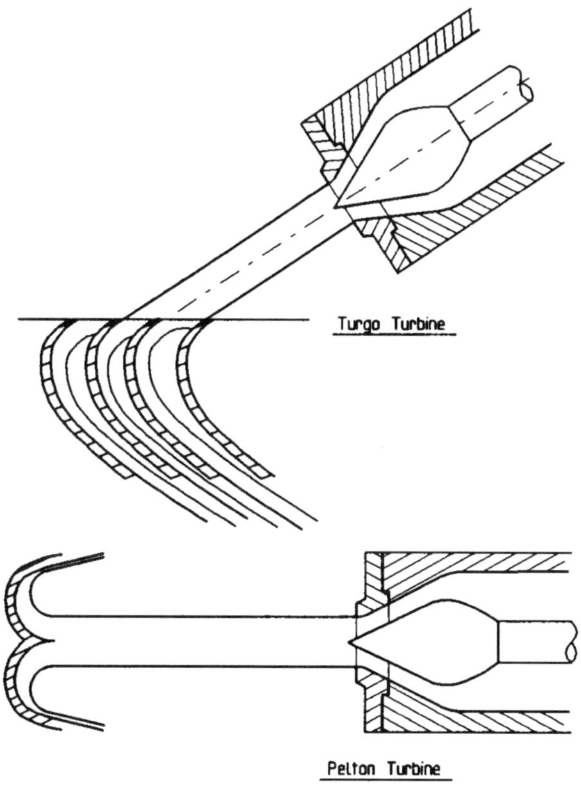

FIGURE 3-16: *Comparison between Turgo and Pelton Impulse Units.* *This figure shows the path of a jet as it strikes a Turgo bucket as compared to a Pelton bucket.*
(Courtesy Gilbert Gilkes & Gordon Ltd.)

Chapter 3

reference diameter was 1 meter and the head was 1 meter (or 1 foot diameter and 1 foot head). An example that uses N_{11} to describe a model test point in metric units follows in Equations 3-8 through 3-12.

EQUATION 3-8 (DERIVED FROM EQUATION 3-5):

$$\frac{DN}{H^{0.5}} = \text{Constant}$$

where:
D=actual model diameter, 0.49403 meter;
N=actual model rpm, 1367.9 rpm; and
H=actual model net head, 61.139 meters.

Equation 3-8 assumes that g is the specific local value and is constant on both sides of the equation.

Inserting the values gives Equation 3-9.

EQUATION 3-9:

$$\frac{DN}{H^{0.5}} = \frac{0.49403 \times 1367.9}{61.139^{0.5}} = \frac{1N_{11}}{1^{0.5}}$$

From this equation, the value of N_{11} is found in Equation 10.

EQUATION 3-10:

$$N_{11} = 86.427$$

Similar coefficients define the flow rate and output as in Equations 3-11 and 3-12.

EQUATION 3-11:

$$Q_{11} = Q/(H^{0.5}D^2)$$

EQUATION 3-12:

$$kW_{11} = kW/(H^{1.5}D^2)$$

where:
Q, H, D, and kW are all raw model data.

One can use dimensionless parameters to present model data. This is less frequently used, but only differs from the above relationships by constants.

Field Testing

Another practical use of these laws can be applied during field testing. If the contract stipulates performance guarantees at a certain net head, but that net head is not attainable at the time the unit is tested, using the P-H-D and Q-H-D laws can determine the field measured output and discharge for the guaranteed head conditions. Limitations for using these equations are in the ASME and IEC test codes.

Specific Speed

A characteristic number, called specific speed, classifies hydraulic turbines. There are six or more different mathematical expressions for specific speed. Two of the commonly used ones are the metric (flow-based) system and the English (power-based) system, as shown in Equations 3-13 and 3-14.

EQUATION 3-13:

$$N_q = \frac{N(Q)^{0.5}}{H^{0.75}} \qquad \text{metric system (flow based)}$$

where:

N_q=specific speed;
N=speed, rpm;
Q=flow, cubic meters/second; and
H=net head, meters.

EQUATION 3-14:

$$N_s = \frac{N(P)^{0.5}}{H^{1.25}} \qquad \text{English system (power based)}$$

where:

N_s=specific speed
N=speed, rpm;
P=power, horsepower; and
H=net head, feet.

The specific speed value defines the approximate range of application for a turbine type. It also provides a means of comparing all types of turbines on the same basis of head and capacity. Sometimes pump-turbines will be referenced according to a flow-based specific speed with flow in gallons per minute.

There are several factors that influence the selection of specific speed for a given application. They are: the head available to the turbine; the desired output (power and annual energy); the rotational speed; unit setting; and the number of units. The consequences of selecting specific speed are efficiency, unit setting, runaway speed, and machine size, weight, and price. The Selection Process section gives more details of these consequences.

Model to Prototype Performance

The performance characteristics of a full-scale turbine can be predicted from the results of tests performed on a homologous scale model in the laboratory. Homologous turbines are geometrically similar turbines that have similar velocity vector diagrams and also have geometrically similar streamlines. From a series of such tests, manufacturers develop a line of turbines that cover the full range of specific speed in increments of design variations. These incremental design variations are chosen to provide an overlapping of performance so that nearly all combinations of head and flow (or power) required by an owner's specification can be accommodated exactly or with minimal deviation from homology.

Initially, the rated net head and planned relative elevations of runner and tailwater help the manufacturer to select the specific speed and turbine type. This selection leads to one or several of applicable tested models.

Since the full-scale turbine will be designed to be homologous to the tested model, the laws of similitude are applicable.

Using the Q-H-D and D-N-H similitude equations developed earlier, Equation 3-15 shows the following.

EQUATION 3-15, Q-H-D:

$$\frac{Q_m}{H_m^{0.5} D_m^{\;2}} = \frac{Q_{11}}{1^{0.5} 1^2} = \frac{Q_p}{H_p^{0.5} D_p^{\;2}}$$

where:

Q=flow;

m=model;

subscript 11 means H=1 meter, D=1 meter; and

p=prototype.

Therefore, Equation 3-16 follows.

EQUATION 3-16:

$$D_p = \left[\frac{Q_p}{H_p^{0.5} Q^1_1} \right]^{0.5}$$

Using the same logic for D-N-H and assuming $g_m = g_p$, Equation 3-17 is as follows.

EQUATION 3-17:

$$\frac{1N_{11}}{1^{0.5}} = \frac{D_p N_p}{H_p^{\;0.5}} \quad or \quad N_p = \frac{N_{11} H_p^{\;0.5}}{D_p}$$

In this example the metric system is used. The English system also could be used.

Scale Effects

From very early in the development of reaction turbines, it has been known that leakage and friction losses are not directly scalable. That is, they are relatively smaller on the full-scale prototype than on the tested model. As progressively larger size ratios came into being, it became apparent that the differences in these losses needed to be quantified so that a more realistic prototype efficiency would be shown.

Many researchers and scientists have developed formulae for scaling, based primarily on the theory for pipe losses. However, testing instruments and methods were not sufficiently accurate to confirm or revise these formulae. Eventually, as many as 17 different formulae were published, but only two of these, Ackeret's and Moody's, found popular use. Ackeret's work was based on comparative Reynolds numbers, while Moody's work was based on the physical scaling of runner diameter. In 1953, S.P. Hutton proposed a modest revision to Ackeret's formula as it is applied to propeller and Kaplan turbines. Until 1991, IEC publications recommended using the Hutton formula. More recent IEC publications recommend using scalable losses based on Reynolds number as in Ackeret's formula for calculating the increase of hydraulic efficiency. Owner specifications normally require efficiency guarantees to be based on a model test and then stepped-up to the full-size turbine using one of these formulae or a fraction of the value derived from one of these formulae. In addition to specifying what formula to use, it is customary to stipulate the method by which the correction is to be applied.

The three principal methods are:

1. Calculate prototype efficiency using the optimal model efficiency point, and then use the ratio of these two efficiencies as a multiplier to step-up all other model efficiencies;
2. Calculate the prototype efficiency as in Method 1, but use the difference between model and prototype efficiency as an adder to step-up all other model efficiencies; and
3. Calculate each model efficiency point separately to determine the equivalent prototype efficiency.

Methods 1 and 2 are the most common. The third method tends to overestimate efficiencies away from the optimum.

It is evident from the definition of efficiency that this scaling of losses, which results in increased prototype efficiency, must also acknowledge either a proportionate increase in output or reduction in flow quantity, or both. Definitive verification of which of these occurs has not been fully tested, and manufacturers choose their own method for plotting prototype performance. It is essential that the purchaser specify, in detail, the method to be used to ensure a true comparison of manufacturers' data.

While the prior discussion has pertained to reaction turbines only, formulae have been developed for scale effects of Pelton turbines by combining the effects of Reynolds number, Froude number, and Weber number criteria. The results of such calculations show small efficiency increases at best and, sometimes, even show a negative effect. Field test results also have not consistently shown a step-up in efficiency. The IEC code acknowledges this experience, and recommends against using scale effect to predict increased Pelton prototype efficiency.

Selection Process

The first step in this process is the determination of site conditions and the anticipated operation scheme. If annual energy is to be used as an evaluation or a selection criterion, then specific site information needs to be well-documented in the specification (see the following section on Annual Energy). If annual energy is not going to be used in the evaluation, a description of the net head range and power ranges is required. The percent of time spent at each of the heads also should be specified for the bidder. If the setting of the plant is not fixed, the cost of excavation and preferred setting should be given. Water temperature and quality, headwater elevation, and tailwater elevation ranges also should be specified. Performance evaluation formulas and powerhouse limitations, such as spiral case width, should be provided. This information needs to be well-defined to avoid leaving too much to be assumed by the bidder. Otherwise, the purchaser may not get correctly optimized solutions, and the purchaser may have difficulty in selecting the best solution.

To optimize performance of the plant, the experience and recommendations from potential equipment suppliers should be used in selecting the following items during the economic evaluation of the plant: type, size, number of units, speed, and setting. The specification of too many of these parameters limits the manufacturer's flexibility to optimize performance.

Satisfactory intake conditions are essential, especially for low-head plants. Also, hydraulic model testing of the intake is often advantageous on low-head units. The turbine designer must assume a relatively smooth velocity distribution for the full flow range entering the unit. It is the owner's responsibility, in cooperation with the plant's designer, to have good intake flow distributions.

Turbine Characteristics

Prototype tests, model tests, or computer analysis coupled with test experience determine the turbine performance characteristics. Efficiency and flow as a function of head and operating speed are typically summarized on a turbine performance hill curve, similar to the one shown in Figure 3-17. The rotational speed of the turbine is constant across the head range. Within the performance range of the turbine, limits of operation exist. Low-head operation is limited by noise, vibration, and potential cavitation damage caused by the fixed cavitation limits, while, at the high head, the limitation is mainly pitting-related or the maximum power of the turbine. Full gate opening, the generator limit, cavitation, or overall surging normally limit operating at high loads. The cavitation-based power limit is a function of the turbine setting with respect to tailwater and flow through the turbine. At this limit, the turbine is operating at the highest flow possible before cavitation begins to adversely affect performance on the machine. Some pitting damage could occur at this condition. Operating at higher flows will result in a reduction in operating performance and/or in damage to the turbine components because of cavitation pitting.

Another limit at high load is the maximum power, or saturation, limit. This limit occurs where, at a given head coefficient, a value of power is reached that cannot be exceeded. This value is typically designated as P_{max}. Operating the tur-

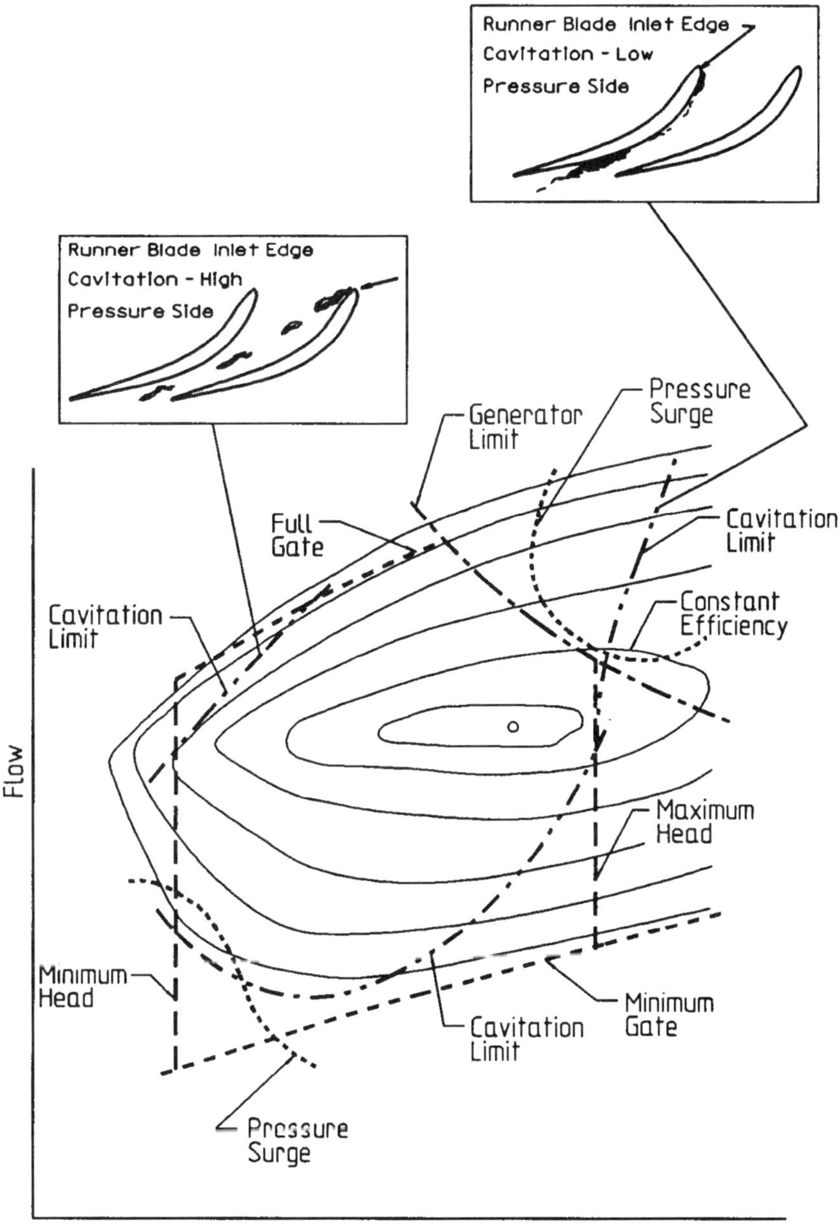

FIGURE 3-17: ***Prototype Turbine Hill Curve.*** *Efficiency as a function of head and flow typically are summarized on a hill curve, similar to the one shown in this figure. Operation is limited by pressure pulsations, potential cavitation damage, and maximum power.*

bine at higher flows than P_{max} is possible but is not practical because of reduced efficiency and because less power is generated than at P_{max}.

At low outputs, part-load pressure pulsations and the potential for cavitation damage caused by part-load vortices typically limit operation. Francis turbines and fixed-blade machines are particularly susceptible to flow-induced vibrations and power swings. During operation at best efficiency, the water leaves the runner with a relatively small rotational component in the direction of the runner rotation. As the flow is decreased from the optimal flow rate, the tangential component of the velocity increases. As a result, a tightly formed, spiral "rope" of cavitation is attached to the runner cone and extends into the draft tube. This large unstable "rope" causes relatively large pressure fluctuations in the draft tube, resulting in an associated change in turbine output, power swings, vibration, and noise.

Whether or not the surging is objectionable to the machine operation depends on the design of the plant. The development of these secondary flows is characteristic of all Francis and fixed-blade turbines and cannot be entirely eliminated by increasing the suction pressure. Although several methods, such as draft tube inserts and admitting compressed or atmospheric air at various locations, have been developed to reduce surge-induced vibrations and power swings, it is best to minimize operation within the area of large pressure fluctuations.

Selection of Machine Type

The specific speed of a machine defines the limits of application for a machine type and forms a basis for maintaining specific design data for each. Table 3-1 can be used as a guideline for ranges of applications for each machine type. While these guidelines define typical ranges, there is overlap between machine types.

Figures 3-2 and 3-18 show the general range of performance parameters versus specific speed and power respectively for each type of turbine. While these figures are typical, manufacturers have their own range of application for their designs that might differ from these figures.

A model test or computerized flow analysis determines the specific speed of a turbine design. Manufacturers rate the specific speed of their machines at the flow (or power), net head, and speed conditions that result in either the peak efficiency point or the rated point (a customer-specified head and output). Most manufacturers use the peak efficiency point.

TABLE 3-1: ***Range of Application for Turbine Types***

	English (Power-Based) System	Metric (Flow-based) System
Pelton Turbines	$N_s < 16$	$N_q < 20$
Francis Turbines	N_s = 16 to 95	N_q = 20 to 120
Axial Flow Turbines	$N_s > 78$	$N_q > 100$

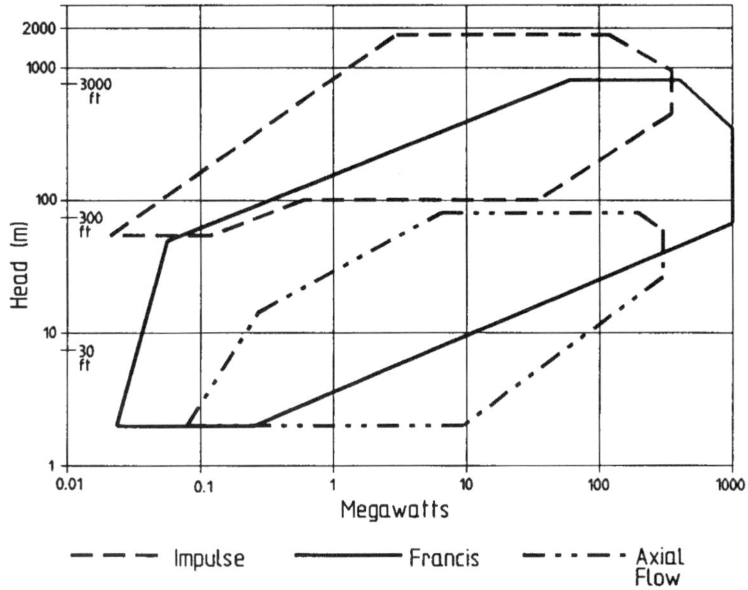

FIGURE 3-18: ***Application Ranges Graph.*** *This figure and Figure 3-2 show the general range of head versus power or specific speed for each type of turbine. While these figures are typical, manufacturers have their own range of application for their designs that might differ from these figures.*

For some sites, it is obvious from the specific speed which type of unit to use. However, for many sites, the specific speed falls in an area where an axial flow/ Francis decision or Francis/Pelton decision must be made. Table 3-2 provides some comparisons to assist in determining types.

A comparison of Francis and Kaplan efficiency levels is shown in Figure 3-19. Typical efficiency versus load characteristics for a low specific speed Francis turbine and a six-jet Pelton turbine with the optimal number of jets in service are compared in Figure 3-20.

If a peak load project is intended, the variations in system demand and the energy availability shown by the flow duration curves, shown in Figure 3-21, will indicate, in a general way, whether one large or several smaller turbines will be needed. An analysis of the head and flow capabilities of the project will help to determine the type of turbine to select. Knowledge of performance characteristics of each turbine type also is needed in reaching a decision.

Figure 3-18 shows the choices of turbine types that are available for a given set of net head and power requirements. For example, a potential site evaluated at 46 meters (150 feet) net head and 50 MW (67,000 hp) would be suitable for axial flow or Francis-type runners. Further, more than one of a type or even a mix (Kaplan and propeller) can be considered. The final selection of a turbine type for a given site is based on a wide variety of factors, including economic, environmental, technical, and operational considerations.

TABLE 3-2: *Comparisons of Axial Flow and Francis Turbines and Francis and Pelton Turbines*

Francis/Kaplan Decision

Advantages of Francis turbine include:

- Higher setting for given size diameter;
- Lower overspeed/synchronous speed ratio; and
- Less expensive turbine.

Advantages of a Kaplan turbine include:

- Smooth operation to low flows;
- Higher specific speed and higher rotational speed and, therefore, smaller generator;
- Higher efficiency is realized over a wide range of head and flow conditions. This is the principal advantage of the Kaplan turbine over a propeller or Francis turbine. A graph illustrating this appears in Figure 3-19; and
- May result in a single unit instead of two Francis or propeller units.

Francis/Pelton Decision

Advantages of a Francis turbine include:

- Higher specific speed and higher rotational speed and, therefore, smaller generator;
- Higher peak efficiency; and
- Higher effective head.

Advantages of a Pelton turbine include:

- Excavation cost will be less;
- Often better for applications with erosive water—easier to maintain;
- Better part-load efficiency—very flat efficiency curve for approximately 25 percent to 100 percent load, particularly with multi-jet units;
- Less sensitive to changes in head;
- Wider operating range—particularly with multi-jet units;
- Rapid operation of deflectors minimizes waterhammer and may, therefore, eliminate the need for a surge tank. A Francis turbine needs longer closing time, resulting in longer and higher overspeeds; and
- It is simple with fewer moving parts and may have lower maintenance costs.

Selection within a Machine Type

After the type has been selected, further refinements in the selection must be made. The specific speed, rotational speed, size, and whether to have adjustable or fixed blades must be determined.

Selection of Synchronous Rotational Speed

There are a finite number of synchronous rotational speeds that may be selected. The system frequency, whether 50 or 60 Hertz (Hz), and the number of poles in the generator determine these speeds. Equation 3-18 shows available rotational speeds for 50 and 60 Hertz.

EQUATION 3-18:

$$N = \frac{f \times 120}{z}$$

where:

N=speed, rpm;
f=system frequency, 50 or 60 Hertz; and
z=number of poles (must be an even number), dimensionless.

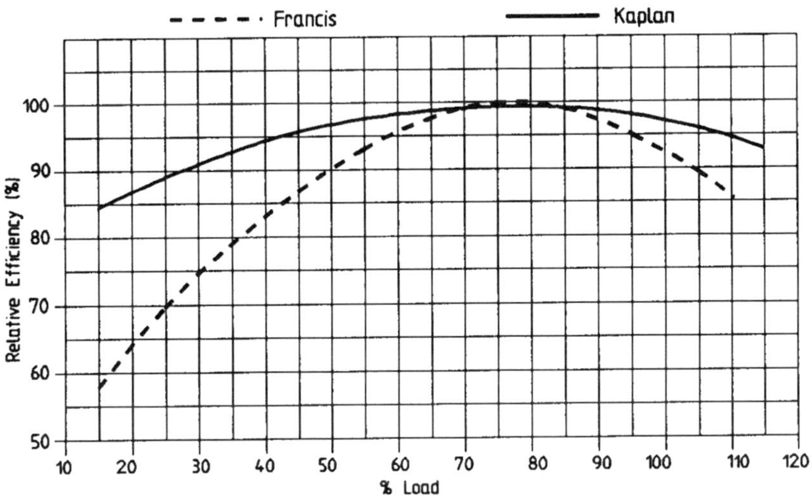

FIGURE 3-19: *Francis versus Kaplan Performance.* *This figure illustrates that a higher efficiency value is realized over a wide range of head and flow conditions and is the principal advantage of the Kaplan turbine over a propeller or Francis (where applicable) turbine. The blades of a Kaplan turbine runner can be rotated about their axes to vary the pitch angle of the blades. This permits a better match to the changing entrance conditions, and results in reduced losses.*

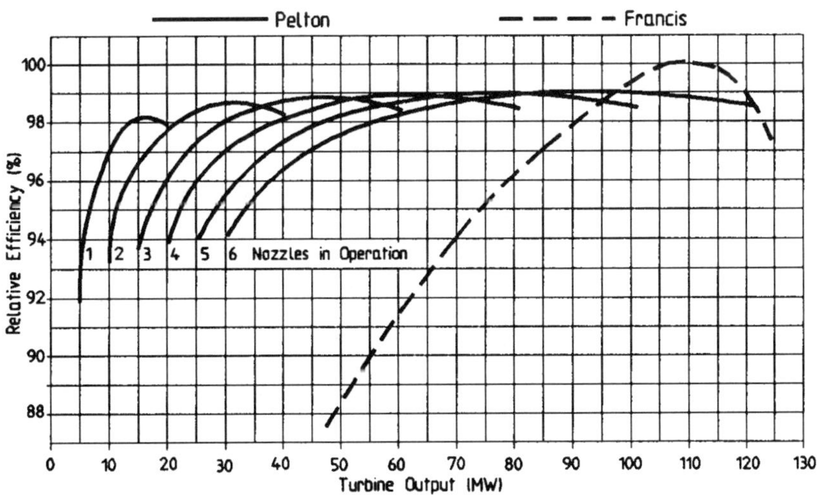

FIGURE 3-20: *Francis versus Pelton Performance.* *This figure compares the typical efficiency versus load characteristics for a low specific speed Francis turbine and a six-jet Pelton turbine with the optimal number of jets in service.*

Chapter 3

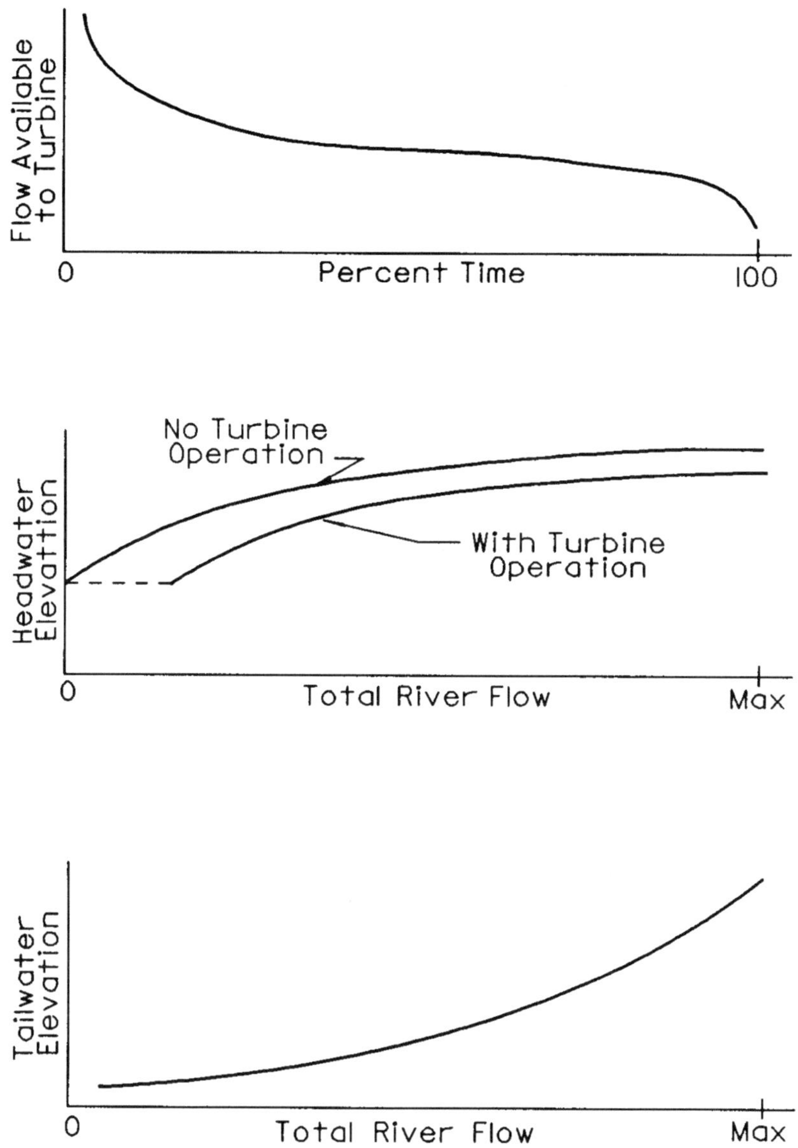

FIGURE 3-21: ***Flow Duration Curves.*** *An annual energy study requires flow rate data as a function of time. Flow rate data can be monthly, weekly, daily, or hourly (for peaking plants with relatively small forebay ponds). Head and tailwater elevations as a function of that flow or as a function of time also are required.*

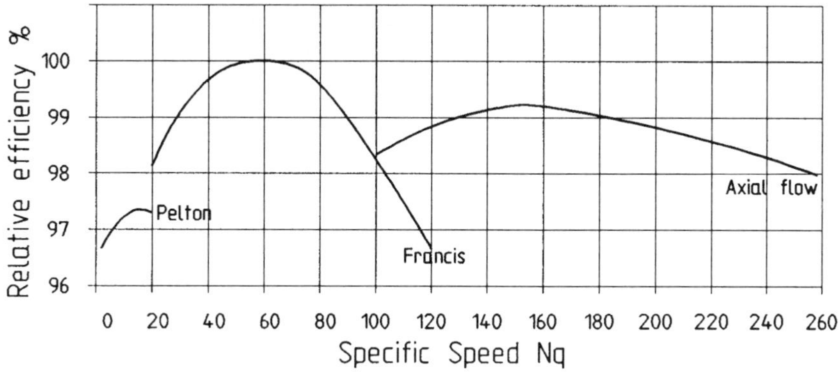

FIGURE 3-22: ***Variation in Efficiency.*** *This figure shows the variation in efficiency with specific speed for Pelton, Francis, and axial flow turbines.*

A small percentage of plants use adjustable-speed generators. While more expensive, these allow the turbine efficiencies to remain relatively high over a wider range of operation.

Specific Speed

For a given application of head and output, the size of the turbine decreases as the specific speed increases. This reduction in machine size tends to reduce the cost of the turbine, but increases the fluid velocities. Increasing the velocity increases the potential for cavitation damage and, therefore, requires the unit to be set at a lower elevation to increase the ambient pressure at which the unit operates. The additional excavation required increases the powerhouse costs. This is not generally applicable for underground powerhouses. The higher rotational speed associated with higher specific speed results in smaller and, therefore, lower cost generators. Finally, auxiliary equipment such as compressors and pony motors can be smaller, thereby reducing costs. Efficiency also is affected by specific speed, with efficiency dropping at both extremes of specific speed.

An increase in the loads in the machine accompanies the use of the same specific speed machines for a higher head. If these resulting loads exceed good design practices, higher strength materials or an increased section thickness may be required—both of which will tend to increase the turbine costs. In the case of increased thickness, it may also adversely affect the hydraulic performance of the machine. There has been a historical trend toward higher specific speeds for a given application. Figure 3-22 shows the variation in efficiency for Pelton, Francis, and axial flow turbines.

Annual Energy

Annual energy is the quantity of energy that a hydro turbine will produce at a given site over a period of one year when operating within the machine's constraints. It is a function of the site head and available flow. It is influenced by the elevation of the machine relative to the tailwater elevation at the site as a function of head and flow.

Mathematically, annual energy (AE) can be expressed as in Equation 3-19.

EQUATION 3-19:

$$AE = \int_{o}^{t} P(t)dt = f(H, Q, N, H_s, P_{max})$$

where:

t=31,536,000 seconds (number of seconds in a year);
P=power;
H=head;
Q=flow;
N=rpm;
H_s=submergence; and
P_{max}=maximum power.

This concept of annual energy is extremely useful in comparing alternate designs and variations among the design parameters. The concept provides a tool for evaluating plant revenue and computing the return on the total investment involved in constructing the plant.

Primary data are required for conducting an annual energy study related to site information. Flow as a function of time through the plant site is required. Figure 3-21 shows flow duration curves averaging over several years. Flow rate data can be monthly, weekly, daily, or hourly (for peaking plants with relatively small forebay ponds). Head and tailwater elevations as a function of that flow or as a function of time also are required. It is very important that the information in Figure 3-21 be generated and specified in tabular form. To do a detailed and accurate annual energy calculation, the user and manufacturer should understand the intended operation of the plant and the method used for energy calculations. This will ensure that:

- The manufacturer fully understands the intended use of the plant;
- Various bids and options are compared on the same basis; and
- The best hardware solution is found.

To be able to evaluate the revenue production and the cost of the installation, dollar values must be assigned to the energy produced. The costs of dam construction and how it is influenced by machine size is required. Unit cost of excavation must be provided to be able to evaluate the additional energy produced for deeper machine settings. Machine cost as a function of size and rotational speed are significant factors in the overall economic analysis.

The turbine performance characteristics described previously also are required to predict annual energy.

The owner or consultant should do the annual energy calculation to maximize consistency. If the manufacturer does this calculation, detailed procedures should be specified.

Computer Modeling of the Annual Energy Calculation

Annual energy prediction is most effectively done by using a computer program. For the number of turbine units available, the operational point is selected that maximizes overall machine efficiency while keeping the point of turbine operation within the performance constraints. For periods of high flow, all of the machines will be operating at maximum power, cavitation, or low-efficiency constraints, and water will be spilled over the spillway. As the flow decreases, the number of machines will be selected to produce the maximum energy from the flow available. The energy produced by all, n, machines is compared to the energy produced by one less machine, (n–1), successively, until the optimal number of machines is obtained. Starting at the lowest allowable flow, one of the machines will operate. As the flow increases, successive machines will be brought on line one at a time to keep the energy production at its maximum value. The energy produced in each time increment is summed to obtain the total annual energy production from the plant.

Chapter 3

Other Considerations

Cavitation. Cavitation is the formation of bubbles from small gas nuclei as the local pressure allows the flow to momentarily enter the vapor phase. These bubbles subsequently collapse as they enter an area of higher pressure. Minute bubbles form and collapse as shown in Figure 3-23. The collapse is characterized as an implosion in which the surface is attacked by high pressure intensity of the

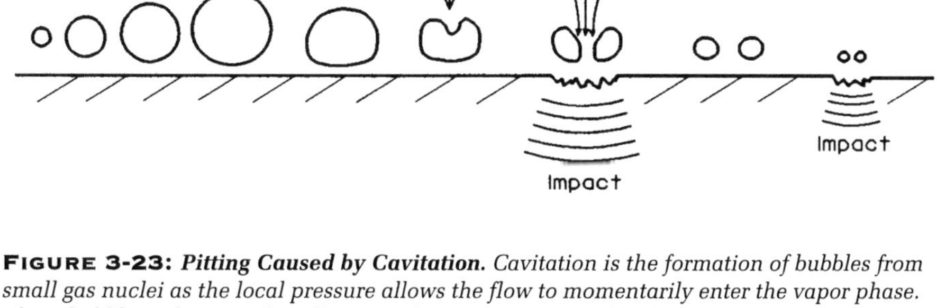

FIGURE 3-23: ***Pitting Caused by Cavitation.*** *Cavitation is the formation of bubbles from small gas nuclei as the local pressure allows the flow to momentarily enter the vapor phase. These bubbles subsequently collapse as they enter an area of higher pressure, as shown in this figure. The collapse acts like an implosion in which the surface is attacked by high pressure intensity of the impinging jet. This pressure alters the micro structure of the metal, causing a flake to peel out. Eventually, these microscopic-sized flakes form a visible pit and, if not repaired, will develop into major damage.*

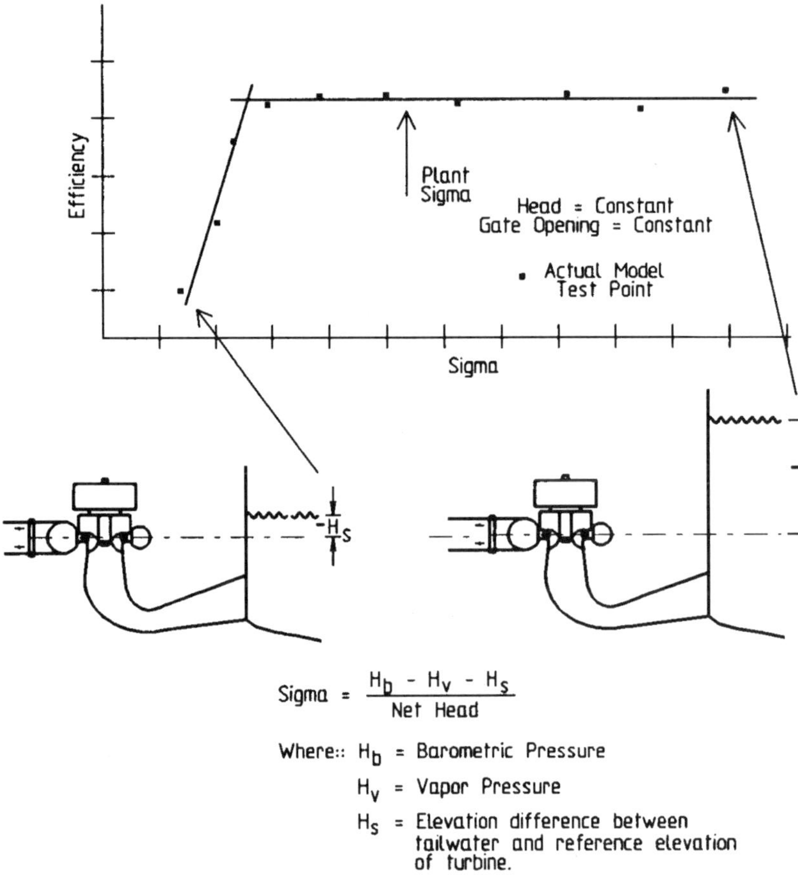

FIGURE 3-24: ***Cavitation Break; Sigma versus Efficiency.*** *As sigma is changed in the laboratory, curves of sigma versus efficiency and sigma versus power are generated. Eventually, as shown in this figure, a sigma value will be reached where a measurable drop in efficiency or output will be noted. The two primary objectives of cavitation testing are to establish the shape of the sigma "break" curve and to do visual studies of cavitation.*

impinging jet. This pressure alters the micro structure of the metal. This causes a flake to peel out. Eventually, these microscopic-sized flakes form a visible pit and, if not repaired, will develop into major damage. Increased noise, vibration, and reduced performance indicate that this phenomenon has occurred.

Dieter Thoma observed that the relationship of the difference between the absolute pressure and the vapor pressure was proportional to the net head. This proportionality, called sigma, is shown on Figure 3-24, along with a diagram illustrating the factors involved. It can be seen that lowering the runner or unit centerline with respect to the tailwater elevation will lessen the risk of cavitation.

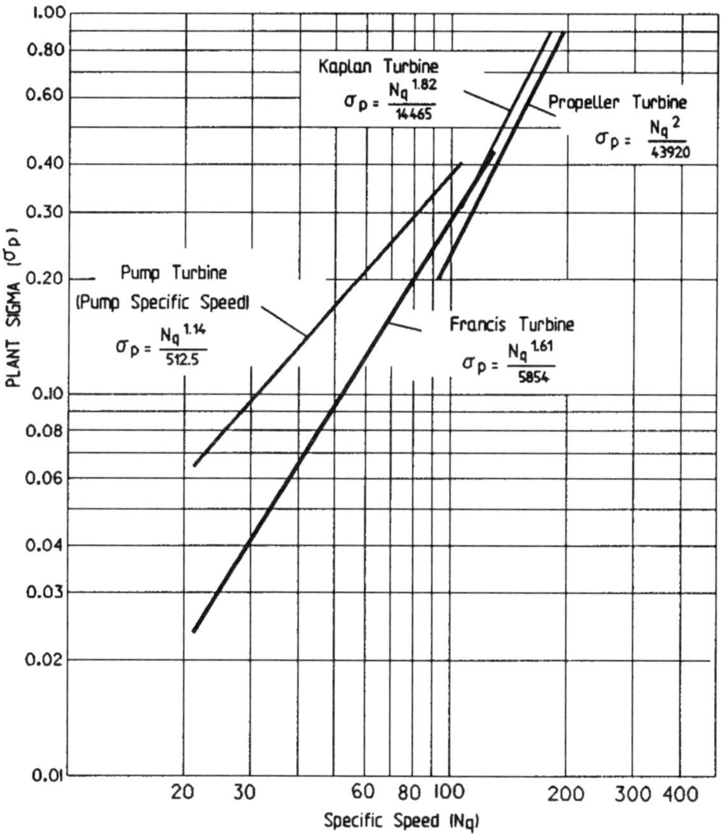

FIGURE 3-25: ***Specific Speed versus Plant Sigma.*** *This figure shows approximate plant sigma values versus specific speed for the typical turbine types.*
(Source: Cavitation Pitting Mitigation in Hydraulic Turbines, *EPRI)*

Figure 3-25 shows approximate plant sigma values versus specific speed for the typical turbine types.

When planning a project, one important influence on cost is the unit centerline elevation. This elevation is selected so that the sigma value for the plant (plant sigma) is higher than that for the tested runner design. The cost of increased excavation required to lower the unit centerline is evaluated against the additional power that becomes available due to the increased cavitation limit. Machine size also becomes a factor as it is related to the cavitation limit.

IEC Publication 609 furnishes a guide to the contractual relationship between the owner and manufacturer for common understanding and guarantees.

It is likely that some amount of pitting will occur on the runner during its operating life. Visual observations during model testing will delineate those zones of the runner that are most susceptible to damage. To minimize future damage, a weld overlay may be made of a more cavitation-resistant material such as stainless steel, cobalt, or other alloys, or the runner might be made entirely of stainless steel. The trend is certainly toward solid stainless steel runners. In summary, pitting is minimized by using pitting-resistant materials, using state-of-the-art hydraulic design techniques, adherence to model/prototype homology, and adherence to operating design limits. A properly designed, manufactured, and operated unit will provide ten or more years of operation before needing cavitation repair. It is, however, very important to perform periodic inspections to confirm that repair is not required.

Metallurgical tests, principally done by ultrasonic vibration of test specimens in water, have determined the relative resistance of various materials to pitting. However, these tests are not used for predicting the quantities of metal removed on prototype units.

Field repairs to pitted areas must be performed with caution to maintain the runner geometry as designed. This involves making templates of the area to be repaired and using the templates to re-establish the correct contour of the welded surface. Original finish, waviness, shape, thickness, angle, and vent (the closest distance between two adjacent vanes) tolerances must be re-established if the original performance is to be restored.

Overspeed and Runaway

An important contributor to the cost of the generator is the maximum speed for which the rotor must be designed. This is typically the maximum runaway speed. The turbine model test furnishes the input to make this determination.

Maximum steady-state runaway speed is the maximum speed a turbine-generator unit will eventually achieve under maximum net head with no load on the generator and wicket gates in the most adverse position. Figure 3-26 shows comparisons of runaway speeds.

Maximum transient runaway speed is that speed reached momentarily during a load rejection. The ratio of transient to steady-state runaway is very site-dependent.

If the unit is operating at full output and the line breaker is opened for some reason, this sudden loss of load, without a change to the input energy, results in a rapid increase in speed. If the governing system is functioning normally, it will begin closing the wicket gates as soon as a few rpm of speed increase have been detected. This reduction in gate opening normally limits the speed rise to less than runaway speed (for Francis and Kaplan, but not pump-turbines). The lower value, called "overspeed," is dependent on several factors, including the rate at which the wicket gates close, specific speed, geometry of water column, and the rotating inertia of the rotating turbine and generator components.

The Kaplan turbine requires special attention during model testing. The Kaplan's governor control keeps the runner blades and wicket gates

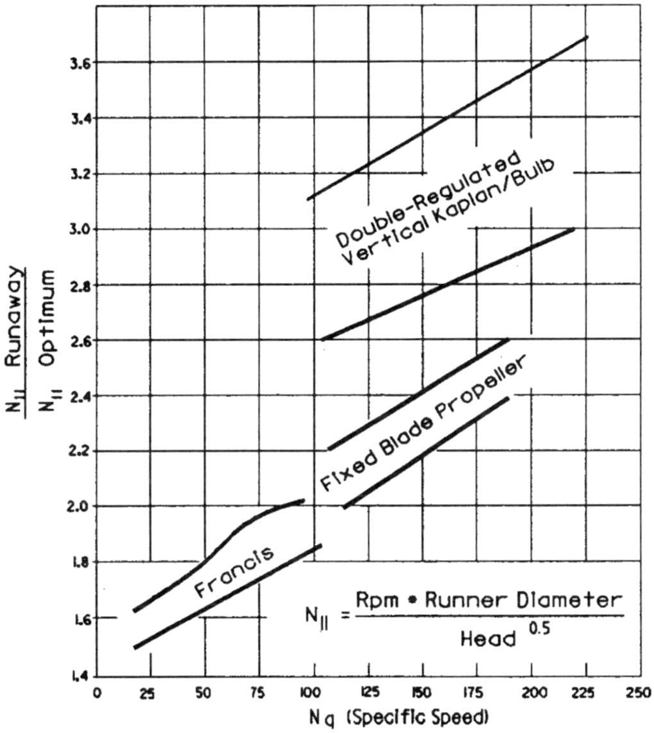

FIGURE 3-26: *Runaway Speed versus Specific Speed.* *This figure shows runaway speeds for different turbines. Maximum steady-state runaway speed is the maximum speed a turbine-generator unit eventually will achieve under maximum net head with no load on the generator and wicket gates fully open. An important cost factor of the generator is the maximum speed for which the rotor must be designed. This is typically the maximum steady-state or transient runaway speed.*

---

synchronized. During a load rejection, this synchronization can no longer be maintained and the turbine blades and gates go into an "off-cam" mode. That is, the blades close too slowly for the amount of speed increase and, consequently, the low pressure side (side toward draft tube) of each blade cavitates. Thus, the pressure difference across the blade from the high pressure side to the low pressure side increases and causes the speed to increase to a value higher than "on-cam." This change reaches a limit when the flow fully separates from the low pressure side, thus stabilizing the pressure difference. If the plant sigma is numerically close to the runner sigma, this process occurs early in the excursion into overspeed. If the plant sigma is significantly higher than the runner sigma, the acceleration due to "off-cam" conditions occurs later and the resulting maximum speed is not quite as high.

Kaplan turbines typically have their maximum runaway speed at a low blade tilt. Therefore, by fixing the low limit of the tilt to an increased value, the runaway speed can be decreased. Limiting of lowest blade tilt may cause a loss of efficiency. Some Kaplan designs have included mechanical devices to move the blade tilt to a high position to minimize runaway.

The "on-cam" and "off-cam" runaway speeds of the prototype are determined from the model test by plotting sigma versus unit speed for no-load conditions with various combinations of fixed blade angle and wicket gate opening. This information permits the project designer to select an overspeed value for generator rotor design.

Hydraulic Vibrations

Hydraulic vibrations are usually greater during operation at part-load and overload than at operation near peak efficiency. In addition, natural frequencies and forced oscillations of pressure, such as the vortex core beneath a Francis turbine runner, vary with load. Hydraulic vibrations should be considered for the complete operating range, including start up and transients. Vibrations are caused by bends in the water conductors, flow patterns in the machines, enlargements in the flow paths, and vortices produced in places such as the draft tube and discharge edges of runner blades and stay vanes. The magnitude of most vibrations will be relatively small. However, the vibrations should be investigated to find whether resonance may occur since the amplitude of pressure surges with resonance may damage the system. To avoid resonance, the natural frequencies of the hydraulic water passages and the machinery in the system should be designed to be at least 20 percent different than the frequencies of harmonics of the forced vibrations. Some of these calculated frequencies will be approximations and may result in an increase in the 20 percent value. The most prevalent forced-vibrations are:

- The oscillations of the pressure waves during the transient;
- Vortex cores in the draft tube beneath Francis turbine runners;
- Disturbances caused by the frequency of the blades (buckets) of the runner passing a non-rotating part such as wicket gates or stay vanes; and
- Resonances with penstock vibration and wave travel frequencies.

The vortex core beneath a Francis turbine runner rotates with the frequency calculated from Equation 3-20.

EQUATION 3-20:

$$f_f = \frac{N}{60k}$$

where:

f_f=forcing frequency, Hertz;
N=unit rotational speed, rpm; and
k=Rheingans number, dimensionless.

The value of k for part load, as first noted by William Rheingans, is usually from 2 to 4. On a given machine, k varies, depending on the operating conditions. For some machines and plants, k may be as high as 8. The value of k for overload is approximately 10.

The vortex core is a region of low pressure. Admitting air may diminish pulsations caused by such vortices. Possible locations for admitting air are at the top of the draft tube, behind the runner band, at the bottom of the runner crown, or at the bottom of the runner cone. It is very important that air is admitted only when needed. Used when needed to reduce pulsations and vibrations, it can increase efficiency by as much as 1 percent at the operating point. Admitting air when not needed may decrease the efficiency by as much as 4 percent. Introducing air lowers the wave propagation speed and hence could be helpful or detrimental. There are techniques, such as draft tube fins and co-axial diffusers, that may reduce pressure pulsations on some machines at some operating conditions.

Example of Machine Selection (Francis)

To achieve maximum prototype efficiency, it is evident that the value of N_{11} that passes through the optimum efficiency of the model test, as shown in Figure 3-17, must be chosen as the starting point. Likewise, the minimum runner size will result from choosing the Q_{11} at the intersection of that N_{11} line and limiting the full gate opening line. This latter line is the manufacturer's choice for limiting performance of that particular model due to power or cavitation limitations.

From the specified net head and flow and the selected model Q_{11}, the prototype diameter can be calculated. Likewise, the speed can be calculated by using that diameter and the selected N_{11}.

Since this speed will not likely be a synchronous speed for the system's electrical frequency, the next closest synchronous speed must be found and the N_{11} recalculated. From this new N_{11}, the value of Q_{11} also will vary slightly and the prototype Q_P must be recalculated. For low-speed applications, a speed increaser may be incorporated to take advantage of a lower cost, higher speed generator. The speed increaser, which is located between the main turbine shaft and generator, provides a suitable step-up ratio from the designed turbine speed to generator speed.

For example, if the selected N_{11} results in N=176 rpm, the theoretical number of poles is calculated by using Equation 3-18. Using this equation for 60 Hz, z=7200/176=40.9. By refining the number of poles to the closest even integer of 40, N becomes 180 rpm and is now synchronous for a 60 Hertz electrical system. In this example, the N_{11} is changed by the ratio of 180/176=1.023.

The trial value(s) of machine speed should be reviewed with generator manufacturers since not all power (kW) and speed (rpm) combinations are practical from a generator perspective.

The performance curve (power versus flow and efficiency) for a given net head can be calculated by picking off the values of Q_{11} and efficiency at the constant N_{11} line that corresponds to the given head.

Use Equation 3-21 to calculate efficiency (η).

EQUATION 3-21:

$$\eta = \frac{P}{\rho g Q H} \; \textit{kilowatts} \qquad \text{or} \qquad \eta = \frac{550P}{\rho g Q H} \; \textit{horsepower}$$

where:

ρ =water density, kilograms/cubic meters (slug/cubic feet);

g= acceleration due to gravity at the specific location, meters/second squared (feet/second squared);

η=efficiency (in decimal form), dimensionless;

P=power kW (horsepower);

H=head, meters (feet); and

Q=flow rate, cubic meters/second (cubic feet/second).

The calculation process is the same for each net head of interest. That is, the appropriate N_{11} can be determined for each new net head using the previously calculated diameter and speed.

HYDRAULIC DESIGN

Sophisticated fluid analysis computer programs, linked with design programs, have been developed to provide good computational tools to evaluate designs. Operation of turbines and pump-turbines generally takes place within a narrow operating range for most of the life of the design. Within this range, the flow characteristics generally are well-behaved. Large head or power ranges create unique design challenges. Analytical models have been developed to predict characteristics of flow in this region with satisfactory accuracy.

The source of cavitation arising from the flow between the bladed passageways generally also can be predicted with good accuracy. In many cases, these tools have been well correlated against model tests. However, model testing remains an important method to confirm the design and predict the performance for all types of turbines. Away from the normal range of operation, the results of mathematical models become less valid, and the calculations become imprecise. Neither steady nor time-varying flow characteristics are predicted accurately. Model testing is required to confirm the behavior of the design when operating in these off-design regimes.

For reaction turbines, flow characteristics can be predicted with reasonable accuracy in the spiral case, between the stay vanes and wicket gates, and through the runner at and near the design point. The flow through the spiral case, while composed primarily of the principal through-flow characteristics, has the secondary flows superimposed upon it that may be neglected in the overall computational scheme. The analytical tools can closely approximate the flow through the wicket gates and stay vanes for well-designed units near the design point, and the characteristics of the flow in the boundary layers can be evaluated with two and three dimensional boundary layer codes. However, for units with high fluid incidence or with high vane loadings, secondary flows that are significant will take place. This results in separated regions. When the flow is separated, the computational assumptions used in the tools may not be valid and the accuracy of the results degrade. In the runner, the major flow calculations near the design point provide excellent correlation with observed model data.

The most difficult flows to compute are in the turbine draft tube. Highly three-dimensional and separated flows are present and are quite typical in turbomachinery draft tubes.

Model Testing

Over the years, model testing has evolved as an essential design element for hydro turbine manufacturers. Turbine manufacturers principally own and operate model testing laboratories. Also, independent laboratories have provided a source for relative performance evaluation between competitive manufacturers. Traditionally, major hydro projects have had proof-of-performance model tests as part of their contract.

Typical Model Test Program

To see how a model test fits into the overall scheme of a project, it is necessary to understand the steps involved in the hydraulic design of a turbine. Figure 3-27 is a flow chart giving the sequence of events that occur from the prototype design through to the manufacturing process. The modeling discussed here involves the wetted surfaces of the turbine. It is not a mechanical modeling of items such as the gate servomotor systems or other mechanical features outside the water passages. However, if a "mechanical feature," such as thickness of stay vanes, influences the hydraulic contours, it is part of the hydraulic modeling.

The first phase of the hydraulic design is to determine the type, specific speed, rpm, submergence, and size of the unit required to meet the specifications. At this point, it can be determined if there is an existing model test that meets the requirements of the job. In some cases, the manufacturer will have an existing design and accompanying model test that suits the requirements of a particular application. Most major manufacturers have a large selection of prior designs and model tests available. The use of existing model test data will reduce the cost and lead time of the project.

In some cases, an existing model cannot be used. This can happen for a number of reasons:

- Constraints imposed by powerhouse construction;
- Constraints imposed by existing designs (for replacement parts);
- Design parameters for existing model test information is not applicable; or
- A model test may not have the most up-to-date design with respect to efficiency levels, optimized mechanical construction, etc.

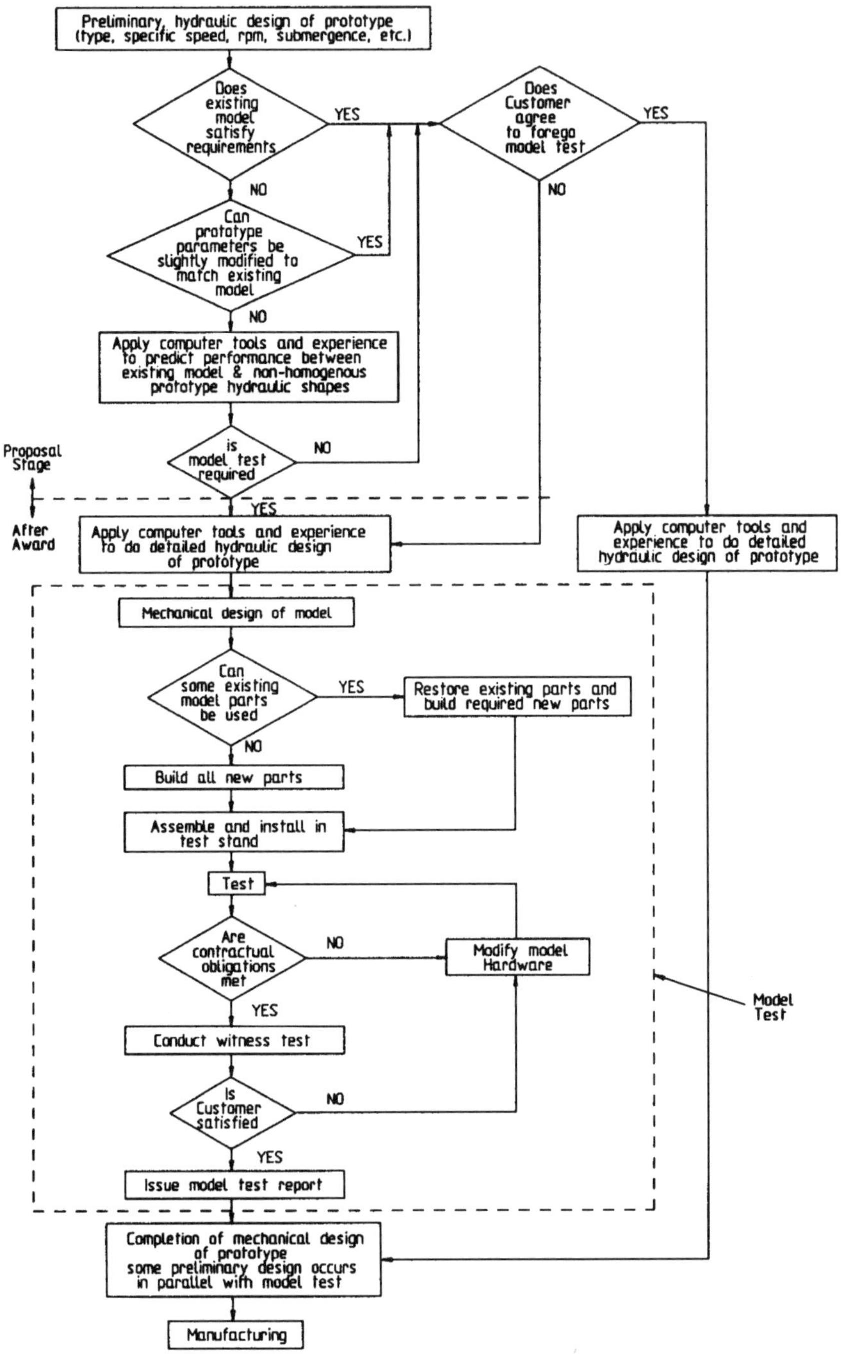

FIGURE 3-27: *Contract Sequence Flow Chart.* *This flow chart gives the sequence of events that occur from the prototype design through to the manufacturing process. To see how a model test fits into the overall scheme of a project, it is necessary to understand the steps involved in the hydraulic design of a turbine.*

Cost-Effectiveness of Testing

The need to specify model tests depends on the overall objectives and economics of the project. Generally, for small machines, it is not economical to conduct a comprehensive model testing program because the model testing may cost as much as the equipment being purchased. In this case, the purchaser must rely on the experience of the manufacturer, results from tests on similar specific speed machines, or on comprehensive calculations to verify the overall operating characteristics of the machine within the range of normal operation.

If the project is large, extensive model testing is well-justified for the following reasons:

- Confirmation of prototype performance provided that dimensional tolerances of IEC 193 code are maintained in the manufacturing of the prototype. The efficiency measured in model testing is turbine efficiency, not plant efficiency. Per IEC 193, it is the efficiency of the portion of the machine shown within the dashed lines on Figure 3-1;
- Measurements are made in the controlled and accurate environment of the laboratory rather than in the field;
- Potential field problems related to transient phenomenon, cavitation, and pressure pulsations can be identified and design changes can be implemented;
- A comprehensive view of operating conditions is possible, so that undesirable regions can be avoided in operating the prototype;
- Loadings for structural design purposes can be provided;
- With model efficiencies at their current levels, it becomes increasingly difficult to achieve higher performance values. Although there are powerful computer-aided design tools available to optimize designs, the fine-tuning for a full optimization is possible only through model testing; and
- The performance characteristics of the turbine are necessary for transient, hydraulic oscillation, and stability analyses. The transient conditions involved in speed increases and flow decrease/increase require study because they contribute to the project's civil design and governor requirements. During the feasibility studies and design phase, these characteristics may be derived from the available data corresponding to plants of similar specific speed. As the design progresses, the actual performance characteristics should be obtained from hydraulic model tests or computer-based analyses.

Competitive Model Testing

On some large contracts, especially multi-unit plants, equipment manufacturers are asked to bid on pre-contract award hydraulic turbine model tests. Generally, these tests are done at an independent hydraulic laboratory at the purchaser's expense. The contract is then awarded based on the price and best performance evaluation.

The main advantage of competitive model testing is independent verification of performance. The test can be especially useful when a small difference in efficiency can alter the ranking of bid proposals.

Disadvantages of competitive model testing include increased costs and additional time required.

Results of Testing

Table 3-3 summarizes, in broad terms, the principal results obtainable from a model test series.

Description of Laboratory

Figure 3-28 shows a typical model test stand including the circuit pumps (used to deliver high pressure water to the test stand), motor-generator, and flow measurement and calibration system. Today's test stands usually have at least the following capabilities:

- Performance testing with efficiency accuracy of approximately ±0.25 percent;
- Complete four-quadrant testing, as shown in Figure 3-29;
- Cavitation testing with viewing capabilities;
- Acoustic monitoring of cavitation;
- Auxiliary data-gathering including pressure pulsation, flow induced stresses, vibrations, and acoustic emissions;
- Determination of direction and magnitude of three-dimensional flow patterns;
- Frequency analysis of dynamic signals;
- Testing turbine or pump arrangements covering a wide range of configurations;
- Qualified to meet or exceed all international and domestic test codes for model turbines and pumps; and
- Calibrating test instruments against in-house standards that come from government standards.

Testing of Civil Features

In addition to the turbine testing, hydraulic model testing of the civil project features is done to determine flow characteristics of the intake and discharge hydraulic conduits of the turbines and pump-turbines. This testing is typically not performed at the turbine manufacturer's laboratory.

Model Performance Testing

For the fixed-blade runners, such as Francis and propeller turbines, the model turbine performance testing is run by setting the model turbine wicket gates to a constant gate opening. Varying the speed while holding the head constant or varying the head and holding the speed constant traverses the N_{11} (speed or head) range. Values of discharge, head, speed, and torque are measured directly, and the results are presented as unit quantities. Additional parameters, such as axial hydraulic thrust and gate torques, also are measured. The cycle is repeated, varying gate opening until the entire range is completed. The exact procedure is laboratory dependent.

TABLE 3-3: ***Data Obtained or Verified through Turbine Model Testing***

Turbine Efficiency	Evaluate Performance/Civil Trade-offs
Turbine Power	Determine if the location of Winter-Kennedy taps for the turbine cycle and the draft tube taps for the pump cycle are at locations that give suitable results. It is not recommended to develop prototype coefficients from model data.
Verify Cavitation (Area)	Loading and Structural Design
Verify Cavitation (Vortex)	Verify/Determine Pressure Loading
Determine Runaway Speeds	Determine Axial Thrust
Determine Four Quadrant Characteristics (P/T)	Determine Radial Thrust
Determine Shut-off Head (P/T)	Determine Wicket Gate Torque
Determine Instability (Pump Cycle P/T)	Determine Blade Torque (Kaplan Turbine)
Determine Cavitation Effect on Instability (Pump Cycle P/T)	Evaluate Flow Distributions
Determine Pressure Pulsations	Design Tool Improvements
Evaluate Air Admission	Determine Component Losses
Determine Optimal (Best) Position of Turbine Components	Correlation with Analytical Design Tools
Determine Impact of Shape Changes on Performance	
Determine Draft Tube Effectiveness	

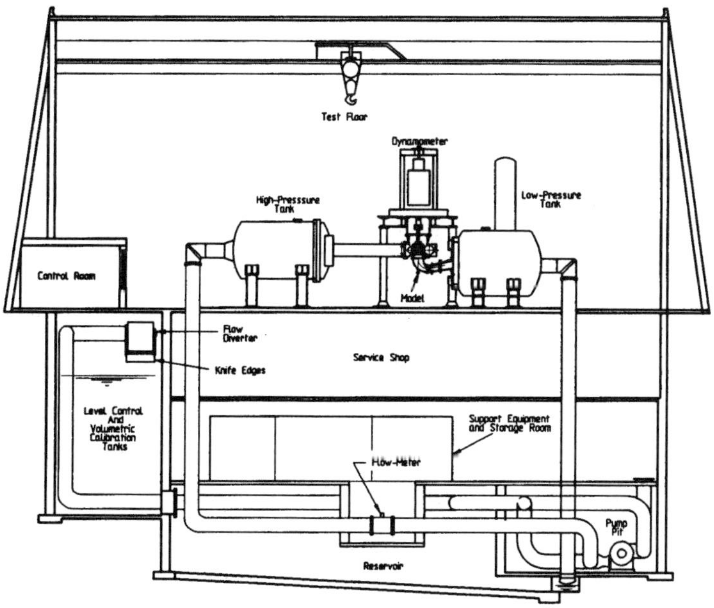

FIGURE 3-28: ***Test Stand.*** *This figure shows a typical model test stand, including the circuit pumps (used to deliver high pressure water to the test stand), motor-generator, and flow measurement and calibration system.*

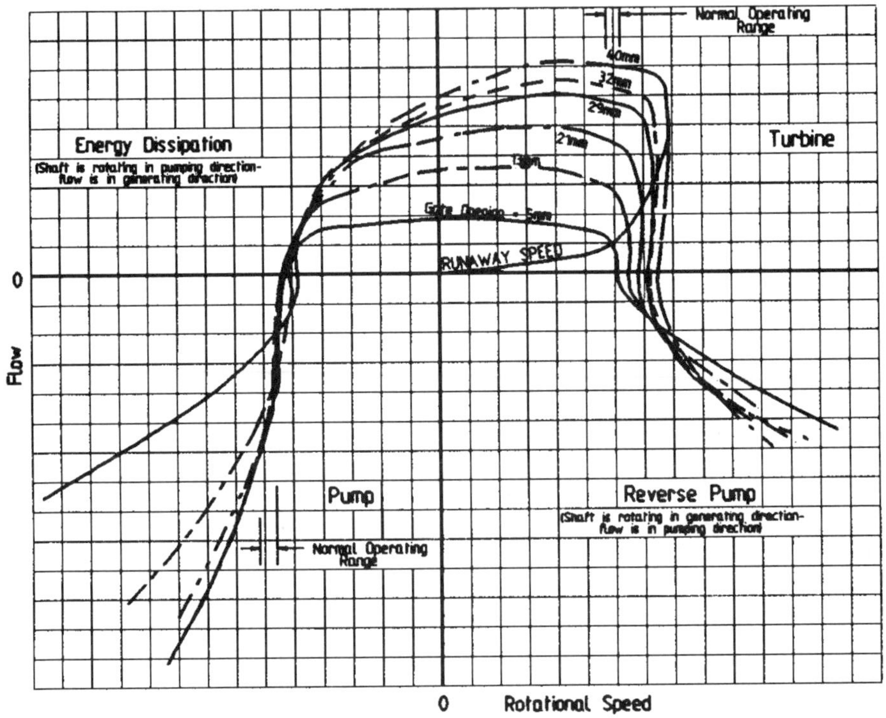

FIGURE 3-29: ***A Typical Four-Quadrant Curve.*** *This graph illustrates a four-quadrant curve for a typical model pump-turbine. Most test stands are capable of doing complete four-quadrant testing.*

For movable blade runners, such as Kaplan and Deriaz, the performance testing is run by setting the blades to a constant tilt and performing a "fixed tilt" test as described in the previous paragraph. The blade tilt is then changed, and the entire process repeated until all blade tilts have been tested.

Model Cavitation Testing

The cavitation coefficient, sigma, is varied during model testing by simultaneously changing the headwater and tailwater pressures by the same amount so that the net head remains the same while the ambient pressure at which the runner operates changes. As sigma is changed in the laboratory, curves of sigma versus efficiency and sigma versus power are generated. Eventually, a sigma value will be reached where a measurable drop in efficiency or output will be noted, as seen in Figure 3-24. This process is repeated for several gate openings at this net head. The full testing program involves multiple net head tests to cover the full range of heads for the prototype. The two primary objectives of cavitation testing are to establish the shape of the sigma "break" curve and to do visual studies of cavitation.

Several different characteristic values of sigma can define the cavitation characteristics of a model turbine. These values are designated based on the observed patterns of cavitation and how these patterns affect the overall performance of the machine. Figure 3-30 illustrates the commonly used definitions of sigma. A family of curves, drawn from this multitude of individual test points, such as Figure 3-31, delineates the cavitation characteristics of the tested model. Sigma is based on a reference elevation in the prototype turbine. Because the change in sigma with location in the turbine is not the same from the model to the prototype, it is important to reference the prototype plant sigma carefully. It is common practice in the turbine industry to define plant sigma near the discharge of the turbine runner during cavitation testing because this is the elevation where the cavitation typically first begins.

Sigma P (σ_p) is the plant sigma corresponding to the head and tailwater conditions and reference elevation of the prototype turbine.

Sigma begin (σ_{beg}) is the beginning of visual cavitation on the surface of the runner bucket.

Sigma 0 (σ_0) is the lowest value of sigma for which efficiency remains unchanged as compared to non-cavitation conditions.

Sigma 1 ($\sigma_{1.0}$) is the lowest value of sigma for which a 1 percent drop in efficiency is attained, compared to non-cavitating efficiency.

Sigma standard (σ_s) is the intersection of the constant efficiency line (non-cavitating) with the strongly sloping portion.

When sigma standard, sigma 1, or sigma 0 are used for determining the runner centerline elevation, a safety margin of additional submergence based on past experience is applied. This assures the owner of adequate protection from pitting resulting from cavitation.

When the project centerline elevation is determined using Sigma begin, usually no safety margin is needed.

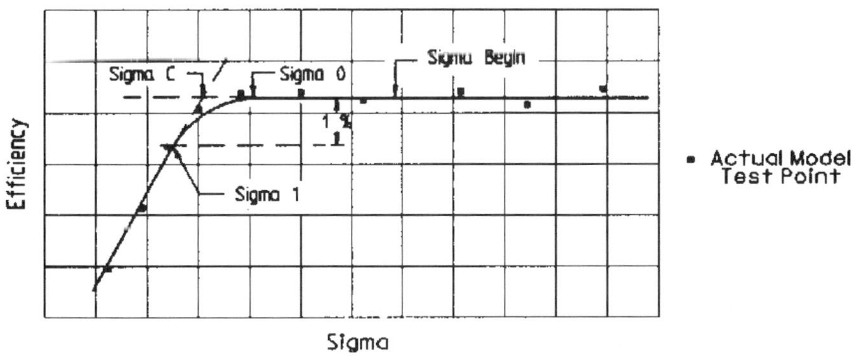

FIGURE 3-30: ***Typical Sigma Break Graph.*** *Several different characteristic values of sigma can define the cavitation characteristics of a model turbine. These values are designated based on the observed patterns of cavitation and how these patterns affect the overall performance of the machine. This figure illustrates the commonly used definitions of sigma.*

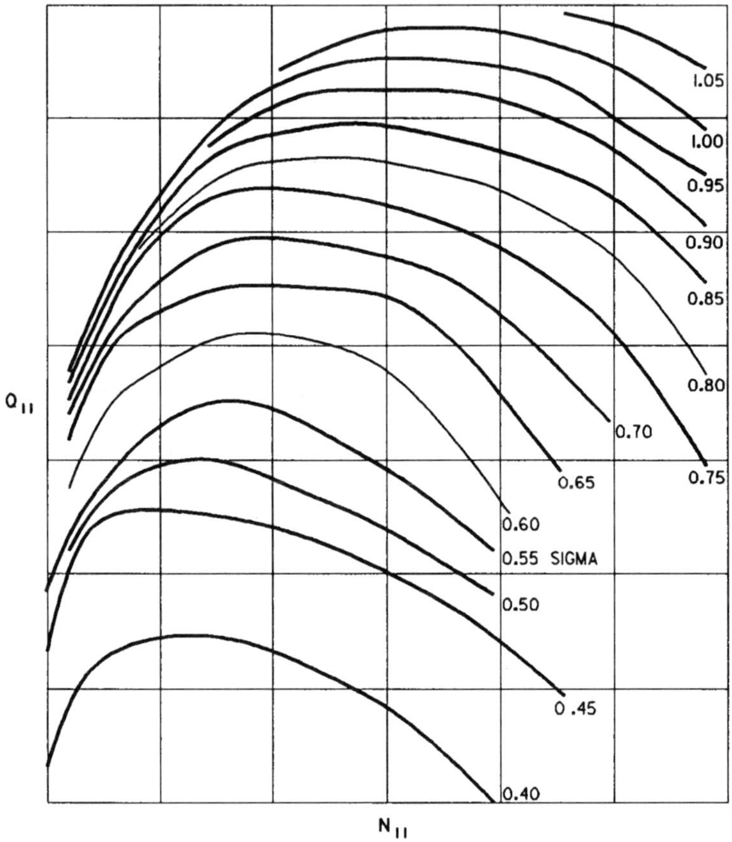

FIGURE 3-31: ***Typical Model Sigma Values.*** *This figure shows a family of curves, drawn from a multitude of individual cavitation breaks. It delineates the cavitation characteristics of the model over the entire operation range. Sigma is based on a reference elevation in the prototype turbine. Because the change in sigma with location in the turbine is not the same from the model to the prototype, it is important to reference the prototype plant sigma carefully. It is common practice in the turbine industry to define plant sigma near the discharge of the turbine runner during cavitation testing.*

Pump-Turbine Four Quadrant

For pump-turbines, measuring torque, rpm, head, flow, and gate opening as the model is run through the following four quadrants of flow and rpm establishes the four-quadrant characteristics:

1. Pump Rotation/Pump Flow
2. Pump Rotation/Turbine Flow
3. Turbine Rotation/Turbine Flow
4. Turbine Rotation/Pump Flow

This data, along with a geometrical description of the penstock system of the plant and the moments of inertia of the rotating parts, are used for predicting transient behavior. (See Figure 3-29.)

MECHANICAL DESIGN

Basic Considerations

For the hydraulic turbine manufacturer, the design of hydraulic turbines requires close cooperation between hydraulic and mechanical designers. Hydraulically desirable shapes and profiles that form the water passages of hydraulic turbines may not be structurally acceptable and, conversely, structurally acceptable water passages may not be hydraulically desirable shapes.

The mechanical design discussion limits itself primarily to vertical Francis turbines and secondarily to vertical axial flow machines. It reviews design considerations for most of the structures associated with water passages, namely, stay rings, spiral cases, draft tube liners, wicket gates, headcovers, discharge rings, and runners. This section also discusses completion parts such as the wicket gate mechanism, turbine guide bearing, and main shaft seal.

Chapter 3

Designs Start with Hydraulics

The hydraulic engineer lays out an outline of hydraulic water passages for the specific turbine design. Performance characteristics and trends provided by scale model testing of many types of turbines covering a wide variety of heads and flows guide the designer who applies the principles of fluid mechanics and dynamics in the design. The design accounts for flow velocities and velocity distributions, angles of attack and discharge at flow obstructions, water passage finish, and waviness. Because of the behavior of optimal fluid flow, the desired hydraulic water passage design for a hydraulic turbine strives for smooth fluid flow stream lines with as few discontinuities as possible.

Hydraulically ideal water passage shapes are often not structurally feasible and sometimes their manufacture is cost-prohibitive. The structural integrity of the unit must be ensured throughout the process of balancing the hydraulic component "manufacturing costs" with "hydraulic performance."

Mechanical Design Considerations

Spiral Case

The mechanical design of the plate steel for the spiral case is based on special adaptation of the thin-walled pressure vessel theory. Because the longitudinal stress is only one-half of the hoop stress, the higher hoop stress dictates the wall thickness. The loading for the design calculations must consider the maximum internal pressure, which is normally taken as the maximum static head above the unit centerline elevation plus the transient over-pressure created by emergency shutdown procedures. Generally, the surrounding concrete is considered to not supply any support and, therefore, is of no assistance to lowering the case stress level.

Some applications require a spiral case plate thickness that is too large for convenient and economical manufacturing. The designer will then select a steel classification with higher mechanical properties to reduce the thickness. At the juncture of this high strength plate with the lesser strength material, typically used for the stay ring, a tapered transition, using high strength material or high strength weld buildup, may be needed to fulfill the maximum allowable stress criterion for the lesser strength material.

The concrete semi-spiral case is a variation, in form, of the spiral case, and is developed for low head applications. Its use has traditionally been limited to under approximately 30 meters (100 feet) of static head. The area of each section downstream of the transverse centerline is reduced proportionately as is done with the full spiral case. Figure 3-4 illustrates such a concrete semi-spiral case. While the figure shows unsymmetrical proportioning above and below the horizontal centerline, symmetrical areas above and below the horizontal centerline may also be used. These options are available to optimize powerhouse designs.

Stay Ring

The stay ring is the structural backbone of the Francis and vertical axial flow type turbine. It transmits hydraulic, equipment, and concrete loads and supports most of the major turbine components through mechanically fastened or welded joints. This component always spans the water passages and transmits loads across the span to the unit's foundation through hydraulically shaped structural members called stay vanes. Generally, the number of stay vanes is the same as the number of wicket gates, varying from 16 to 24. However, under some conditions, such as low-head units, only half as many stay vanes as wicket gates might be used.

Physically, the stay ring is subjected to tensile, compressive, and bending loading. The membrane-type loading, contributed by the attachment of the spiral case, introduces a tensile force in the leading edge of the stay vane and a bending moment in the ring portions. The superimposed concrete mass above, the generator stator weight, and the thrust bearing load contribute to the compressive forces applied to the stay vanes. Additional tensile forces and bending moments result from the vertical pull of the headcover at its attachment to the stay ring.

In the special case of not having an attached spiral case, such as a propeller or Kaplan concrete semi-spiral case, the compressive loading makes the potential for vane buckling more critical and, therefore, usually predominates over the tensile loading.

Design and structural calculations for stay rings proceed based on the previously mentioned loads.

In the hydraulic design, the objective is to reduce frictional losses or optimize the angle of incidence by using thinner, shorter stay vanes. This is in direct conflict with the mechanical need to ensure the rigidity of the stay ring. The design of a stay ring that has a spiral case attached to it illustrates the optimization that is performed to provide the best stay vane hydraulic shape within structural requirements. This optimization is possible when consideration is given to the location of the spiral case membrane load that affects the bending moment (a force acting at a distance from a point of interest) on the stay ring crown section. This moment is then balanced with moments on the same crown section that result from powerhouse loads and headcover or discharge ring loads. This is done to reduce the overall net moment on the stay ring crowns and, therefore, the moment applied to the stay vanes. In combination with the moment optimization, the spiral case attachment point on the stay ring crowns and the headcover and the discharge ring connection points are located away from the leading and trailing edge locations of the stay vane. This reduces the concentration of the transmitted loads at these areas of the stay vanes, allowing the stay vane edges to be somewhat thinner. The thinner edges allow a more desirable hydraulic profile to be presented to the water flow.

Wicket Gates and Operating Mechanism

The primary function of wicket gates, or guide vanes, is to control and direct the flow of water to the turbine runner. Wicket gates are generally supported in two or three bearings or bushings. Usually, Francis and vertical Kaplan turbines are three-bearing arrangements.

Model testing provides information for plotting a hydraulic loading curve of gate torque versus gate position. When sizing the wicket gate servomotors, the mechanical friction in the operating mechanism is considered along with the hydraulic torque on all of the gates.

Some designs incorporate individually operating servomotors for each wicket gate; however, the conventional arrangement uses a gate operating ring and two gate servomotors. The design assumes that the servo effort is uniformly distributed to all wicket gates. However, there is the possibility that a foreign object may become lodged between two gates. To prevent the gates from breaking or stalling, a protective device is often provided. This device can be a breaking link or shear pin designed to fail at a load about two times greater than normal. At this failure load, the combined stress in the gate itself and the rest of the mechanism is permitted generally to approach 70 percent of the yield point. A representative breaking link or shear pin sample must be tested to destruction to determine that it breaks as designed. Some designs use friction connections rather than shear pins or links. Also, some pit or bulb units have a spring mechanism for this function.

After the protective device has broken and the gate is separated from the gate operating mechanism, it could oscillate violently because of hydraulic torques acting on the asynchronous gate. To avoid this, a gate-restraining mechanism may be specified to interact, usually by friction, between the wicket gate and its operating mechanism or linkage. Gate-restraining mechanisms often are specified for pump-turbines and large hydraulic turbines where high hydraulic torques and inertial forces are anticipated for an asynchronous wicket gate.

The major mechanical design considerations for wicket gates center around the following areas:

Deflections

• Leaf (gate hydraulic profile) deflection with respect to sealing off water leakage at the gate-to-gate contact surface; and

• Leaf and stem deflections with respect to gate binding in the bushings. Gate stem slopes need to be compared to headcover and discharge ring bushing slopes under loaded conditions to evaluate binding potential.

Stress

• Torsional stress, particularly at the gate stem to the leaf junction. This must be held within limits for the maximum torque applied throughout the opening and closing stroke of the gate including the closed gate (squeezed) condition and shear pin break condition; and

• Combined bending and torsion must be evaluated at the same junction of stem and leaf.

Natural Frequency

• When wicket gates are in close proximity to the runner bucket or vane periphery (for example, low specific speed machines), it is advisable to calculate the natural frequency of the gate to ensure that its frequency is sufficiently different from the anticipated forcing frequencies to avoid resonance.

Fatigue

• For machines that undergo vibrations associated with transient operational changeover conditions (for example, pump-turbines) or spend a large part of their operating life at substantially off-design conditions, fatigue analyses of the wicket gates would be recommended.

Headcover

The function of the headcover is pressure containment and support for various other turbine components and systems, most notably, the wicket gates, the wicket gate operating system, the turbine guide bearing, and, in some applications, the unit's thrust bearing.

Headcover mechanical design considerations to be addressed are:

Geometry/Configuration

• The headcover is typically removed vertically from the turbine pit and through the stator bore. Often it is necessary to split the headcover into pieces to accomplish this. Other designs incorporate an additional intermediate shaft between the generator and turbine shafts that can be removed sideways from the turbine pit through an equipment removal passageway. Once the intermediate shaft is removed, the headcover is raised and also removed laterally through the same passageway using special handling equipment.

Deflections

• Sufficient stiffness must be provided in the headcover construction to minimize the angular rotation of the upper deck of the headcover and, in so doing,

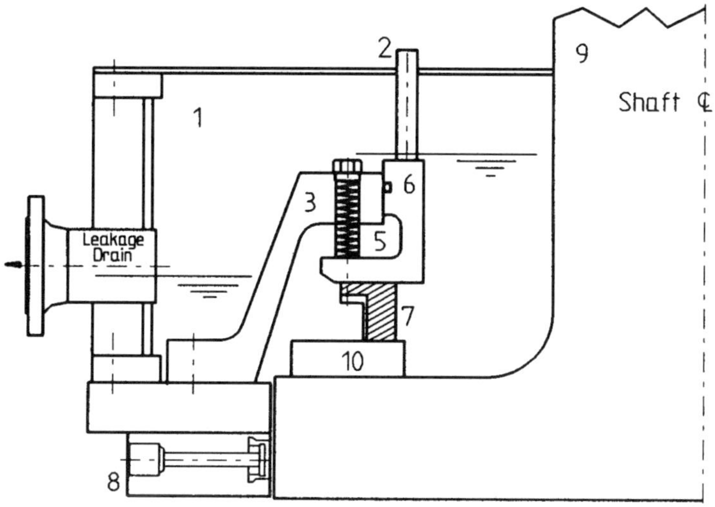

Hydrodynamic Seal

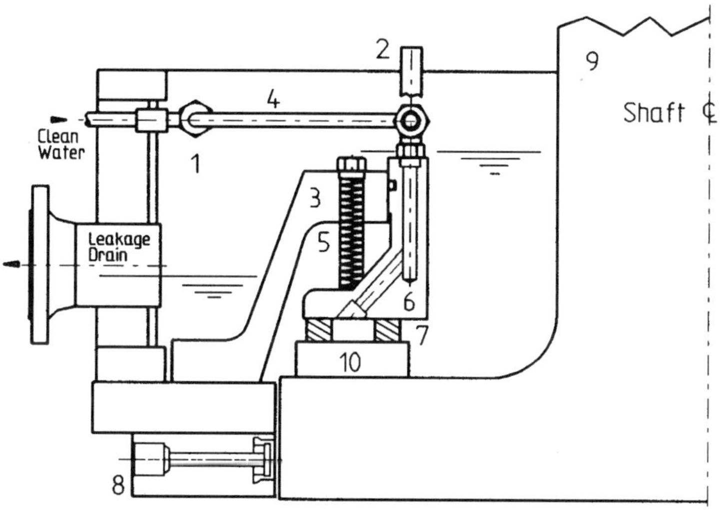

Hydrostatic Seal

Key:

1	*Water Collecting Pan*	*6*	*Seal Ring*
2	*Retaining Pin and Indicating Pin*	*7*	*Seal Ring-Insert*
3	*Casing*	*8*	*Maintenance Seal*
4	*Seal Water Pipe*	*9*	*Shaft*
5	*Compression Spring*	*10*	*Sliding Ring*

FIGURE 3-32: *Mechanical Design Features of Shaft Seals.* *This figure shows the two versions of the axial face seals: hydrostatic (bottom), which requires clean water injection into the sealing interface, and hydrodynamic (top), which generates its own interface liquid film by using the fluid being sealed.*

also reduce the slope at the gate stem bore locations. The ribs between the upper and lower decks of the headcover must transfer sufficient shear load to minimize the offset between the upper and lower headcover wicket gate's stem bore locations.

Stress

• Stress is generally not a problem with headcovers except at areas of abrupt discontinuity between ribs and axisymmetric decks and barrels. These discontinuities must be avoided when possible or adequate stress transition provided when unavoidable.

Shaft Seal

The shaft seal provides the interface between the pressurized turbine water passage and atmospheric pressure at the location of the turbine shaft. The shaft controls the leakage of the water passage medium, whether it is water or air, and must, in some cases, also operate if a vacuum is present at the water passage side of the shaft seal.

Hydraulic turbine shaft seals usually fall into two main types: mechanical seals and packing, or stuffing, boxes.

Mechanical seals consist of two different configurations: radial face seals and axial face seals. Each of these types of seals gets its name from the orientation of the "sealing interface" with respect to the turbine shaft axis. "Radial" face seals provide a sealing interface in a plane parallel to the shaft's rotating axis. "Axial" face seals provide a sealing interface in a plane perpendicular to the shaft's rotating axis.

Two versions of the basic "axial" face seal concept have been used: hydrostatic, which requires clean water injection into the sealing interface, and hydrodynamic, which generates its own interface liquid film by using the fluid being sealed. Figure 3-32 shows both versions.

The use of hydrostatic seals is especially advantageous when the following special operation requirements exist:

- Lowest possible face wear rate;
- Contaminated operating water, particularly when the contaminants are abrasive solids; and
- Synchronous condenser operation (with runner operating in air).

As an example, cooling water must be provided to the seal in a pump-turbine installed with a high submergence and required to run in air as a synchronous condenser.

Radial face seals incorporate segmented sealing elements that are mechanically held against the shaft to provide the sealing interface pressure and compensate for seal element wear.

The packing-box type of main shaft seal is the simplest configuration of the mentioned shaft seals. The sealing element is generally a series of square-woven packing compressed by a packing gland to provide sufficient packing pressure on the main shaft to provide controlled leakage through the packing box. It should be pointed out that packing-box seals require a small amount of leakage to cool the seal properly.

For shaft seals where the seal remains submerged even at minimum tailwater elevation, a separate maintenance seal is normally specified. This maintenance seal, when actuated, will allow exchanging the main sealing elements of the shaft seal without dewatering the draft tube.

Bearings

The bearing system for hydraulic turbine-generator units typically consists of two or three guide bearings and a thrust bearing. The guide bearings are arranged to meet the requirements of the shafting system's mechanical design. The number of guide bearings and their spacing are important factors in determining the critical speed of the shafting system.

The turbine guide bearing must be located as close to the runner as can be physically accomplished and still provide access to the shaft seal. The close proximity to the runner is required since the runner and its radial hydraulic loads are essentially "overhung" loads, and the amount of overhang must be as small as possible to optimize the shafting system characteristics. These characteristics relate to shaft stress, deflection, and the shaft system's critical speed.

The critical speed is that speed at which the number of revolutions per second of the shaft is equal to the frequency of its natural lateral vibration.

The rigidity of the bearing supports also will have an effect on critical speed, and, thus, this must be considered when taking into account the margin of safety to use. The lowest critical speed should be at least 25 percent above the maximum runaway speed of the machine.

Most vertical guide bearings used on hydraulic turbines are oil-lubricated. These bearings are generally of two design types: shell, or journal, and adjustable shoe, or segment. For the shell or journal type, the oil is fed to the bearing surface by pressure or gravity from a separate circulating system or is pumped from a revolving oil pan or shaft collar that is immersed in an oil bath. The adjustable shoe or segment type bearing is self-lubricating since it uses a self-generated hydrodynamic fluid pressure distribution enhanced by the pivoting capability of the shoe or segment.

The type of guide bearing that offers adjustability, good combined (oil and structural) stiffness, and the closest load support to the runner is the adjustable shoe or segment bearing. The resulting close proximity of the guide bearing to the runner, and, therefore, the shaft seal, limits space and precludes the use of a large oil basin necessary for a self-pumping bearing. Therefore, the benefits of this bearing arrangement are fully realized when the lubricating oil is pumped from an external sump that includes cooling coils (or heat exchangers), controls, and redundant pumps and motors.

A thrust bearing supports the weight of the rotating elements in the turbine-generator and hydraulic thrust loads. Normally, thrust bearings are designed so that the rotating element builds a supporting oil film against a series of thrust shoes that are rigidly or elastically supported. The thrust bearing components are housed in an oil basin and flooded with oil for lubrication. Oil coolers in the oil basin or an external lubrication system equipped with heat exchangers, controls, and pumps removes the heat. Many thrust bearings use a hydrostatic lift system to pump oil into the bearing surface for starting and stopping to minimize wear at low speed operation. The high pressure lift system is used often during maintenance to reduce bearing friction during manual rotation of the turbine-generator.

The bearing shoes should have temperature detectors to monitor the shoe temperatures during operation. Any temperatures greater than the established limit indicates a potential problem that requires investigating and, normally, stopping the machine for inspection.

Thrust bearings for vertical machines are mounted on structural supports that provide good axial stiffness and are located either above the generator, directly below the generator, or mounted on the turbine headcover. A thrust bearing mounted on the turbine headcover generally can reduce the overall height of the turbine-generator unit and, therefore, potentially can reduce the powerhouse volume required to contain the units. This is an important cost factor, particularly for underground powerhouse construction.

Discharge Ring

The discharge ring's function is to transmit pressure containment loads and portions of the powerhouse loads above it to the foundations that support the unit. The discharge ring, or the bottom ring when not integral with the discharge ring, supports the lower wicket gate stem bushings, forms a portion of the water passage adjacent to the wicket gates, and forms a tight clearance with the ends of the wicket gates. For Francis-type machines, it also contains the stationary portion of the lower wearing ring that provides a close running clearance with the rotating runner. For axial flow machines, the discharge ring forms the primary water passage adjacent to the periphery of the runner blades.

Sufficient stiffness must be provided to ensure that the discharge ring will not undergo unacceptable distortions in these critical areas during anchorage pre-load, embedment in concrete, grouting operations, and transient operating conditions. To provide protection against cavitation pitting, consideration should be given to specifying stainless steel material for portions of the discharge rings adjacent to the runner in axial flow machines.

Runner

The runner is the energy conversion element of the turbine, and, therefore, it transmits hydraulic loads through blades, or vanes, to a drive disk, known as the runner crown. The shroud that forms the lower/outer portion of the water passage is attached to the blades, or vanes, and is known as the runner band.

The runner for axial flow turbines has multiple blades that transmit hydraulic loads to a central component, known as the hub. Axial flow machines do not have a shroud or band fixed to the blade at their outer edge as Francis turbines do. Kaplan runners are axial flow turbines that have movable, controlled, adjustable blades. Propeller runners are axial flow turbines that have blades similar to Kaplan turbines, but the blades are fixed to the runner hub, and therefore, are not adjustable. A hybrid of the Francis and propeller runner is the mixed flow runner that resembles a high specific speed (low head) Francis runner except that it does not have a shroud or band attached to the blades.

The material selected for new project, replacement, or upgrade constructions of hydraulic turbine runners has, over the years, become, almost exclusively, stainless steel. Martensitic grades of stainless steel (13 to 16 percent Chrome, 3 to 5 percent Nickel) have been used to a large degree because these materials offer a combination of good weldability, high strength, and cavitation resistance. Austenitic stainless steels offer excellent cavitation resistance and weldability, but tend to have lower strength characteristics. Austenitics (18 to 20 percent Chrome, 3 to 12 percent Nickel) inherently have better corrosion characteristics than martensitics due to the higher chrome content. With any stainless, steps should be taken to avoid contamination during manufacturing and handling.

Francis runner designs typically use stainless steel blades, bands, and crowns with some designs using carbon steel crowns instead of stainless steel. Kaplan runner designs typically use stainless steel blades and carbon steel hubs.

Francis turbine runners provide close-running clearance seals located so as to limit leakage losses around the runner, minimize friction losses at the seal, and limit hydraulic thrust. Removable and replaceable stainless steel or aluminum bronze wearing rings often form these seal areas. When stainless steel crowns and bands are used, the seals often are machined as an integral part of these components.

The major design considerations for runners include:

Deflection

• The expansion of the runner crown and band (primarily, the band) for all normal operating conditions must be checked to ensure that the tight running-seal clearances at the wearing rings will not contact. This must be checked in combination with the shafting system deflections.

Stress

• The crown, band, and blade, or vane, thicknesses should be compatible, particularly at the discharge edge junction of the blades, or vanes with the crown and band. This allows each component to carry its share of loads and moments without unnecessary constraint; and

• Adequate stress transition must be provided between blade, or vane, discharge edges and crown and, to a lesser extent, the band.

Natural Frequency

• For some applications, the runner's natural frequency should be calculated to ensure that it is sufficiently separated from the normal operating frequencies of the unit to avoid resonance.

Draft Tube Liner

A plate steel draft tube liner is used below the discharge ring and extending to the region where the maximum velocity, averaged across the section, is about 5.5 meters/second (18 feet per second). The steel liner protects the concrete walls from being eroded by the higher velocity flow. Some unit configurations provide for the top portion of the draft tube liner to be removable. This allows for the runner to be removed from below after the draft tube section has been removed. The advantage of this arrangement is that runner repair or replacement can be accomplished without having to disassemble the turbine stationary completion parts (for example, headcover, bearings, and gate mechanism or other rotating parts (such as the shaft or generator).

For inspection and maintenance purposes, there typically is a man-door (access door) in the draft tube liner, just below the runner. For safety, all man-doors should have a valve or petcock at an elevation below the bottom of the door to test the level of the water behind the door.

The two main loading conditions for the mechanical design are an assumed vacuum (usually one-half atmosphere) in the draft tube during operation and the pressure of the fluid concrete poured on the outside of the liner during its embedment. Liner thicknesses of 13 to 19 millimeters (0.5 to 0.75 inch) are quite common but require sufficient external ribbing and anchorage to provide the stiffness needed to resist the loading. Frequently, internal bracing is required during the pouring operation.

The first-stage concrete is blocked out for this liner and fitted with anchorage for aligning and securing it during placement of the second stage concrete. If only one or two small units are involved, the full length of the draft tube may be fabricated from plate steel. This eliminates the need for any form work. However, the conical inlet position is still blocked out and attached later to avoid alignment problems with the remaining embedded parts.

Structural Analysis

The major structural considerations for hydraulic turbines are stress, deflection, and vibrational amplitude and frequency. Hydraulic turbines' continued reputation as prime movers with long life expectancy depends on an accurate determination of loads and stresses.

The performance and long-term reliability of hydraulic turbines are influenced also by several close-running clearances; therefore, the turbine design's success also depends upon an accurate analysis of deflection.

Deflection is assessed by analyzing the interaction of the stay ring/spiral case, headcover, and discharge ring as an integral unit. Hydraulic forces for each of the various operating conditions are combined and the resulting load is distributed throughout the unit. The analysis yields the deformation of critical areas such as the headcover O-rings, wearing ring/seals, bearing, and gate stem bushings. The deformation of the stationary parts is then examined relative to the deflected rotating assembly. Clearances are evaluated for manufacturability, performance, and long-term reliability.

Hydraulic turbines are analyzed also to avoid adverse vibrational amplification caused by structural resonance. Higher head and capacity increase the amplitude of the forcing frequencies. Additionally, reducing clearances between the runner blade periphery and the wicket gates can improve performance, but can increase the amplitude of the forcing frequencies. The rotating system—the generator-motor and hydraulic turbine or pump-turbine—is analyzed for its interaction with stationary components, then the analysis is correlated with prototype and model data. Table 3-4 lists the known forcing or exciting frequencies that hydraulic turbine designers need to compare to the turbine components' natural frequencies to avoid structural resonance.

The structural design and analysis of individual components and the combined assembly of hydraulic turbines use advanced computer techniques. The principal analysis methods available to the designer are:

- The finite element method for stress, deflection, and vibration analysis; and
- Fatigue and fracture mechanics methods for predicting potential operating life of the equipment as well as failure analysis.

Chapter 3

Finite Element Method

The finite element method is a modern method used to determine the stress, deflections, and the complex interaction of major components of turbine and

TABLE 3-4: ***Expected Frequencies in Hydro Equipment***

Frequency (Hertz)	Remarks
RPS = RPM/60	One per Revolution
RPS/3.6	Rheingans number of 3.6 varies from 2.0 to 4.0 (could be as high as 8)
RPS x (Number of Runner Blades)	Hydraulic Forces from Runner Blades
RPS x (Number of Wicket Gates)	Hydraulic Forces from Gates
$\frac{RPS \times Runner\ Blades \times Gates}{(Gates - Vanes)}$	Interaction of Wicket Gates with Runner Blades
RPS x (Number of Bearing Pads)	Normally Noticeable in Shafting and Bearing Data
$\frac{a}{4L}n \begin{cases} n = 2, 4, 6, \ldots \\ n = 1, 3, 5, \ldots \end{cases}$ *where:* a=sonic velocity L=length	Natural Frequency of Pipe a/4L or a/2L Normally Most Predominant and Can be Estimated for Very Simple Systems Only
RPS x (Number of Generator Poles)	Generator Associated
RPS x (Number of Stator Slots)	Frequencies

pump-turbine distributor sections. The design optimization of major components such as the stay ring, spiral case, runners, headcovers, wicket gates, and discharge rings use the finite element method. Detailed analyses are completed to generate the optimal shapes to minimize stress in the transition or fillet areas of major components.

The accurate stress and strain results calculated in local areas can be used in the structure fatigue life analysis that is very sensitive to the values of input stress and strain. Resonant frequencies and mode shapes are calculated for components subject to dynamic loading such as wicket gates, runners, and headcovers.

Fatigue and Fracture Mechanics

Significant advancement in fatigue analysis technology has led to its increased use in the hydro turbine industry. It is being applied routinely at the design phase for structural life prediction, material selection, and to define material flaw acceptance criteria. It also provides insight into fatigue failures of older equipment and has been successfully used in the structural redesign of this older equipment to correct fatigue-related problems. Fatigue analysis is particularly important for designing pump-turbines that undergo repeated mode changes and significant dynamic loads, therefore, increasing the number of stress cycles.

The process of fatigue failure consists of a crack initiation stage, followed by a controlled crack propagation stage, after which sudden failure occurs. Analyzing the crack initiation phase of fatigue failure relies heavily on the strain-amplitude versus number of reversals to failure and the material stress-strain properties.

The crack propagation analysis requires the crack growth ratio versus stress intensity data of the material. Crack initiation and propagation fatigue properties for most materials can be found in available literature. Laboratory tests may be required for materials not found in the literature or if the material will be operating in a hostile environment that is substantially different from the laboratory-tested environment.

Material Selection

Unit availability is extremely important to the project's owner. For this reason, material selection must be keyed to the need for low maintenance and maximum reliability for each component. In addition to the usual chemical and physical property considerations for materials typically used for hydraulic turbines, the following additional factors must be considered:

- Corrosion and the possibility of galvanic action;
- Abrasion resistance (surface hardness);
- Cavitation resistance;
- Resilience (to absorb energy within the elastic range);
- Toughness (to absorb energy within the plastic range);
- Compatibility (to avoid galling of close-running surfaces or during reassembly of parts);

• Weldability and repairability;
• Fatigue properties; and
• Machinability.

Manufacturing methods also must be considered in material selection. The behavior of the material during the manufacturing process needs to be understood. Materials that become embrittled when welded or develop a non-reversible work hardening during the machining process will not be suitable for some components.

Foundations, Embedment, and Unit Assembly

Foundations

The powerhouse's first stage concrete placement establishes the foundation upon which the turbine and generator will be built. For medium-sized plants, a large mass of concrete forms the base for the pedestals that support the spiral case and stay ring, forms the blockouts for the embedded parts, and contains the anchors for the tie-down rods and any embedded piping necessary for auxiliary services.

The support pedestals are designed to carry the weight of the stay ring and spiral case, generally assuming that the spiral case is filled with water. Anchors attached to the turbine equipment are used for alignment and securing components from "floating" during embedment in concrete.

The draft tube elbow and downstream portion are frequently included in the first stage concrete pours. In a multi-unit project where it is economical, the form work for these can be made from wood in a convenient number of sections to permit dismantling and reassembling into subsequent units. If the discharge end of the foot is too wide for economical support of the roof or for a reasonably sized exit gate, a pier can be used effectively to reduce the span without harmful changes to the flow characteristics.

The stay ring, spiral case, draft tube liner, discharge ring, and pit liner constitute the embedded parts because they are normally anchored to, and cast into, the large mass of concrete that forms the powerhouse. While all other parts of the turbine are considered removable, these, with the occasional exception of the discharge ring and draft tube liner cone, are not.

The Embedment Process

Depending on the design used, the stay ring or the discharge ring establishes the unit centerline elevation, the vertical centerline location, and the level or plumbness of the entire assembly. It is set upon adjustable jacks or wedges placed on columns that are usually concrete. These adjusting devices position the ring for elevation and level. Depending on the type of unit, adjustable tie rods, attached at one end to the anchors in the first stage concrete and at the other end to the stay ring or discharge ring, position the ring laterally and restrain it from moving during embedment.

Multi-arm or adjustable spider braces are placed into the upper and lower rings of the stay and discharge rings to preserve their roundness during the subsequent assembly work.

Those spiral case plates that could not be attached in the shop because of shipping limitations can now be welded into place to complete the spiral case.

If the spiral case is to be hydrostatically pressure tested in the field, a cylindrical barrel must be placed inside the stay ring and sealed against leakage. This barrel must be very rigid to resist deflections due to the high external pressure. Small deflections can cause a loss of seal and an aborted test. Likewise, a dished test head must be welded or bolted onto the spiral case entrance and provided with the means for admitting and releasing the pressurized water for the test.

Initially, the pressure is gradually brought to 50 percent or higher above maximum design pressure and held there for a specified period of time. Meanwhile, all of the weld seams and bolted joints are examined for leakage. When this holding period expires, the pressure is reduced to normal to complete one cycle. This cycle may be repeated depending on specific requirements.

After the last cycle, with the case still at normal pressure, the full assembly is checked for accuracy of position, elevation, level, and roundness. Even though the assembly is usually massive, typical level and roundness tolerances of 0.1 millimeter per meter (0.001 inch per foot) of diameter are achieved. Any final adjustments to the jacks, tie rods, and bracing must be completed before proceeding.

The second stage concrete pours are normally made while the case is held at or near maximum static head pressure.

Instead of embedding the spiral case under hydrostatic pressure, some end users, consultants, and contractors prefer to place a resilient mat over the top half of the spiral case, extending down to the outside horizontal centerline, or in some cases, covering the entire spiral case. After the concrete embeds the spiral case, this resilient mat provides room for it to expand under the internal operating pressure variations without transferring major loads to the concrete. This does not imply that all spiral cases are embedded. Sometimes, for smaller turbines, the spiral case is anchored to the concrete but not embedded.

A third method, used for smaller spiral case turbines, is to install the spiral case in a non-embedded condition. There are widely varying opinions as to the best approach; however, all have been used successfully.

Concrete exerts large buoyancy and lateral displacement forces in the fluid state. To help prevent distorting or moving the turbine parts, restrictions are placed on how the concrete is poured. Normally, a maximum depth of 1 meter (3 feet) is allowed in any pour. Further, during the pour, the fluid material is limited to a 0.3-meter (1 foot) difference in height around the part being embedded.

It is not possible to force concrete into all of the recesses formed by the underside of the draft tube liner, stay ring, spiral case, and discharge ring. To achieve a complete filling of these recesses, or pockets, a thin non-shrink concrete or epoxy grout is forced into those recesses that are known to provide air pockets during concrete pours. Two holes are made in the region to be grouted—the lower one for forcing the grout in and the upper one for releasing the trapped air. A low pressure is used to insert the grout so that equipment and anchorage are not unduly stressed and movement does not occur.

The final embedded part to be considered is the pit liner. The primary concern during field assembly of the pit liner is to ensure that the servomotor mounting flanges are set at the correct elevation and distance from the unit's centerline, as well as in the correct plane. If so, the installed servomotors will connect properly with the gate operating ring.

Unit Assembly

The turbine stationary completion parts are those that form, support, or seal the flow control surfaces, or water passages, and are not embedded parts. These consist of the headcover, bottom ring (if used), wicket gates and their operating mechanism, the turbine guide bearing, and the main shaft seal or packing box. These components are removable for access or extended maintenance. As such, they are also the components that, along with the rotating parts like the main shaft and runner, are involved in the final installation step of unit assembly.

The rotating and completion parts assembly sequence for a vertical Francis turbine generally proceeds as follows:

1. Install maintenance platform in the draft tube liner;
2. Clean the discharge ring, stay ring, etc., for completion parts assembly;
3. Install the discharge ring stationary wearing ring;
4. Install bottom ring (if provided separate from discharge ring);
5. Install wicket gates;
6. Lower runner into place and set down on the discharge ring ledge. The runner is set exactly level and centered in the bottom/discharge ring wearing ring. The runner needs to be set on shims lower than its final elevation. This will permit the runner with shaft to be set completely clear of the spigot fit with the generator shaft when the generator shaft is installed and resting at its final elevation on the thrust bearing;
7. Install the headcover into the turbine pit. It is sometimes required that the headcover be installed in two or more sections and bolted together in the turbine pit before lowering it over the wicket gates;
8. Set the headcover in place on the stay ring. Orient the gate stem bores (by rotating the headcover) to obtain the optimal vertical position of the wicket gates. Rotate the gates to prove that no binding exists. Center the headcover by establishing the proper clearances of the wearing rings between the runner and the headcover. Bolt the headcover to the stay ring (temporarily with eight bolts), drill, ream, and install the permanent dowels and the rest of the bolts between the headcover and stay ring. Matchmark the headcover to stay ring (Matchmarking is the process of scribing lines across connected surfaces in an assembly of parts in the factory to aid in the reassembly of those parts at the site);

9. Lower the turbine shaft into the turbine pit and set the shaft, in proper orientation, on runner;
10. Bolt the shaft to the runner using studs and nuts;
11. Check the concentricity of the runner with the discharge ring at the lower wearing ring gap. Check the level on the top surface of the upper shaft flange;
12. Install the shaft seal support, install dowels and bolts;
13. Install the shaft seal onto the headcover. Install the oil and water deflectors;
14. Install and center the outer bearing housing. Install the oil basin;
15. Install the gate servomotors in the pit liner;
16. Install the gate operating ring and all the gate linkage, gate arms, and other components. At this point the unit is "shaft free" and ready to begin the generator installation;
17. Install the turbine guide bearing with the oil basins and covers; and
18. Install the bearing oil pumping unit, if required, and all the necessary piping in the turbine pit for oil, air, and water. Install the turbine pit walkways.

Throughout the assembly process, critical measurements should be recorded for future reference and maintenance purposes.

The preceding sequence is general only. A specific alignment and assembly procedure with acceptance tolerances should be supplied by the manufacturer.

Prototype Operation

To maximize the performance, efficiency, power, flow, or annual energy of a plant, it is necessary to know the machine characteristics over the entire operating range.

The performance characteristics of a Francis turbine can be summarized as shown in Figure 3-17 where lines of constant efficiency have been plotted as a function of head and flow. This "hill" is developed from the model curves described previously. When operating the plant requires operation at extremes of power or head, the ramifications of such operation need to be known by the operator.

Superimposed on Figure 3-17 are typical operation limits. Operating at high loads is normally limited by full gate opening, the generator limit, the turbine setting (cavitation), or overall surging. At low outputs, operation is limited typically by part-load pressure pulsation and the potential for cavitation damage caused by part-load vortices. Low-head operation is limited by noise, vibration and potential cavitation pitting caused by the fixed cavitation limits while at the high-head the limitation is mainly pitting related.

The runner losses may be quite large when the turbine is operated at head or flow conditions that differ from those used for the runner design. The incoming flow velocity and entrance angle will be different than those values for which the runner was designed. Since the Francis and fixed-blade propeller turbines cannot alter their shapes to match these new conditions, the losses will increase and the efficiency will be proportionately lower.

Tailwater Depression/Synchronous Condensing

A hydro project is ideal for supplying low-cost, reactive power to the system and for voltage regulation by driving the generator as a motor carrying no load (synchronous condenser operation). To do this at minimum power input, it is necessary to lower the tailwater inside the draft tube (or wheel pit for a Pelton turbine) to some level well below the bottom of the runner so that it spins in air. For the same lower input power reason, tailwater depression is used for pump-mode starts on reversible pump-turbine units.

Typical tailwater depression systems include compressors, storage tank(s), large and small admission valves, water-level sensors, and controls.

To operate a unit as a synchronous condenser, the wicket gates are opened to achieve synchronous speed, at which time the line breaker is closed and the gates are returned to the closed position. When fully closed, compressed air from the storage tank or blowers depresses the tailwater to a preset low level. As this level is reached, the air admission is shut off. The runner is now rotating in air at synchronous speed and being driven by input from the system. The tailwater may rise slowly because of leakage and entrainment of air in the water. When an upper preset level is reached, the controls open the air admission to return the level to its lower preset value. While this cycle is being repeated, the compressor(s) recharges the storage tank(s) to prepare for depressing the next unit, if desired. When in condenser operation, the seal water should be supplied to the runner seals at all times.

Opening the wicket gates to admit the full penstock flow that sweeps the air into the tailrace can return a turbine unit to full generating service from tailwater depression. However, this must never be done on a pump-turbine unit when operating depressed in the pump mode. The large volume of air would rush up the penstock, growing larger and accelerating as it approached the surface. The water volume pushed ahead of this air mass has the potential for causing severe damage to the forebay intake structure.

To allow a lower setting with greater use of head than otherwise would be possible when high tailwater exists, a tailwater depression system can be applied to a Pelton turbine. Such a system supplies compressed air, preferably near the center of the runner, to lower the tailwater in the wheel pit, thus allowing the runner to rotate in air when the water in the tailrace is above the elevation of the bottom of the buckets.

BIBLIOGRAPHY

ASME Power Test Code, PTC18, American Society of Mechanical Engineers, New York, New York, 1992.

Avallone, Eugene A., and Theodore Baumeister III, *Marks' Standard Handbook for Mechanical Engineers,* Ninth Edition, McGraw-Hill, Inc., New York, 1978.

Brekke, Hermod, "A Discussion of Impulse Turbines versus Francis Turbines for High Head Plants," Paper Presented at the Joint ASME/ASCE/IAHR Symposium, Fort Collins, Colorado, 1978.

Brekke, Hermod, "Recent Trends in the Design and Layout of Pelton Turbines," *International Water Power & Dam Construction,* November 1987.

Brown, J. Guthrie, *Hydroelectric Design Practice,* 3 Volumes, Second Edition, Blackie and Son Limited, 1964.

Buckingham, E., "On Model Experiments and the Form of Empirical Equation," *Transactions of the American Society of Mechanical Engineers,* Volume 37, 1917, pages 3-10.

"Cavitation," Movie #2160, Encyclopedia Educational Corporation, Chicago, Illinois.

Cavitation Pitting Mitigation in Hydraulic Turbines, Volume 1: Guidelines and Recommendations, EPRI AP-4719, Palo Alto, California, 1986.

Dawson, P.B., "Multi Jet Impulse Turbines," 64-WA/FE-24, American Society of Mechanical Engineers, November 1964.

de Siervo, F., and F. de Leva, "Modern Trends in Selecting and Designing Francis Turbines," *International Water Power & Dam Construction,* August 1976, pages 28-35.

de Siervo, F., and A. Lugaresi, "Modern Trends in Selecting and Designing Impulse Turbines," *International Water Power and Dam Construction,* December 1978, pages 40-48.

Fischer, J.H., and G.J. Russell, "Equipment Selection and Site Optimization Based on Energy Considerations," *Waterpower '87: Proceedings of the International Conference on Hydropower,* Portland, Oregon, August 1987.

Grein, H., D. Klicov, and W. Wieser, "Efficiency Scale Effects in Pelton Turbines," *International Water Power & Dam Construction,* May 1988, pages 32-36.

Grein, H., J. Meier, and D. Klicor, "Efficiency Scale Effects in Pelton Turbines," Presented at the International Association of Hydraulic Research Symposium, Montreal, 1986.

Grein, H., R. Angehrn, M. Lorenz, and A. Bezinge, *Inspection Periods of Pelton Runners,* Sulzer Escher-Wyss, Zurich, Switzerland.

Guide for Hydraulic Machinery Model Testing, EPRI AP-5876, Palo Alto, California, 1988.

Hutton, S.P., "Component Losses in Kaplan Turbines and the Prediction of Efficiency from Model Tests," *Proceedings of the Institute of Mechanical Engineers,* London, 1954.

Ida, T., *An Analysis of Scale Effects on Water Turbine Characteristics,* Report No. 18, Kanagawa University, March 1980.
IEC Publication 41, International Electrotechnical Commission, Geneva, Switzerland, 1991.
**IEC Publication 193 (with amendment 1), International Electrotechnical Com*mission, Geneva, Switzerland, 1965.
**IEC Publication 497,* International Electrotechnical Commission, Geneva, Switzerland, 1976.
**IEC Publication 609,* International Electrotechnical Commission, Geneva, Switzerland, 1978.
**IEC Publication 995,* International Electrotechnical Commission, Geneva, Switzerland, 1991.
Kent's Mechanical Engineers' Handbook, 12th Edition, John Wiley, 1950.
Kovalev, N.N., *Hydroturbines, Design and Construction,* Translated from Russian, Israel Program for Scientific Translation, Jerusalem, 1965.
Krueger, R.E., *Selecting Hydraulic Reaction Turbines,* Engineering Monograph No. 20, U.S. Bureau of Reclamation, 1976.
Kubota, T., "Observations of Jet Interference in Six-Nozzle Pelton Turbines," *Journal of Hydraulic Research,* Volume 27, No. 6, 1989.
Miller, D.S., *Internal Flow Systems,* British Hydromechanics Research Association, 1978.
Nechleba, Miroslav, and Artia Prague, *Hydraulic Turbines, Their Design and Equipment,* 1957.
Raabe, Joachim, *Hydropower: The Design, Use and Function of Hydromechanical, Hydraulic and Electrical Equipment,* VDI-Verlag, 1985.
Raabe, J., "The Negative Efficiency Scale Effect in Impulse Turbines and its Causes," Paper Presented at the International Association for Hydraulic Research Symposium, Tokyo, Japan, 1980.
Sinclair, J.P., and P.R. Rodrique, *Cavitation Pitting Mitigation in Hydraulic Turbines,* EPRI Report AP-4719, Palo Alto, California, August 1986.
The Value of Competitive Model Testing in the Bid Evaluation Process for Hydroelectric Turbomachinery, EPRI EM-4174, Palo Alto, California, 1985.
Vladislavev, L.A., *Vibration of Hydro Units in Hydroelectric Power Plants,* Second Edition, Amerind Publishing Co., 1979.
Warnick, C.C., *Hydropower Engineering,* Prentice Hall, 1984.

* Tentatively to be superceded by new IEC code 41.

In addition to the foregoing documents, numerous papers available from the American Society of Mechanical Engineers, the International Association of Hydraulic Research, the Waterpower Conference series, and the HydroVision Conference series are useful references.

Checklist for Chapter 3

This checklist includes information needed from the customer for the manufacturer to provide the optimal turbine design and manufacture.

1. Definition of scope of work.
2. Site conditions including:
   - Gross head range as well as sufficient information concerning the inlet, penstock, and tailrace design to calculate head loss;
   - Net head range;
   - Available flow;
   - Headwater and tailwater elevation ranges;
   - Flow data as a percentage of time, flow versus headwater elevation, and flow versus tailwater elevation (data should be supplied in tabular form);
   - Water temperature, water quality, and powerhouse elevation;
   - Powerhouse layout and unit rotational direction; and
   - Preferred centerline elevation.
3. Intended operational use such as peaking load, or run-of-river. Also included should be any constraints and/or requirements for special applications, such as irrigation.
4. Environmental constraints.
5. Powerhouse and/or geometry constraints.
6. Customer requirements (if any) of turbine type, configuration, size, number of turbines, and rotational speed.
7. Performance evaluation criteria and penalties. Include those for efficiency, power, metal removal, and excavation.
8. Testing requirements for both model testing and field testing.
9. Codes and Standards for design, manufacture, and testing of turbines.
10. Mechanical design requirements.
11. Sufficient penstock detail for transient analysis.
12. Delivery schedules.

For replacement units, the above information plus the following is needed:

1. Existing geometry.
2. Existing generator limitations including maximum capacity and synchronous and runaway speed.
3. Mechanical limitations such as servomotor stroke, thrust bearing capacity, clearances, and shaft stresses.

Chapter 4
Governors

Varying the water flow through a hydraulic turbine controls its speed or power output. A governor adjusts the water flow to maintain the generating unit's speed and the system frequency. Also, it maintains a balance between the water input to the turbine and the power requirements of the system. In general, this chapter describes the construction and operation of governors for hydroelectric units. For detailed information on governors, ANSI/IEEE standard 125 is a valuable reference.[1]

The primary functions of the hydraulic turbine governor when controlling a unit directly connected to an AC system are:

- To maintain and adjust unit speed before the unit goes on line;
- To maintain system frequency after synchronization by adjustments to the output of the turbine;
- When operating on a system in parallel with other units, to share load changes with the other units in a planned manner in response to system frequency error;
- To adjust the output of the unit in response to operator or other supervisory commands; and
- To protect the unit from uncontrolled runaway following sudden isolation from the electrical load or to initiate a unit shutdown in response to an abnormal condition.

The governor is a key component in an installation that uses an automated start-up and shutdown sequence. Operation of the governor usually is coordinated with protective relays, lubrication pumps, excitation control, generator braking, the synchronizer, and the main generator breaker.

With an isolated system and a single generating unit, the governor controls the frequency. When there is a large system, the governor may be used to match the frequency of the system for synchronizing, to regulate the load on the unit, and may contribute to system frequency control.

Some generating units are so small in comparison to the system they are electrically connected so that they cannot contribute significantly to system frequency control. In this case, a governor may not be required. Instead, a gate actuator may be used to position the flow-controlling devices to produce the desired output. If a synchronizer is used, it produces commands to raise and lower the actuator to control speed before and during synchronizing.

Basic Governor Principles

A governor has three major sections, shown schematically in Figure 4-1. These are:

- A speed sensing element that senses turbine speed and provides an output that is proportional to speed;
- A control element that compares the turbine speed to the desired speed set point and provides an output signal that represents the required control action; and
- A power amplification element that produces the mechanical force needed to position the water flow-controlling devices (wicket gates, blades, needles, or deflectors) in response to the control element's output signal.

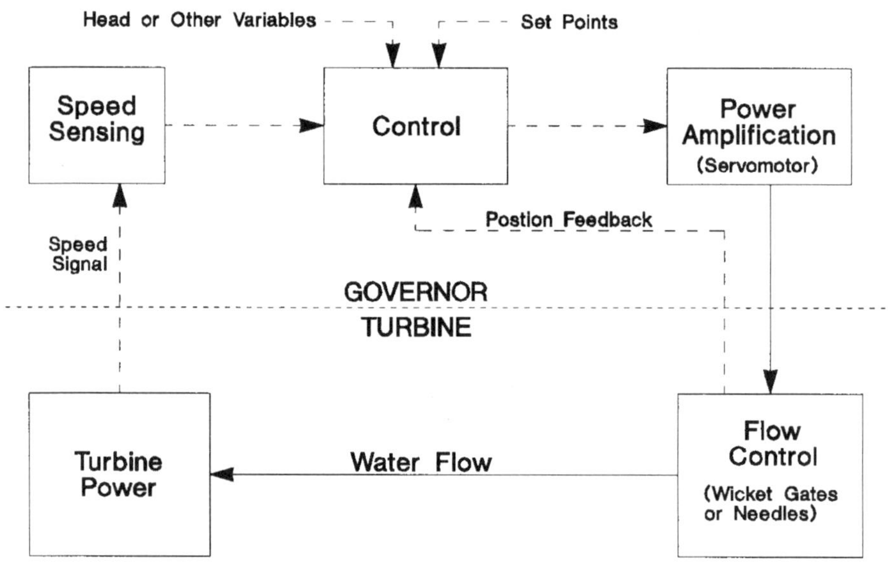

FIGURE 4-1: ***Basic Block Diagram of a Governor.*** *This figure shows the three major sections of a governor. Every governor contains a speed sensing element that senses turbine speed and provides an output proportional to speed; a control element that compares the turbine speed to the desired speed set point, and provides an output signal for the required control action; and a power amplification element that produces the mechanical force to position the water flow-controlling devices in response to the control element.*

Reliability

Any governing system should be designed with a careful analysis of failure consequences and fail-safe operation. This may include a cost-benefit analysis of redundant components, secondary automatic or manual systems, and engineering designs that have proven reliability. Also, this review will determine the type and number of spare components that should be kept on hand. Whenever possible, the entire system should be designed and reviewed so that if a failure should occur, the consequences of the failure have been planned for and appropriate remedial actions can be taken quickly.

Types of Governors

In the early twentieth century, engineers developed mechanical-hydraulic governors. These governors use hydraulic amplification, as a function of sensed speed, to provide the force that controls the water flow to the turbine. When the governor senses that the desired speed is decreasing, wicket gates are opened to increase the turbine power. For increases above the desired speed, the gates are closed to reduce the turbine power. This positions the wicket gates proportionally to the changes in speed caused by changes in load. When a unit is connected to a power grid, the resulting changes in turbine output contribute to the overall system frequency stability. Some utilities prefer a mechanical governor because of long-term parts availability, proven reliability, low maintenance costs, the lower level of technology necessary for maintenance and adjustment, and personnel familiarity with maintenance and repair.

In the 1960s, analog electronic governors were developed that use analog electronics to perform the control functions done mechanically on earlier governors. These governors provide electronic PID (Proportional-Integral-Derivative) action for improved control without the levers and dashpots used in mechanical governors.

In the 1980s, digital electronic governors were designed that use microprocessors, programmable logic controllers (PLCs), and software programs to perform the control functions. These governors provide control of speed, load, flow, water level, or any system parameter using a variety of algorithms.

Owners considering using an electronic governor should evaluate the need for additional spare parts (such as printed circuit boards) and the need for additional skills and training for maintenance personnel. This may be a particular concern for plants located in remote areas.

The governor may be packaged in various arrangements. Three configurations are common: the gate-shaft type governor; actuator-type governor; and the cabinet-type governor.

Gate-shaft type governor. The gate-shaft type governor is a configuration developed early in the twentieth century. These governors are of the mechanical-hydraulic type and use large castings for the gate-shaft bearing, servomotor, and

Chapter 4

FIGURE 4-2: ***Gate Shaft-Type Governor.*** *The gate shaft-type governor was developed early in the twentieth century. These are mechanical-hydraulic type governors and use large castings for the gate shaft bearing, servomotor, and governor housing. They have separate pump units and pressure tanks. Many gate shaft governors are still in service, but are not used for new construction today.*

(Courtesy Woodward Governor Company Turbomachinery Controls)

governor housing. These governors have separate pump units and pressure tanks. Many gate-shaft governors are still in service, but are not used for new construction today.

Figure 4-2 shows an example of a gate-shaft type governor. The control valve, servomotor, hand-control mechanism, and other associated parts of the governor-control mechanism are mounted on a single base. The turbine-control mechanism is operated through a gate shaft.

Actuator-type governor. With large units, the force requirements on the control linkage prevent the servomotors from being located remotely from the water-regulating devices. Therefore, actuator-type governors are used primarily for larger units. This type of governor provides for additional flexibility in powerhouse design because the servomotor(s) are mounted directly at the turbine. In an actuator-type governor, the control valve, hand-control mechanism, and associated parts of the governor-control mechanism are mounted on a single base or stand. The servomotor(s) are mounted remotely from the governor and adjacent to or integral with the hydraulic turbine. If the governor is either an analog or digital governor, the electronics usually are mounted in a separate cabinet. The pressure system may be separate from or may be integrated with the actuator assembly.

Cabinet-type governor. Almost all cabinet-type governors have remotely mounted servomotors and can be referred to as cabinet actuators. The cabinet-type governor consists of a base on which the governor components are mounted. The base and the components are enclosed in a cabinet with operating controls on the cabinet face. When the base is used as the oil sump, the pump assemblies are mounted atop the base. A section of the base or the upper cabinet housing commonly is used as the governor terminal enclosure. The governor electronics may be in an electrical compartment or cubicle that may be attached to, or part of, this cabinet.

The governor structure for the cabinet-type governor shown in Figure 4-3 includes a cabinet that encloses specified governor-control or hydraulic-pressure supply mechanisms.

Electronic governors may be supplied in a variety of arrangements or configurations. The electronic components may be located in a regulated environment in the control room with the electric-hydraulic transducer and control valve located on the pressure supply system. If a suitable environment exists in the plant, the electronics, electric-hydraulic transducer, and control valve may be combined in a configuration similar to the cabinet- or actuator-type governor. For small units, the electric-hydraulic transducer may be used as the control valve.

The physical size of the governor depends upon the volume of the servomotors, the operating pressure, and the required servomotor timing. The requirements for auxiliary equipment and control panel instruments also will affect the governor size. If the governor base is used as the oil sump, the required sump volume and pump assembly size(s) must be considered.

FIGURE 4-3: ***Cabinet-Type Governor.*** *Almost all cabinet-type governors have remotely mounted servomotors and can be referred to as cabinet actuators. The cabinet-type governor consists of a base on which the governor components are mounted. The base and the components are enclosed in a cabinet with operating controls on the cabinet face. The governor electronics may be in an electrical compartment or cubicle that may be attached to, or part of, this cabinet.*

(Courtesy Woodward Governor Company Turbomachinery Controls)

Speed Sensing

The ballhead mechanism of mechanical governors is the oldest speed sensing device. In its best known form, the ballhead consists of a pair of weights at the end of two arms. These arms are pivoted near the axis of rotation in such a way that the flyweights can move radially in a plane through the axis. Over the years, the centrifugal ballhead has progressed through a great number of design configurations to the more modern form seen in Figure 4-4. As shown, an increase in system speed will increase the centrifugal force on the flyweights and lift the speeder rod upward.

Most speed sensors today use electrical methods. One type uses a small DC generator with a permanent magnet field for which the output voltage is proportional to speed. Eliminating commutators and brushes, a variation uses a small permanent magnet alternator whose output is rectified with the DC voltage used to generate a speed sensing signal. Methods that result in a frequency output that

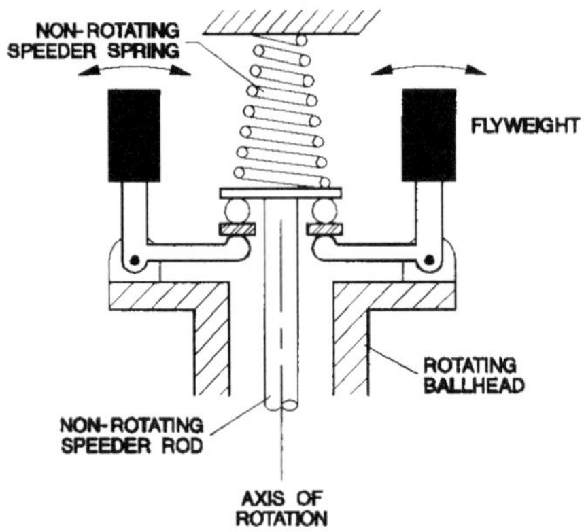

FIGURE 4-4: ***Modern Ballhead.*** *The ballhead mechanism of mechanical governors is the oldest speed sensing device. In its best known form, the ballhead consists of a pair of weights at the end of two arms. As shown in the figure, these arms are pivoted near the axis of rotation in such a way that the flyweights can move radially in a plane through the axis. Over the years, the centrifugal ballhead has progressed through a great number of design configurations to the more modern form seen in this figure.*

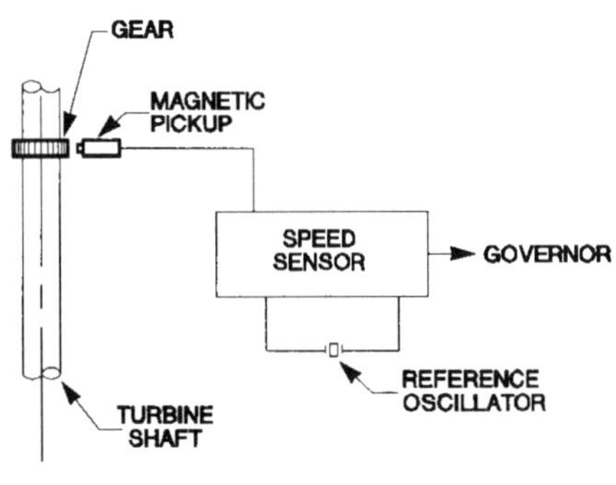

FIGURE 4-5: ***Electronic Speed Sensing.*** *A speed-sensing method commonly used today requires using a toothed wheel and a magnetic pickup. The toothed wheel may either encircle the turbine shaft or be driven by an auxiliary shaft and coupling. The output frequency of the pickup is measured by the speed sensor to determine the speed of the shaft.*

is proportional to speed are suited to analog or digital governors. Among these methods are:

• Either a permanent magnet alternator or generator output voltage feeds its output into a frequency sensitive network that provides a signal approximately proportional to the deviation in speed from an equilibrium value; or

• A speed-sensing method commonly used today requires using a toothed wheel and a magnetic pickup. The toothed wheel may either encircle the turbine shaft or be driven by an auxiliary shaft and coupling. The output frequency of the pickup is measured by the speed sensor to determine the speed of the shaft, as seen in Figure 4-5.

Control

The control element of a governor consists of those components that receive and act upon external signals such as speed, wicket gate position feedback, power, and pond level to produce appropriate commands to the governor power amplification element. The major differences between mechanical-hydraulic, analog, and digital governors occur in the control elements employed.

Mechanical-hydraulic control. The simple mechanical-hydraulic governor provides an easily visualized example of governor control principles. The ballhead is driven at a speed proportional to that of the unit and the servo is connected to operate the wicket gates, as seen in Figure 4-6. The simple combination of ballhead and directly connected pilot valve has only one equilibrium position: the position in which the valve is closed—neither admitting oil to nor discharging oil from the servo cylinder. This is known as isochronous governing.

Unfortunately, such a system is also inherently unstable. This is because the turbine speed does not instantly assume a value proportional to the wicket gate position due to the inertia of the rotating mass and the momentum of the water in the penstock. Therefore, if the turbine speed is below the governor speed setting, the pilot valve is positioned to allow oil flow to move the servo to increase the turbine gate opening. By the time the speed has increased to the setting so that the valve is centered and the servo stopped, the flow of water has already been increased too much and the turbine continues to speed up. Then, the pilot valve opens the other way and water flow begins to decrease. As before, when the speed gets to the right value, the servo has traveled too far, the turbine underspeeds, and the whole cycle is repeated. Some means for stabilizing such a system is required to secure a satisfactory governor.

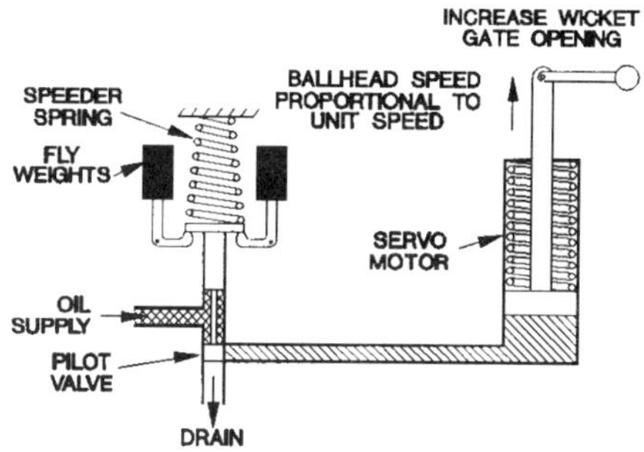

FIGURE 4-6: *Mechanical Speed Sensor and Servo.* *As seen in this figure, a ballhead is driven at a speed proportional to that of the unit, and the servo is connected to operate the wicket gates. The simple combination of ballhead and directly connected pilot valve has only one equilibrium position: the position in which the valve is closed—neither admitting oil to nor discharging oil from the servo cylinder.*

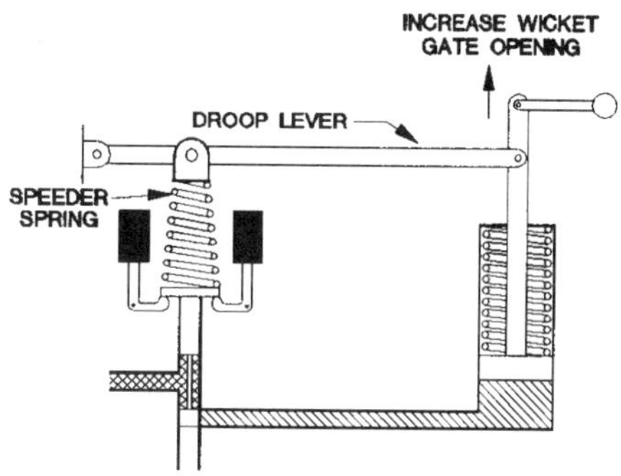

FIGURE 4-7: *Mechanical Speed Sensor and Servo with Speed Droop.* *Speed droop is a governor feature that requires a decrease in speed to produce an increase in gate. Since an increase in gate is required if the turbine is to carry more load, it follows that increased load results in decreased speed. As seen in this figure, the simplest method of securing stability is to add a way that will provide speed droop in the governor.*

The simplest method of securing stability (avoiding hunting) in the system described is to add means that will provide speed droop in the governor, as seen in Figure 4-7 (see droop lever). Speed droop is a governor feature that requires a decrease in speed to produce an increase in gate opening. Since an increase in gate opening is required if the turbine is to carry more load, it follows that increased load results in decreased speed. Speed droop may be a nonlinear relationship. For a complete definition of speed droop, refer to the ANSI/IEEE standard 125.[1]

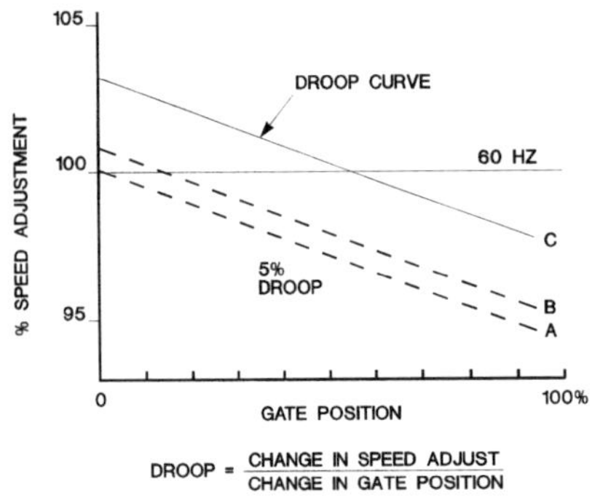

FIGURE 4-8: ***Speed Adjustment versus Gate Position Showing Droop.*** *For each position of the manual speed adjustment, there will be a corresponding sloping line. A line sloping or "drooping" downward to indicate decreased speed with movement of the servo in the "increase water flow" direction may represent the equilibrium relationship between speed and servo position for such a system. (See text for more details about the example illustrated with this figure.)*

A mechanical interconnection between servo (and, therefore, wicket gate) movement and the speeder spring can provide speed droop in a simple governor so that as water flow is increased, the speed is decreased. A line sloping or "drooping" downward to indicate decreased speed with movement of the servo in the "increase water flow" direction may represent the equilibrium relationship between speed and gate position for such a system. For each position of the manual speed adjustment there will correspond a sloping line, such as those shown in Figure 4-8.

In the example shown in Figure 4-8, assume that the unit is not yet paralleled to the system and the speed adjustment is set to 100 percent (line "A"). Assume, also, that the unit requires a 10 percent wicket gate opening to reach synchronous speed at no load. The intersection of the "A" line and the 10 percent gate position establishes the actual speed of about 99.5 percent. The speed adjustment must be raised to 100.5 percent (line "B") to bring the actual speed to 100 percent. After matching the voltage and phase angle, the unit now can be paralleled to the system. Once paralleled, the system will force the generator to stay synchronized with it. Now, raising the speed adjustment actually opens the wicket gates and loads the unit (line "C"). If the system frequency were to decrease, all connected generators would open their wicket gates in proportion to their individual droop settings. This combined increase in generation will help to restore system frequency. A common droop setting is 5 percent, although an adjustment range of 0 to 10 percent is commonly specified.

What has been described above is called permanent droop. As a practical matter, a governor system with only permanent droop would have an unacceptably large variation in speed over its load range. To avoid this problem, temporary speed droop, or compensation, is used. This produces a larger temporary readjustment of speed setting with servo movement to produce the stabilizing speed droop characteristic. This is then followed by a relatively slow return of speed to the speed droop line shown in Figure 4-8.

The temporary droop feature can be accomplished in a number of ways. Mechanical governors use a variety of schemes. One solution, shown in Figure

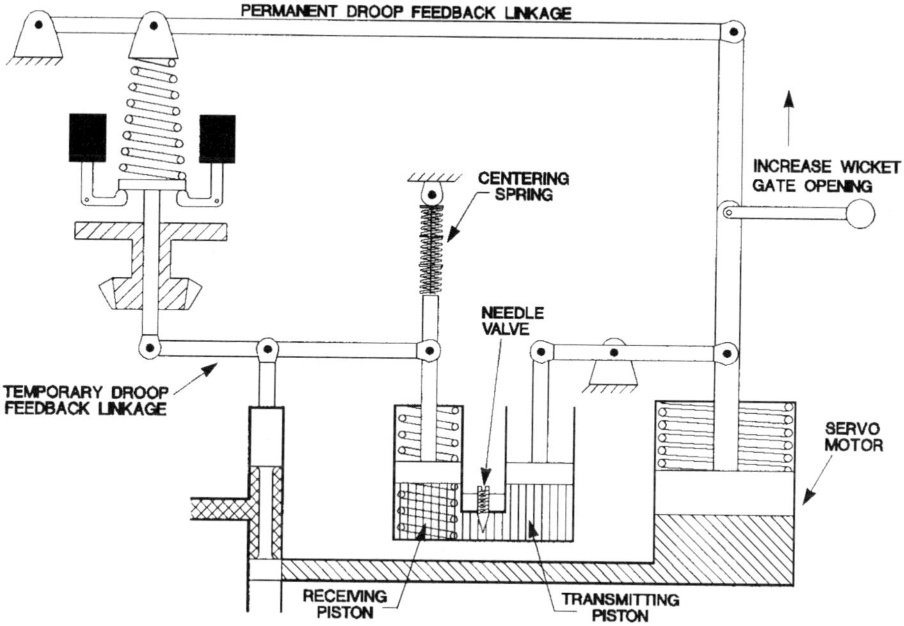

FIGURE 4-9: *Governor with both Temporary and Permanent Droop.* *This mechanical governor has a temporary droop dashpot to provide a quick response to system disturbances, and a permanent droop feedback lever so that the unit will operate on a fixed droop curve, as see in Figure 4-8.*

4-9, involves feedback linkages, transmitting piston, receiving piston, and needle valve. The receiving piston is centered by spring force to an equilibrium position. As long as the receiving piston is in this equilibrium position and there is no change in unit speed, there is no servomotor movement. For a fixed speeder spring setting, this means that the ballhead must, for final equilibrium, be running always at the same speed. Any servomotor movement will move the transmitting piston and initiate an oil flow. This oil flow displaces the receiving piston from its equilibrium position. Thus, if there were no leakage from this compensating hydraulic system, the receiving piston would move as though rigidly connected to the servo, and permanent speed droop would result because the centering spring was ineffective. However, if an adjustable leak in the form of a needle valve is provided between the compensating hydraulic system and an oil sump, the centering spring can return the receiving piston slowly to its initial position. This is done after a disturbance by forcing oil out or drawing oil in through the needle valve as required. As this occurs, the speed setting of the governor slowly returns to its original value although the servo is permitted to remain at the new, different load position. Such a governor is isochronous, although provided with the necessary temporary droop for stability.

Figure 4-9 shows that it is possible to combine temporary and permanent

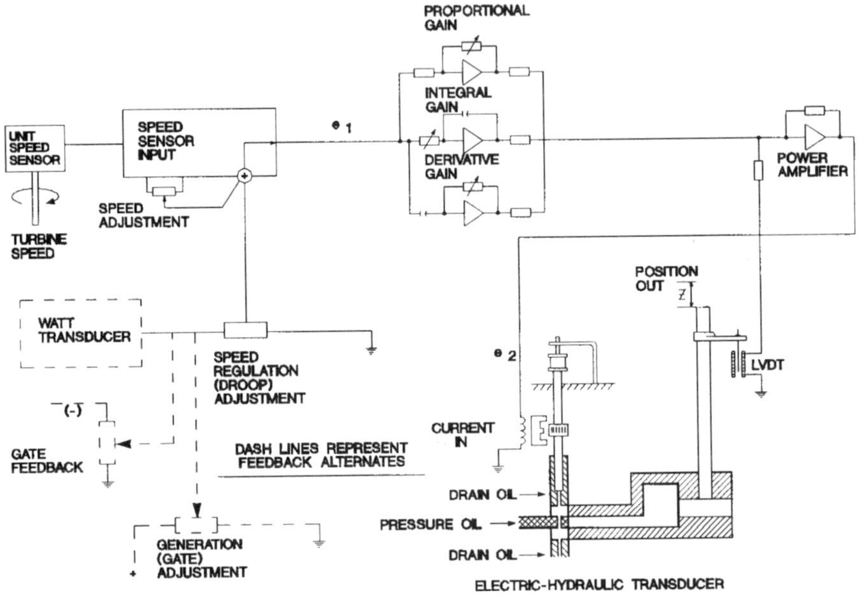

FIGURE 4-10: ***Block Diagram of a Typical Analog Governor.*** *The turbine speed-sensing device in this block diagram produces an output frequency or voltage that is proportional to turbine speed. The speed-sensing circuit compares this frequency or voltage with a reference. Any difference creates an analog speed error voltage. With an analog governor, it is possible to use the generator output instead of gate position as the control feedback. The resulting error is processed by a PID gain circuit, and sent to a power amplifier. The output current is used in the electric-hydraulic transducer as the first stage of the hydraulics system.*

droop in a single governor. This is done in many hydropower units where parallel operation requires a responsive droop governor.

Analog control. Figure 4-10 shows a diagram of a typical analog governor. The turbine speed-sensing device produces an output frequency or voltage that is proportional to turbine speed. The speed-sensing circuit compares this frequency or voltage with a reference; any difference creates an analog speed error voltage (e_1). Various other functions also can be applied to this error voltage.

An important difference between mechanical governors and electronic governors is the possibility of using generator output (true power) instead of gate position as a control feedback. This results in a linear relationship between speed setting and unit load and is called speed regulation. In speed regulation, unit speed is compared to unit generation; whereas, with speed droop, unit speed is compared to turbine servomotor position. The main advantage of speed regulation is the greater accuracy of generation feedback. At constant speed, generation feedback keeps unit output constant regardless of changes in net head or restrictions on water flow. Therefore, using speed regulation may be desirable for electrical dispatch control but is less responsive as it requires lower gains to maintain stability. Generation feedback is undesirable for isolated operation. A governing system may control unit generation or unit frequency, but it cannot do both at the same time.

During steady-state conditions, the error voltage (shown as e_1 on Figure 4-10) is zero. A non-zero error signal will lead to wicket gate movement by the following process. A summing amplifier adds the signals representing speed error, speed adjustment, and speed regulation (a generation feedback signal). The resulting signal is the input for the PID amplifiers (proportional, integral, and derivative). Proportional and integral action are analogous to the temporary droop and dashpot time constant of the mechanical governor. One advantage of analog control is the derivative function that is not available on a mechanical governor. The derivative can increase the speed of response to a change. A power amplifier combines the output signals from the three PID amplifiers. The output from the power amplifier operates the electric-hydraulic transducer, which produces an output position (z) proportional to the input current (e_2). This output is then amplified through other hydraulic stages to move the servomotor.

Thus, the electronic analog governor maintains a zero error voltage by a rapid response to adjust the gate position. This produces the desired balance between unit speed and generation or gate position depending upon the control strategy. PID gain adjustments can modify the speed of this response and governor stability.

Chapter 4

Digital control. A digitally controlled governor can control speed, unit output, flow, water level, or any system parameter that can be sensed by a transducer, switches, or contacts. The digital governor's main advantage is its versatility and the ease with which changes can be made. It also enables easy interfacing with other computer systems.

A digital governor can perform many additional control functions not done previously by mechanical governors and some not done by analog governors. Functions that digital governors can provide include:

- Speed and Power Control;
- Generator Control: Some digital governors include voltage and reactive power (VAR) control;
- Pond Control: There are many different control requirements for headwater control or tail water control;
- Flow Control: Water supply requirements or environmental concerns may dictate minimum flows;
- Turbine Creep Detection: Using zero velocity speed sensors and software logic, digital governors can detect turbine creep;
- Control Sequencing: Governors using relays typically do control sequencing of governor start up, braking, and shutdown. Digital control expands the sequencing capabilities and allows ease of changes for such applications as sequencing of multiple units or unit start or stop based on time to control ponding, irrigation, or scenic water flow requirements;

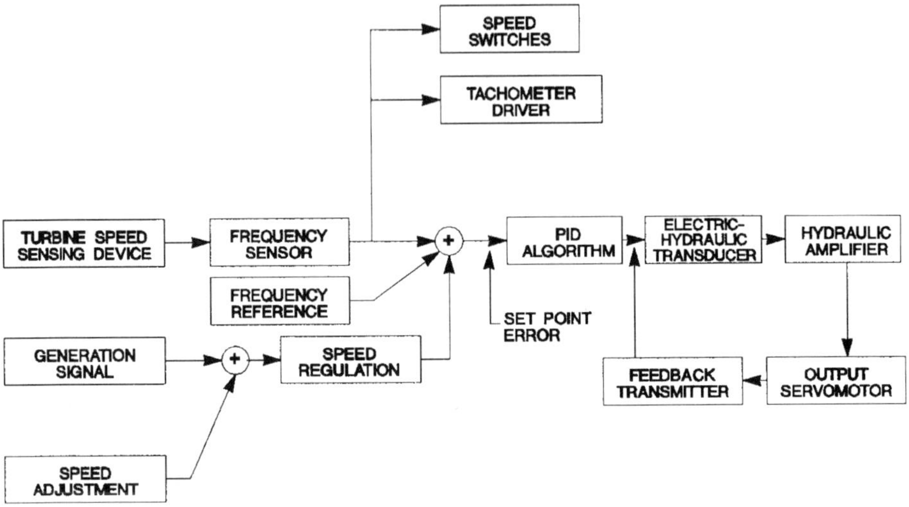

FIGURE 4-11: ***Block Diagram of a Typical Digital Governor.*** *Whatever type of turbine speed-sensing device is used, as seen in this block diagram, its signal provides the input for the digital control unit's frequency sensor and provides signals for the required speed switches or tachometers. Speed adjustment provides another input. Generation feedback combines with these inputs to provide regulation. This regulating signal will drive the electric-hydraulic transducer that operates the control valve to move the gate servomotors.*

- Remote Control of the Unit: Digital governors can start, stop, and change set points and governor parameters. Also, they can send warnings and log data over a communications link;
- Loading Optimization Schemes: With a digital control, the software can be written to maintain higher efficiency regions of the unit operating range and avoid rough zones of operation. Turbine data can be programmed into the governor to make automatic adjustments to avoid cavitation; and
- Dual-Regulated Turbine Optimization: For Kaplan and Pelton turbines, there are optimum regulation requirements for the secondary control servo. In the case of Kaplan turbines, gross head and gate positions are used as inputs to the control to determine optimum blade position. For Pelton turbines, increased efficiency can be gained with needle sequencing. For example, at a half-rated capacity, a six-needle machine has a higher efficiency with three needles in service than with six.

In a digital governor, software typically performs functions done in mechanical and analog governors by devices such as control motors, solenoids, and position switches. Digital controls can be designed in any configuration because software changes or additions can be made easily without requiring additional hardware.

Figure 4-11 shows a block diagram of a typical digital governor system.

Whatever type of turbine speed-sensing device is used, its signal provides the input for the digital control unit's frequency sensor and provides signals for whatever speed switches or tachometers are required. Speed adjustment provides another input. Generation feedback combines with these inputs to provide regulation. A PID algorithm evaluates the error signal and produces an electrical signal to the electric-hydraulic transducer.

An electric-hydraulic transducer is the first control valve. The transducer's force motor converts the electrical signal into a hydraulic flow between the pilot valve and the distributing valve to move the servomotors. Instead of mechanical feedback, a transmitter monitors servomotor position and provides a feedback signal to the system electronics. Thus, the digital control performs the same electronic functions as the analog electronic governor but typically does it with fewer hardware components. The digital control provides the same stability and rapid response, but offers more design enhancements through software or program features.

With new designs of electronic governors, proper attention should be paid to reliability of power supplies, and the transducers that are used in the control system. Input verification logic should be incorporated into the digital equipment to confirm that control values are within range. A "watch dog" feature should verify the correct operation of the control and initiate appropriate fail-safe measures taken in response to a failure.

Power Amplification

The output force of the mechanical governor ballhead or analog or digital governor transducer is not sufficient to move the hydraulic turbine water control devices. With large turbines, considerable force is required (sometimes in the range of millions of Newtons), and a servomotor is used to exert the required force. There may be several stages of hydraulic amplification between the transducer and the servomotor.

The oil hydraulic servo offers a relatively inexpensive, lightweight, and compact means of providing power for operating flow control devices. With accumulators, power can be supplied to make one or two full servo strokes—at a very high force level without requiring a very large pump. This is important where pumps are typically electric motor driven and a limited volume of oil creates a heating problem. Hydraulic energy also can be stored for an emergency or a "blackout" condition. In other cases, small units use spring-opposed cylinders or weight-operated devices to eliminate the need for the shutdown volume of oil. A piston-type valve (the control valve) very readily controls a hydraulic servo with little expenditure of energy. The hydraulic amplification system of a governor comprises the largest components of the governor: control valves, the oil pressure system, and pressure piping. Except for small units, the turbine manufacturer usually supplies servomotor(s).

Servomotor rating. A commonly used simple method of rating governor size is by stall work capacity. This rating method is simply the product of operating pres-

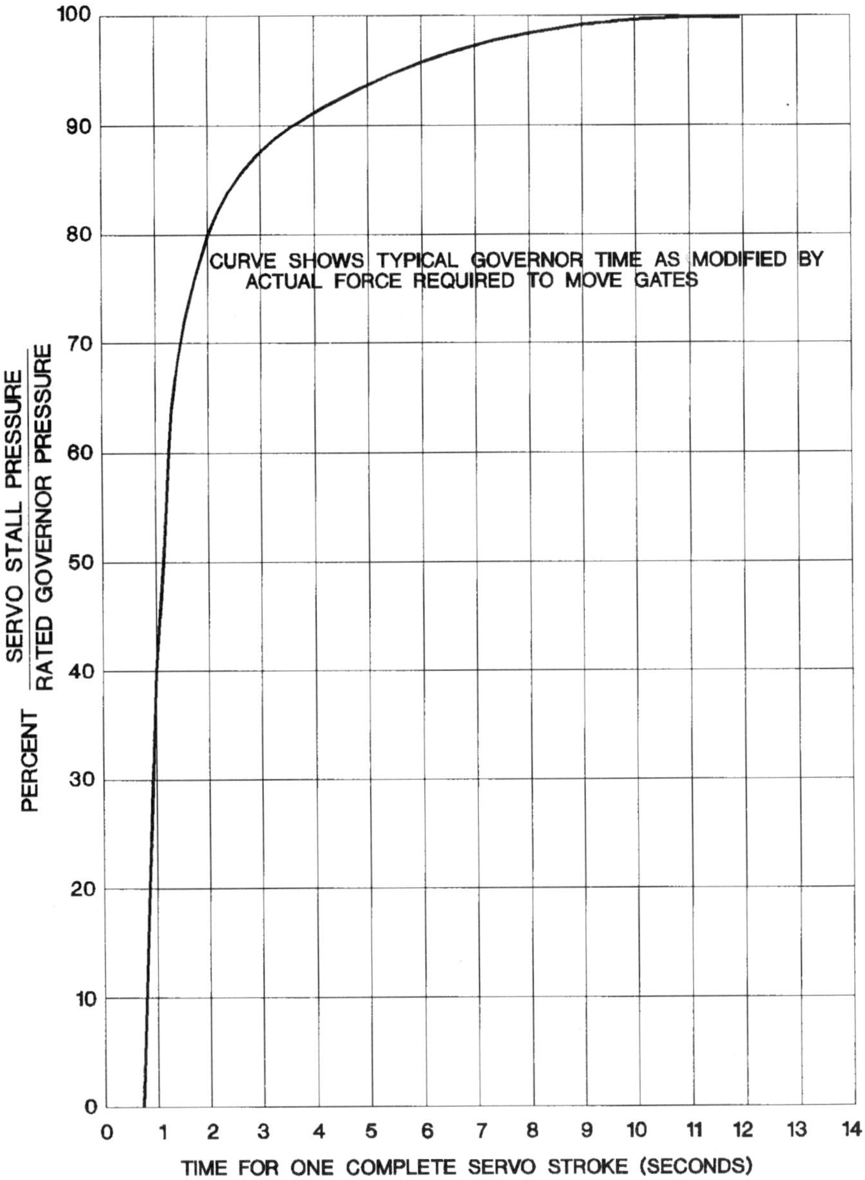

FIGURE 4-12: *Governor Rating versus Governor Time.* *Stall work capacity is a simple method of rating governor size. This rating method is the product of operating pressure, servomotor area, and servomotor stroke—all in suitable units. However, this rating omits the desired time to make a servomotor stroke. This graph is a sample curve showing how rated governor pressure should be increased over servo stall pressure as the time for one complete servo stroke decreases. To allow for pressure drops occurring in the control mechanism, the minimum operating pressure must exceed the maximum stall pressure by at least 25 percent. For example, to achieve a two-second servo stroke, the effective servo stall pressure will be 80 percent of the rated governor pressure, or the rated governor pressure must be 1.25 times the servo stall pressure.*

sure, servomotor area, and servomotor stroke—all in suitable units. However, this rating omits the desired time to make a servomotor stroke. For this reason, a governor must be derated to provide for satisfactory operation.

For example, a governor may be rated at 80,000 Joules (60,000 foot-pounds) and may be used on a turbine that has a stall work capacity of 40,000 to 60,000 Joules (30,000 to 45,000 foot-pounds). Figure 4-12 depicts a sample curve showing governor timing as it varies with actual effort required to move the gates.

To allow for pressure drops occurring in the control mechanism, the minimum operating pressure must exceed the maximum stall pressure by at least 25 percent. Depending upon the length and the design of the piping runs, the value may need to be higher.

Oil pressure system. Oil pumps provide the hydraulic pressure required to operate the servomotors. Older governor designs operated at relatively low pressure, such as at 1,000 to 2,000 kilopascals, kPa (150 to 300 pounds per square inch, psi). Governors operating at these pressures are still used. However, higher pressure units of 7,000 kPa (1,000 psi) allow using smaller valves, pressure tanks, piping, and servomotors to obtain the same effort and, therefore, may be less expensive. Additional care is required with connections to avoid leakage. Also, pressure systems over 7,000 kPa (1,000 psi) must be carefully designed to avoid

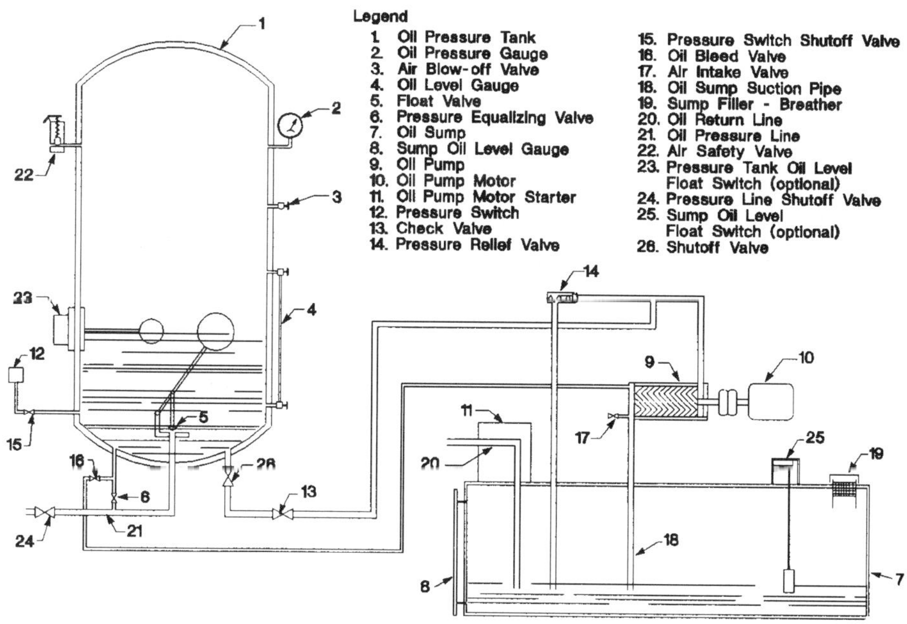

FIGURE 4-13: ***Example of a Typical Governor Pressure System.*** *This figure shows the pressure tank, sump, pump, and associated valves and indicators in a typical system.*

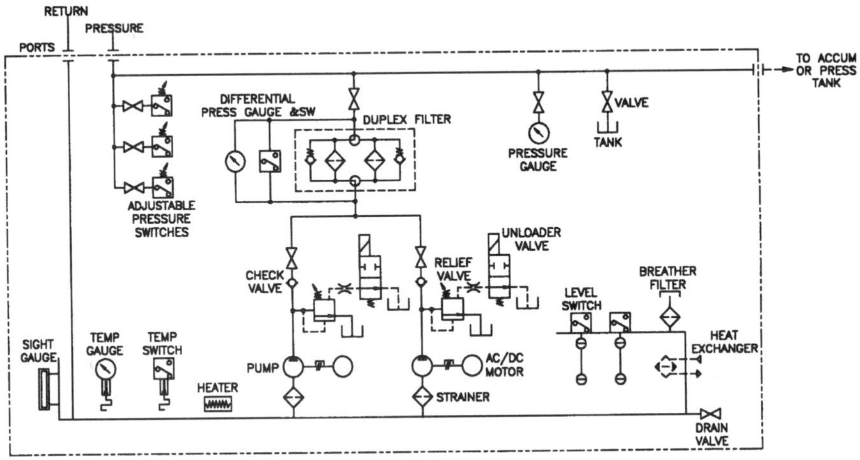

FIGURE 4-14: ***Governor Pressure System.*** *This figure is a typical representation of an AC/DC pump pressurization circuit. This technique uses filters and strainers for oil cleanliness, level, pressure, and temperature switches for remote alarm and shutdown capabilities. Also supplied are local gauges for aiding maintenance (i.e., an optional heat exchanger for maintaining consistent oil temperatures).*

dieseling. If air is entrained in the oil, it can combust during the pumping cycle.

As seen in Figure 4-13, a typical pump system includes oil relief valves, gauges, valves, and monitoring devices necessary to achieve normal operation. The governor oil sump tank serves as a reservoir for the oil pumps; it also may be designed as a dead oil storage area. Figures 4-14 and 4-15 are examples of pressure systems using standard hydraulic symbols.

Typically, the oil sump is designed so that 110 percent of all system oil may be returned to this reservoir. Also, it may require a heating or cooling system to maintain a usable oil temperature range for normal governor operation. Level switches provide indications of high or low oil levels in the sump. Typically, two pumps are used, one as lead and the second as backup. Generally, when pumps greater than 40 liters/minute (10 gallons/minute) are started or stopped, they are provided with a "flow unloader." This device reduces the mechanical and electrical stresses on the pressure system by loading and unloading the oil pumps gradually.

The air-over-oil pressure tank is the pressure oil accumulator used for all large units. An air compressor rated above the operating system oil pressure is needed to maintain the air head within the pressure tank. For smaller units, commercially available piston- or bladder-type accumulators may be used in place of a pressure tank. For operation over 7,000 kPa (1,000 psi), a nitrogen-charged bladder-type accumulator may be required to avoid dissolved air in the oil. All pressure tanks

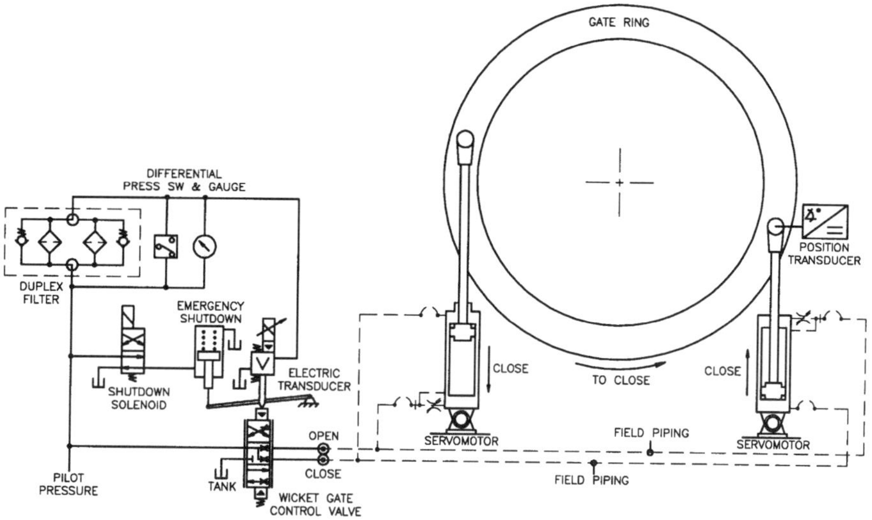

FIGURE 4-15: ***Servomotor Control Hydraulic Schematic.*** *Pressure supplied from the governor pressure system through a three-position throttling valve allows control of the wicket gate ring. Additional fine filtering for feedback and emergency operation ensures turbine protection. Also shown is a position transducer for aiding turbine performance.*

---

or accumulators should be designed to ASME Pressure Vessel Code.[2] The pressure tank allows for governor operation with the pump not running. This allows for intermittent operation of the pumps. Generally, a unit pressure tank is sized to provide a close-open-close servomotor operation with the pump not operative—starting at pump turn on pressure/level. Another consideration with the pressure tank sizing is the pump cycle time of the main pumps. Historically, pump cycle time has kept pumps below ten starts per hour.

An additional requirement may include operating or starting the unit during a “blackout” condition. To meet this need, an auxiliary DC motor or water driven pump(s) may be used. For small units, a hand wheel or hand pump may be installed.

In a hydraulic system with an air-oil accumulator, air from the accumulator can enter the control system and result in uncontrolled, rapid operation of the servomotor system. To prevent this, some manufacturers use a float-operated, disc-type valve that closes whenever the oil level becomes dangerously low. Others use external shut-off valves activated by oil level switch contacts. Prior to this occurrence, other safety features will have shut down the generating unit.

Governor piping. The maximum allowable oil velocity commonly determines the governor piping size. Normal values for the maximum velocity are 5 meters/second (15 feet/second). The total piping pressure losses can prevent the governor from achieving the desired servomotor timing.

Oil Cleanliness

The governor oil system must be free of any contaminants that would hamper its operation or corrode system components. Governor oil is selected based on operating conditions, system pressures, and environmental considerations. Oil usage typically is split between working fluid used for moving the servomotors and a much lesser amount of control oil used in the governor transducer. The tight tolerances in the transducer require a cleaner oil than that sent through the distributing valve to the servomotor. To provide this, a separate transducer oil filter assembly is used. Higher pressure systems require increased oil cleanliness owing to reduced clearances in the hydraulic system components.

SYSTEM CONSIDERATIONS

The turbine's dynamic characteristics, the mechanical inertia of the turbine-generator, the inertia of the column of water, and the system load characteristics influence the governor's performance. A change in any of these system components may affect the governor's ability to control. Changes to these parameters may require reevaluation of governor performance.

Turbine Dynamic Characteristics

The control system requires different designs with different types of turbines. With a Francis turbine, the governor is controlling the position of the turbine wicket gates. However, in the case of a Kaplan turbine, the blade pitch must be controlled in addition to the wicket gates.

Also, the net head is required to provide the input to the Kaplan blade control mechanism. For a unit that experiences a significant change in net head, a three-dimensional relationship positions the blades to increase overall efficiency. Control problems can result from the dual regulation required with a turbine of this type. Therefore, the operating time of each blade must be coordinated with the turbine and the governor designers. If the blade timing is close to the timing of the gates, the movement of the blade may, in turn, result in a movement of the flow control point, and instability may result.

With a Pelton turbine, separate control mechanisms may be required for each needle of a multi-needle machine. Also, the deflector needs a control mechanism. With a turbine of this type, the needle position may be used as a secondary control—the primary control being the deflector. In this mode, called "water wasting," some efficiency is sacrificed to achieve faster control response. Isolated systems often require a Pelton turbine to operate in water wasting mode in order to provide system frequency control.

Mechanical Inertia

An important parameter of the turbine-generator is the rotational inertia of the mass of the turbine-generator combination. It is commonly expressed as GD^2, kilograms-meters squared (WR^2, pounds-feet squared), and is called the "flywheel effect." The magnitude of GD^2 (WR^2) of the machine affects the magnitude of the temporary speed change for a change in load. When the load on the unit is changed, the kinetic energy of rotation changes. The kinetic energy is proportional to the GD^2 (WR^2) of the rotating mass times the square of the speed of rotation. Therefore, the larger the GD^2 (WR^2), the smaller the speed change necessary to absorb a given variation in energy.

The greater GD^2 (WR^2) allows more time for the water regulating devices, such as the wicket gates or needles, to move in the corrective direction and permits maintaining speed with less deviation. Generally, the speed of the unit, generator kVA (kilovolt-ampere) rating, and the turbine type fix the amount of GD^2 (WR^2). Modest increases in GD^2 (WR^2) can be specified with minimal increase in cost. However, a large increase in GD^2 (WR^2) may be cost prohibitive. The term used to measure the flywheel effect on system dynamics is the mechanical starting time, usually expressed as T_m. A way to visualize T_m is the time required to accelerate a machine from standstill to rated speed with the application of rated torque at rated horsepower.

Equations 4-1 and 4-2 give the general formulae for the mechanical starting time in metric and English units respectively.

EQUATION 4-1:

$$T_m = \frac{GD^2 \times RPM^2}{3.65 \times 10^5 \times kW}$$

where:

GD^2=flywheel effect, kilograms-meters squared;
rpm=speed, rpm; and
kW=rated power, kilowatts.

EQUATION 4-2:

$$T_m = \frac{WR^2 \times RPM^2}{1.61 \times 10^6 \times HP}$$

where:

WR^2=flywheel effect, pound-feet squared;
rpm=speed, rpm;
HP=rated power, horsepower.

Inertia of the Column of Water

The length of the water column and the volume of water also affect governor operation. When the governor calls for a change in position of the water regulating devices, time is required to develop the torque at this corresponding new requirement. The term used for this quantity is called the water starting time, T_w. The water's inertia has a limited force available to accelerate the water to a new flow rate. The gross head determines this force.

Therefore, if load is suddenly added, it does no good and may do harm to open the gates faster than the water can accelerate. During this period of deficiency in torque, the turbine will underspeed. These considerations are further complicated when load is reduced because, as the gate opening is decreased, there is initially little change in flow rate due to the time lag. Therefore, the velocity of the water on the runner momentarily increases with a resulting increase, instead of a decrease, in torque.

The water starting time, T_w, is defined as the time required for the water in the water conduit system to accelerate from zero to rated velocity at rated head. Equation 4-3 gives the general formula for water starting time. Additional information can be found in Chapter 12, Hydraulic Transients.

EQUATION 4-3:

$$T_w = \frac{LV}{gH}$$

where:

L=length of water column from upstream free water surface to downstream free water surface, meters (feet);
V=rated velocity, meters/second (feet/second);
g=acceleration due to gravity, meters/second squared (feet/second squared); and
H=rated head, meters (feet).

Because the water conduit system may consist of many sections of varying cross sectional areas, the standard formula for T_w becomes Equation 4-4.

EQUATION 4-4:

$$T_w = \frac{\sum_{i=1}^{N}\left(\frac{L_i Q}{A_i}\right)}{gH}$$

where:

N=the number of sections in the water conduit system;
L_i=length of each section, meters (feet);
A_i=cross-sectional area of each section, square meters (square feet);
Q=water flow, cubic meters/second (cubic feet/second);
g=acceleration due to gravity, meters/second squared (feet/second squared); and
H=rated head, meters (feet).

Figure 4-16 gives an example of the influence of mechanical and hydraulic inertias on governor stability. The graph shows a series of responses to small step

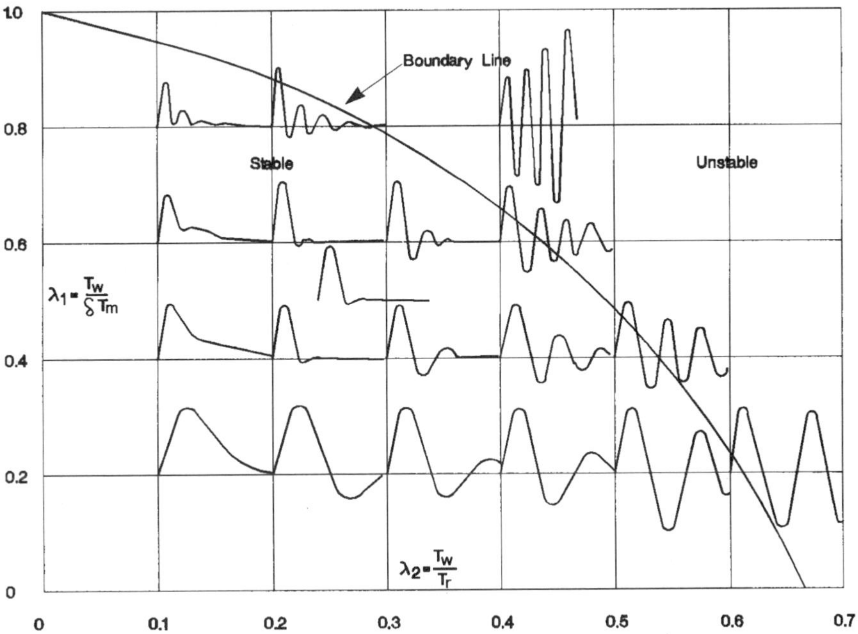

FIGURE 4-16: ***Example of Mechanical and Hydraulic Inertias Defining Governor Stability.*** *This graph shows a series of responses to small step changes in speed setting or load for a turbine-generator operating isolated from other units. The x-axis, λ_2, equals the ratio of T_w (water start time) divided by T_r (dashpot time constant for a mechanical governor). The y-axis, λ_1, equals the ratio of T_w divided by the quantity of δ (the temporary droop for a mechanical governor or reciprocal of proportional gain for electronic governors) times T_m (the mechanical start time). Values inside the boundary line produce a stable system response.*

changes in speed setting or load for a turbine-generator operating isolated from other units. The parameter, δ , is the temporary droop of a mechanical governor or the reciprocal of proportional gain for an electronic PID governor. T_r is the dashpot time constant of a mechanical governor.

Practical values of temporary droop range from 0.3 to 0.8, and the dashpot time constant typically ranges from 0 to 30 seconds. Combinations of λ_1 and λ_2 (as defined on Figure 4-16) that lie below the boundary line are stable, while those outside the line are unstable. The sample response curves shown depict the degree of stability and the response to a step change.

These curves are based on a mechanical governor and idealized turbine and load characteristics. Electronic governors with derivative action or other control strategies may improve response.

A T_m/T_w ratio above 2 is reasonable for a unit on an interconnected system. As the T_m/T_w ratio decreases, the unit will be slow to change generation level in response to a change in set point. New governing algorithms have been developed that can improve the response of systems with lower T_m/T_w ratios. System studies are recommended for any units that are tied to isolated power grids or are major generation contributors.

Load Characteristics

The load characteristics of the system that is being supplied power are important in determining the overall results of system stability. Unfortunately, nothing usually can be done about them; they must be accepted and taken into account in the control design. For instance, consider the type of load typically in motor-driven fans or blowers. As an approximation, the torque required by such a load varies directly with the square of the speed. Such a load is considered to have high positive damping (i.e., helpful to the governor in controlling speed with load); in fact, about the highest damping normally encountered. The connection of such a load to a hydraulic turbine, which also has positive damping, results in a very stable system with a high degree of "self-regulation."

A generator with constant excitation generates a voltage directly proportional to speed. Such a generator supplying a resistance load would have a power output proportional to the square of the speed and a required torque directly proportional to the speed. This load would have positive damping, also. However, modern generators have fast voltage regulators to compensate for the changes in speed and to maintain constant system voltage. In this case the power output would remain constant since the voltage is not permitted to change. Since power is proportional to speed times torque, a decrease in torque would accompany an increase in speed, and negative damping would result. If driven by a unit with little damping, an unstable system would result, and the controlling governor would have to exert a much greater stabilizing influence than in the case of the motor-driven fan. This example illustrates another frequently overlooked fact: that other controls such as voltage and pressure regulators operating on the same controlled system can, and often do, have a definite unstabilizing influence on the speed governing system.

The hydraulic turbine governor is one element in a hierarchy of control elements acting in a multi-loop control system that consists of voltage regulators, power system stabilizers, and so on. In some cases, one of the control systems operating by itself is perfectly satisfactory and completely stable, but combined operation is impossible. The usual solution to this problem is to select the quantity for which the most accurate control is desired, such as the speed. Make its control as fast and accurate as possible, assuming all other quantities constant. Then the controls of the secondary variables deliberately are made sufficiently slow so that the assumption of constant value is substantially true during the time of the speed transient.

A change of torque with speed characterizes the resistance load described previously, but it contains no equivalent to the synchronizing torque that exists in an alternator supplying a synchronous load. In this case, torque equilibrium is not re-established upon return to synchronous speed only. There is an additional torque that is dependent upon the position of the alternator rotor relative to the synchronous field—as though the load were driven through a torsion spring. This type of load demonstrates that, because of this synchronizing or "spring" torque and the rotating mass, a natural frequency of oscillation about the equilibrium synchronous speed position exists.

This makes an entirely different problem: controlling a turbine that drives an alternator on synchronous load compared to the same machine off the line or supplying a non-synchronous load, such as heaters or lamps. Even with the latter load, if other alternators are paralleled with the unit in question, a synchronizing torque exists between them. Analyzing the governor and its effect upon the power system are beyond the scope of this chapter. However, it is important that such an analysis be made to ensure proper operation of the unit on the interconnected power system.

Performance and Testing

Performance and Testing Standards

There are several accepted test standards written by various technical committees. Commonly used standards include ASME PTC 29 and IEC 308.[3,4]

Shop Testing

Shop testing requirements normally include testing and calibrating all electrical circuits, mechanically calibrating stroke and all linkages, and pre-adjusting all position, speed, and pressure switches and timing relays. On digital governors, the input/output (I/O) structure of the program is tested along with sequence testing. In addition, shop test requirements may include dead band, dead time, and stability tests as defined by an accepted standard.

Field Testing

Field testing, at a minimum, normally confirms the calibration of electrical devices and the proper mechanical and hydraulic operation of the governor. Site testing also includes requirements for isolated load tests, load rejection tests, and servomotor differential pressure tests. Also, tests are performed that can only be done with the entire unit in operation, such as setting governor gain adjustments to achieve optimum stability or verification of the ability of the pressure tank to support a close-open-close operation without the pump in service.

References

[1]ANSI/IEEE Standard 125-1988, *IEEE Recommended Practice for Preparation of Equipment Specifications for Speed-Governing of Hydraulic Turbines Intended to Drive Electric Generators,* The Institute of Electrical and Electronic Engineers, New York, New York, 1988.

[2]ASME Boiler and Pressure Vessel Code, Section VIII, Division 1, American Society of Mechanical Engineers, New York, New York, 1995.

[3]ASME Power Test Code, Number 29, *Speed-Governing Systems for Hydraulic Turbine-Generator Units,* American Society of Mechanical Engineers, New York, New York, 1965.

[4]IEC Publication 308 *International code for testing of speed governing systems for hydraulic turbines,* First Edition, International Electrotechnical Commission, 1970.

Bibliography

Benko, G.B., "Stability Criteria Vs. Step Load Handling Capability of Single Units in Isolated Systems," Underground Hydropower Plants Conference, Oslo, Norway, June 22-25, 1987.

Carson, Paul T., "Optimum Frequency Regulation of Isolated Hydroelectric Projects," *Waterpower '83 International Conference on Hydropower, Conference Proceedings,* Volume II, The University of Tennessee, September 1983, Knoxville, Tennessee, pages 1129-1139.

Hovey, L.M., "Optimum Adjustment of Hydro Governors on Manitoba Hydro System," *AIEE Transactions,* Volume 81, Part III, December 1962, pages 581-587.

Paynter, H.M., "Methods and Results from M.I.T. Studies in Unsteady Flow," *A Palimpsest on the Electronic Analog Art,* G.A. Philbrick Researches, Inc., Boston, Massachusetts, 1952, pages 224-227.

Schleif, F.R., *Governor Characteristics for Large Hydraulic Turbines,* USBR REC-ERC-71-14, February 1971.

Schleif, F.R., and A.B. Wilbor, "The Coordination of Hydraulic Turbine Governors for Power System Operation," *IEEE Transactions,* PAS 85, July 1966, pages 750-758.

Working Group on Prime Mover and Energy Supply for System Dynamics Performance Studies, "Hydraulic Turbine and Turbine Control Models for System Dynamic Studies," *IEEE Transactions on Power Systems,* Volume 7, No. 1, February 1992.

Wozniak, L., "Graphical Approach to Hydrogenerator Governor Tuning," *IEEE Transactions on Energy Conversion,* Volume 5, No. 3, September 1990.

CHECKLIST FOR CHAPTER 4

1. Does the governor have to regulate system load?
2. Does the governor have to maintain the upper reservoir water surface elevation?
3. What are the floor space and height requirements for the governor cabinet and pressure tank?
4. What electrical control signals must the governor relay to other equipment in the power plant?
5. What type of oil will be used in the governor? (Include specifications such as viscosity, foam inhibitors, and so forth.)
6. Has the rating of the governor been specified (flow, force, timing requirements, etc.)
7. What is the minimum oil reserve capacity after electrical power failure that is required of the governor?
8. What electrical voltage will be available for the governor to operate on?
9. Have the following unit parameters been supplied to the governor manufacturer?
    - Unit speed;
    - Turbine rated power;
    - Turbine rated head;
    - GD^2 (WR^2) of the unit rotating parts;
    - Length of the water column from upstream free water surface to downstream free water surface;
    - Cross-sectional dimensions of the water column passageway;
    - Turbine servomotor bore and stroke; and
    - Desired flow control timing.

Chapter 5
Major Valves

Valves are mechanical devices that control flow in conduits or piping systems. Major valves are valves used on the major hydraulic conduits in hydropower plants. Using smaller valves on common mechanical piping systems is similar to other applications and is not unique to hydropower plants. Therefore, this chapter does not discuss in detail valves for general mechanical systems. Instead, it focuses on major valve applications used in hydro plants. Additionally, see Chapter 7, Gates, for information about hydropower plant gates that can be used for some applications in place of major valves.

Major valve applications at hydropower facilities can be categorized by the type of service: closure, energy dissipating, pressure control, and flow control. Closure valves (also known as shut-off valves) provide a positive conduit closure against flow. Flow-control valves throttle flow both to regulate the discharge pressure and vary the flow rate. Energy-dissipation valves are specialized flow-control valves that reduce the head on water releases. Pressure-control valves can limit the pressure rise in the conduit or piping system. Various valve designs have been developed to satisfy these service needs. The following paragraphs describe the most common valve types and typical service applications; however, some valves can be used in both closure and flow-control service.

Closure Valves

Hydraulic conduits and penstocks typically are provided with devices to close the upstream low-pressure and downstream high-pressure ends for dewatering to allow inspection and maintenance. Typically, gates are at the upstream intake on large conduits; however, valves also can be used. On long penstocks, the downstream, or high-pressure, end typically has a closure valve to shut off the water supply to the equipment. This allows turbine dewatering for inspection and maintenance without dewatering the penstock. This feature is desirable, particularly when two or more units share a common penstock.

Closure valves installed at the spiral case inlet also are used to shut off flow to the turbine and stop the unit if the turbine wicket gates fail to close. These closure valves are called inlet valves or turbine shut-off valves. Conduits for underground pumped-storage projects typically have valves on both the upstream and downstream side of the pump-turbine.

Turbine closure valves normally operate under zero flow conditions and have balance piping to equalize the pressure across the valve before opening and during closing. Closure valves used for emergency closure at the spiral case inlet should, however, be able to close against maximum flow. The maximum flow may occur at maximum turbine output, turbine runaway speed, or more severe discharge conditions such as penstock rupture. While most closure valves can throttle flow, they normally are not designed for continuous throttling service. Furthermore, throttling of the inlet valves will reduce the available head and also can result in turbulent flow into the turbine.

The design pressure rating of closure valves, as in all valves, must include the effects of waterhammer. The worst-case design condition is often full-load rejection and emergency wicket gate closure. Penstock elongation occurs during closure against flow and during normal shut-off, resulting in an axial valve movement. Closure valves designed for turbine inlets also should compensate for this axial movement. As a result, the valve supports and operating mechanism should be designed for sliding, and expansion joints typically are provided downstream of the valve.

A hydraulic servomotor usually operates closure valves. This servomotor should have mechanical locks that can maintain the valve in the full-open and full-closed positions for personnel safety during maintenance. The locking system design should consider the maximum hydraulic force under maximum hydraulic pressure. The controls and power sources for closure valves have to be reliable to ensure that they are available for emergency operation. Both oil and penstock water hydraulic systems have been used successfully to operate or control closure valves; however, oil systems are more common. Hydraulic systems usually have a double-acting hydraulic cylinder to open and close the valve. Alternatively, a single-acting cylinder and counterweight can be used.

Factory tests of inlet valves should include both hydraulic pressure and leak tests. Operational tests of the valve actuator and valve seals should be performed in the shop, if possible, but in some cases these tests may have to be performed in the field.

The weight of the closure valve often determines the powerhouse crane capacity and service requirements. The following sections discuss the three most common types of closure valves: the butterfly, spherical, and cone valves. Needle valves once were used for closure service at spiral case inlets; however, they are no longer recommended because it is difficult to assure safe closure by securely positioning the needle.

Butterfly Valves

Because of their rugged, compact, simple design and low cost, butterfly valves are commonly used for closure service involving heads up to 122 meters (400 feet) and inlet diameters up to 9 meters (30 feet). While small butterfly valves have been built for heads in excess of 305 meters (1,000 feet), spherical valves usually are selected for higher head applications.

Butterfly valves generally consist of a disc or lattice mounted on a shaft that rotates in a cylindrical body as shown in Figure 5-1. Usually, the disc and lattice profile is contoured in the flow direction to provide a smooth hydraulic flow and balance the forces on the valve. The disc is oriented parallel to the flow to minimize any restriction when open or is positioned at a right angle to the flow to provide full closure. When opening or closing butterfly valves against flow, the torque on the valve trunnion, head loss, and discharge coefficients vary over the valve stroke. Additionally, the torque on the valve disc generally is designed to have a closing valve tendency to facilitate emergency closure.

FIGURE 5-1: ***Butterfly Valve.*** *This isometric cutaway view with the penstock partially removed shows the major components of the butterfly valve. In this view, the disc is in the closed position. (Courtesy Kvaerner Energy)*

While there is essentially no head loss in free flow-spherical and plug valves, butterfly valves cause a head loss in the flow through the valve. Usually, butterfly valves are used when their lower cost offsets the resultant head loss. To reduce the head loss associated with the valve disc, "lattice-type" disc designs have been developed using two thin parallel plates connected by streamlined lattice structures. The butterfly valve head loss increases as design pressure or head increases because of the need for a thicker disc to withstand the pressure—suggesting spherical valves for higher head applications. To offset the head loss, butterfly valves typically are larger than the turbine inlet diameter. This offsets the cross section flow loss caused by the disc. While head can be significantly reduced across partially open butterfly valves, prolonged throttling operation is not recommended and can result in cavitation damage to the disc, body, or seal.

Valve seals are on the disc circumference or in the contact portion of the valve body. Packing glands seal the rotating shaft trunnions. Valve seals can be either resilient material (i.e., rubber or nylon) or metal (i.e., stainless steel or bronze) while the mating stationary sealing surfaces should be stainless steel or other erosion-resistant material. Resilient seals generally are used for lower heads and have less leakage when new. Metal-to-metal seals are generally used for higher heads, are more durable, and last longer. Butterfly valve seals should be adjustable and replaceable without removing the disc from the valve.

The body and discs of larger valves can be sectionalized by field joints to facilitate shipment and field assembly. Typically, valves have flanged connections and spool pieces to facilitate dismantling; however, less expensive, welded connections also are available.

Spherical Valves

Spherical valves are used in applications for heads up to 1,200 meters (4,000 feet) and diameters up to 4.6 meters (15 feet). Spherical valves have a body shaped like a hollow sphere with flanges or other connections for mounting the valve in a piping system. An example of a spherical valve is shown in Figure 5-2.

The body is made in two or more sections with flanged-bolted connections for assembly and disassembly of the rotor. A rotor, shaped like a ball, has a cylindrical hole through its center at right angles to two support shafts located on each side of the valve. In the open position, with the rotor opening parallel to the flow direction, the valve offers an unrestricted water passage with minimal disturbance to the flow path. To close the valve, the valve rotor is turned 90 degrees from the axis of the rotor opening. When opening and closing the valve against flow, the torque on the valve trunnion, and head loss and discharge coefficients vary over the valve stroke. Spherical valves generally have a closing tendency for positions less than 50 percent open, which facilitates emergency closure.

Movable seals reduce valve leakage when the valve is closed. The valve seals are retracted before rotation and should be operated by penstock or water hydraulic pressure to minimize the potential for oil entering the water passage. Valve opening and closing sequencing controls should preclude seal damage by valve rotation when the seals are extended. It is recommended that spherical valves have both upstream and downstream seals. The upstream seal should be used as the maintenance or emergency seal, and the downstream seal as the service or working seal. For applications when erosive particles are in the water, valve seals should be engaged in both the open and closed positions to preclude ero-

FIGURE 5-2: ***Spherical Valve.*** *This isometric cutaway view shows the major components of the spherical valve in the closed position. (Courtesy Kvaerner Energy)*

sion of the valve trunnions and seals.

When the valve is closed under full pressure, the upstream maintenance seal allows replacement or maintenance of the working seal without dewatering the penstock. The upstream maintenance seal should have a positive mechanical locking system such as mechanical screws on the seals to prevent accidental opening while working on the downstream seal. Powerhouses have been flooded when upstream valve seals accidentally opened while the spiral case man-door was open for turbine and valve maintenance.

Valve seal hydraulic systems that use penstock pressure should be designed to avoid "auto-oscillation" of the penstock water column. The purchaser should review the relevant literature about this when specifying the seal hydraulic system.[2]

Cone Valves

Cone valves consist of a conical plug that rotates in a valve body, as shown on Figure 5-3. This is similar to a spherical valve. In the full-open position, a full circular port allows unrestricted flow through the valve, and, when closed, the valve plug is 90 degrees to the flow direction. Metal-to-metal circular seats are on both sides of the valve plug, which is seated in both seats in the full-closed position. For applications that have erosive particles in the water, valve seals should be provided and engaged in both the open and closed positions to preclude erosion of

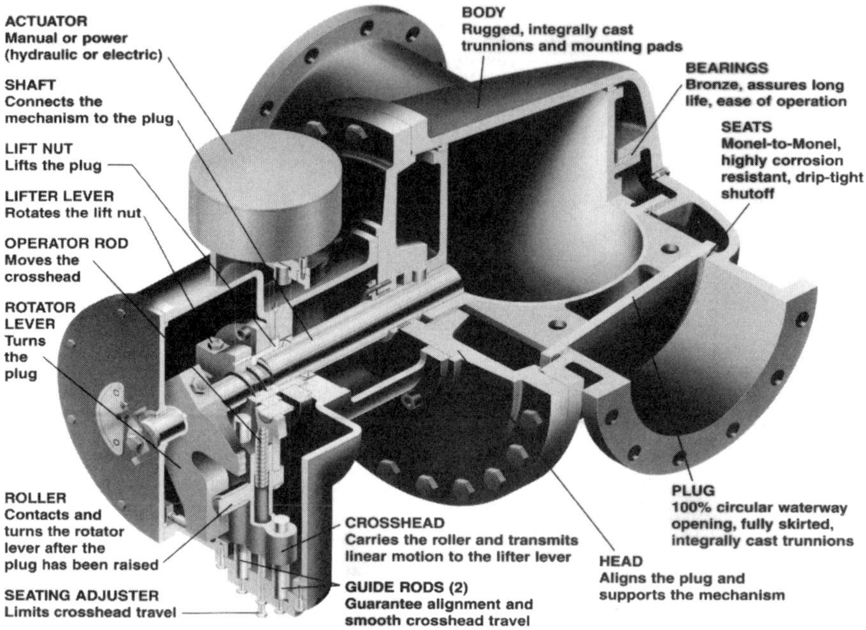

FIGURE 5-3: ***Cone Valve.*** *This isometric cutaway view identifies the major components of the cone valve.* *(Courtesy Rodney Hunt Company)*

---

the valve trunnions and seals.

During operation of some plug valves, the plug moves axially to disengage the valve seats prior to the plug's rotation. After the plug is rotated 90 degrees with the seals disengaged, the plug moves axially to re-engage the seals. The plug is not rotated while the axial-seating motion takes place.

The advantages of the cone valve are its simplified seal and operating design. Cone valves are commercially available in diameters up to 2.1 meters (84 inches) and heads up to 1,750 meters (5,700 feet), depending on the diameter.

Energy-Dissipating Valves

Water is stored in reservoirs to generate hydropower, maintain minimum streamflows, meet downstream water demands, avoid changes in streamflow, and control reservoir levels. To release water not used for generating power requires methods that must dissipate the available potential energy. Therefore, several methods and valves have been developed for these other applications.

Most dams have spillways to pass flood discharges and protect the dam from overtopping. Low head spillways (less than 30 meters (100 feet)) have been used successfully to continuously release water without damage; however, using high head spillways to release water continuously can cause spillway damage and require continual maintenance. Therefore, outlet works facilities and specialized control equipment were developed to provide continuous discharges at medium- to high-head dams.

Initially, gates were used to control reservoir releases and were used successfully for heads below 30 meters (100 feet). Gates used for discharge releases for heads above 30 meters (100 feet) were susceptible to cavitation damage. While modern gate designs and materials have overcome some early cavitation problems, needle valves were developed to regulate reservoir discharges.

Both needle valves and gates used for water releases do not dissipate the energy from the water released, and thus require plunge pool protection to control erosion of the riverbed or banks. Therefore, lower cost, energy-dissipation valves were developed for this purpose. These valves have improved operating characteristics and less stringent stilling basin requirements. Properly designed, the energy-dissipation valves perform their function without sustaining cavitation damage.

Fixed-Cone Dispersion Valves

The fixed-cone dispersion valve, invented by Messrs. Howell and Bunger of the U.S. Bureau of Reclamation, has gained wide acceptance for controlling discharges. Because the valve discharge is sprayed into the air, fixed-cone dispersion valves sometimes are referred to as free discharge valves. The major valve components are shown in Figure 5-4. Moving the cylinder gate sleeve upstream over the valve body opens the valve. The fixed internal cone spreads the conical discharge jet in such a way that the surface area of the jet and area that entrains air increase rapidly. When closed, the movable cylinder seals against the outer edge of the fixed cone. The fixed-cone valve discharge for different-sized valves for various heads are shown in the graph in Figure 5-4. When the valve seals are designed properly and back pressure is maintained on the fixed vanes, a fixed-cone valve can be operated between 2 and 100 percent of maximum discharge without cavitation damage.

The following methods are used to operate the movable cylinder:

- Manual or motor-driven mechanical screw;
- Hydraulic cylinders; or
- Operating mechanism directly coupled to the turbine.

Installations of fixed-cone valves can allow the expanding conical discharge to spray into the air (where water spray creates no problems), but there are many instances in which it is desirable to restrict full dispersion of the jet to avoid icing or other problems caused by the dispersed spray. One way to avoid dispersion is to put a hood, usually a steel shell, around the valve downstream from the discharge of the cylindrical, movable sleeve. The engineer must take great care to design the hood properly, so air admission is adequate and uniform around the jet periphery. An inadequate design will cause pulsations and pressure changes as discussed in Chapter 11, Equipment Layout Considerations. If the hood is a cylinder, rather than a flaring conical shape, the hood will not drain well. In cold climates, minor leakage from the valve may freeze, building up a rim and subsequently interfering with opening the sleeve. In general, unless there are over-

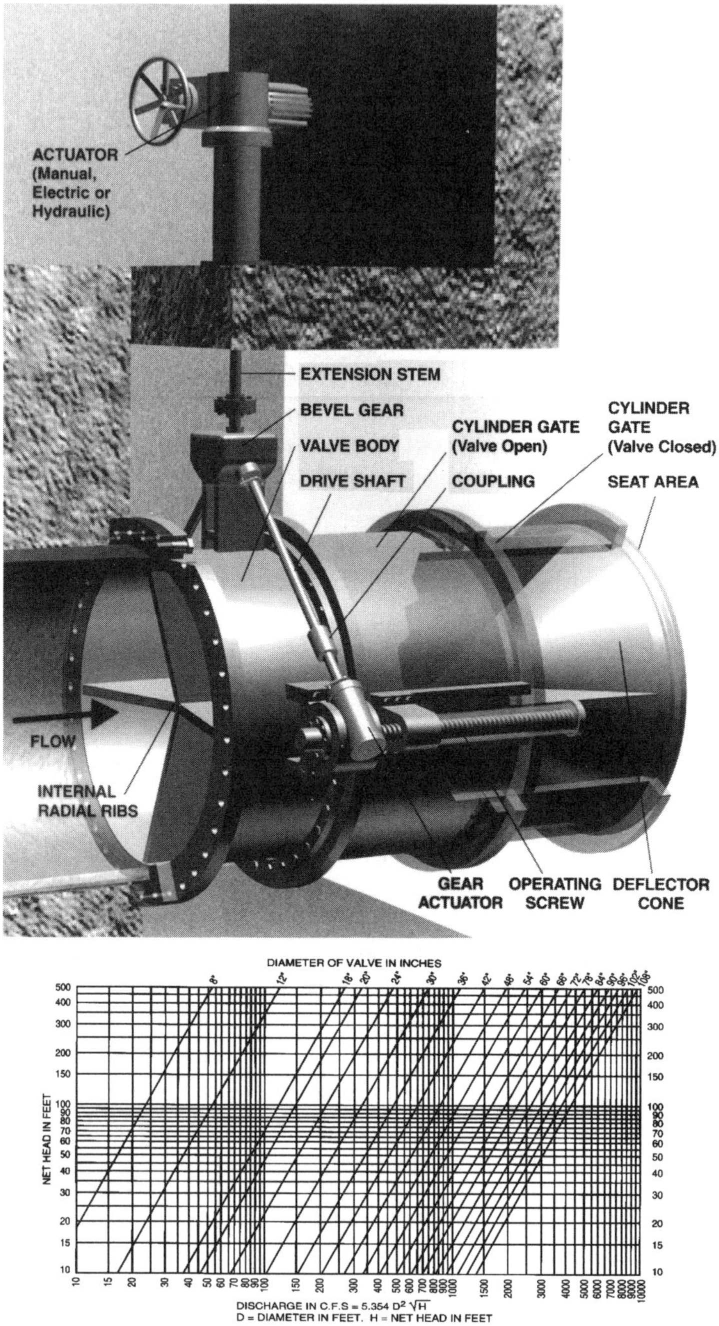

FIGURE 5-4: ***Fixed-Cone Dispersion Valve.*** *The isometric cutaway view (top) identifies the major components of the fixed-cone dispersion valve, invented by Messrs. Howell and Bunger of the U.S. Bureau of Reclamation. The graph (bottom) indicates the discharge for different-sized fixed-cone dispersion valves for various heads.*

(Courtesy Rodney Hunt Company)

riding reasons of great importance, a fixed-cone dispersion valve should not be hooded or set to discharge into a confined space.

The design, location, and alignment of fixed-cone valves should consider the effects of spray on the powerhouse ventilation system and ice buildup potential in the tailrace and on the transmission equipment. Chapter 11 also discusses the proper location for fixed-cone valves.

While most fixed-cone valves are installed above water, they can be operated underwater if there is sufficient water depth and air admission. For submerged applications, special designs and model testing may be necessary. Fixed-cone valves are used in sizes up to 3 meters (9 feet) in diameter and operating heads up to 300 meters (1,000 feet).

The major advantages of fixed-cone valves are the reduced valve cost and reduced plunge pool protection required. Problems with fixed-cone valves can include support cone failure, cavitation, vibration, and closure mechanism problems that may be attributed in part to the relatively light structural design. A straight pipe section that is at least 10 pipe diameters in length should precede the valve to reduce turbulence, pressure pulsation, and problems in fixed-cone valves. Designs using low-strength steel and heavy-valve plate thickness lower the valve's natural frequency and reduce vibration. In spite of past problems, fixed-cone dispersion valves can be successfully used in free discharge applications. In all cases, it is recommended that a guard valve or gate be provided, at least 10 diameters upstream of a fixed cone or other discharge valve to allow valve maintenance without lowering the reservoir.

Hollow-Jet Valves

In the past, the U.S. Bureau of Reclamation widely used the hollow-jet valve, invented by Messrs. Staats and Hornsby of the U.S. Bureau of Reclamation. These valves are not commonly used today because of their high cost. The hollow-jet valve has a movable needle located inside a diffuser that seals in the upstream direction. Either mechanical or hydraulic-operating mechanisms can position the needle. The valve needle is hydraulically smooth and contoured to minimize cavitation and direct the flow. The outer body of the valve directs the water in a controlled stream and supports the inner needle closing mechanism. Radial supports are used to provide air admission to aerate the inner circumference of the jet, stabilize the flow stream, and reduce cavitation. The energy dissipating capability of a hollow-jet valve is less than that of a fixed-cone valve, and a natural rock- or concrete-lined plunge pool and stilling basin usually are required.

The hollow-jet valve is capable of operation down to 5 percent of maximum discharge and is not used for submerged operation.

Sleeve Valves

Sleeve valves as shown in Figure 5-5 have been developed as an alternative to free discharge valves to dissipate the head in a closed system without cavitation damage. Sleeve valves are commercially available in sizes up to 1.2 meters (48 inches) for use at heads up to 30 meters (100 feet).

Multiple small diameter orifices in a cylindrical pipe section are used to throttle the pressure. Sleeve valves are used in relatively clean water conditions so that the small openings do not become plugged. Sliding either a closure sleeve or the orifice section to open portions of the orifice plate area controls the valve opening and throttling. Early models discharged the flow from the inside of the valve to the outside in a stilling well or chamber. Later in-line designs, as shown in Figure 5-5, and angle designs have been developed with inward flows that impinge the individual flow streams to facilitate energy dissipation. The in-line and angle designs are best suited as pressure-reducing valves where some outlet pressure is maintained. Typically, these valves are installed directly in piping systems and do not need stilling basins. Sleeve valves are limited by size and cost considerations and sleeve-type valves are most often used in water distribution and supply systems.

PRESSURE-CONTROL VALVES

There are distinctions among the three types of pressure-control valves: pressure-relief valves, pressure-regulating valves, and pressure-regulator valves as used in the hydroelectric industry.

Pressure-Relief Valves

Pressure-relief valves open when the pressure acting on them reaches a preset value. Often they are used as safety devices on air pressure tanks and on governor

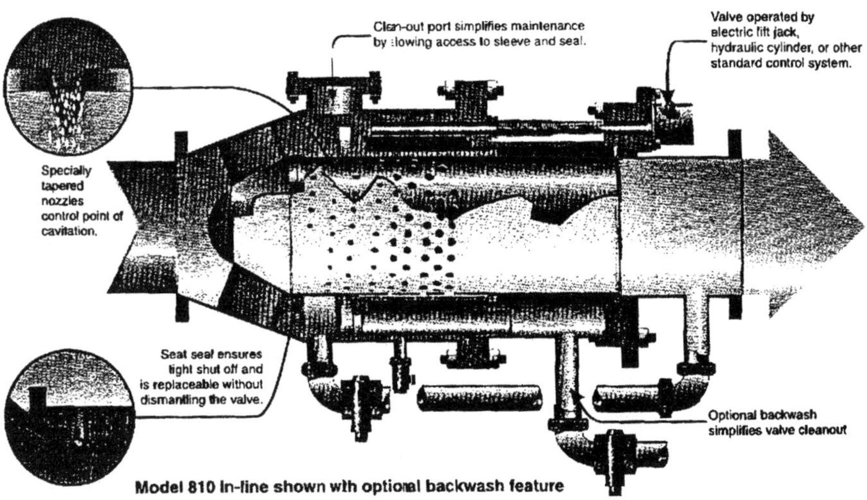

FIGURE 5-5: ***Sleeve Valve.*** *This cutaway view shows the major components of an "in line" sleeve valve. (Courtesy Bailey Division of CMB Industries)*

pressure set accumulators. They may be either self-closing after the pressure reduces, or they may lock open and require manual resetting. Pressure-relief valves are not recommended for controlling hydraulic transients because a significant pressure rise must occur first to initiate valve opening.

A typical angle pressure-relief valve contains a valve disc that is held against the valve seat and system pressure by compression springs on the shaft. When the preset pressure is exceeded, the disc and unrestrained shaft moves to the open position. A one-way port in the piston on the shaft allows oil to enter the lower portion of the hydraulic cylinder. A control valve between the cylinder and oil reservoir controls the rate of closure to provide a smooth controlled shut-off.

Most applications for relief valves for new pressure tanks in the United States should meet the requirements of Section VIII of the American Society of Mechanical Engineers (ASME) Boiler and Pressure Vessel Code.[1]

Pressure-Regulating Valves

Pressure-regulating valves maintain a preset downstream pressure when connected to an upstream higher pressure source. They often are used to provide a constant (regulated) pressure source of air, oil, or water in hydro facilities. Pressure-regulating valves actually reduce their opening as upstream pressure rises and, therefore, are not suitable for minimizing hydraulic transient pressures. When the penstock or unit inlet is the source of the powerhouse cooling water, pressure-regulating valves often are used to reduce the inlet pressure to the required cooling water pressure.

While there are many valve designs for specific applications, most pressure regulators maintain a preset downstream pressure by adjusting the valve opening and, thereby, changing the flow and the pressure loss across the valve. A sensing line, typically a length of small diameter tubing, admits the downstream pressure to one side of the actuator diaphragm. The pressure exerted on the diaphragm counteracts the closing force provided by a calibrated spring to control the flow as required to maintain the desired valve outlet pressure. Changing the compression of the spring that operates the valve mechanism adjusts the pressure setting.

Pressure-Regulator Valves

Hydraulic-pressure transients (waterhammer and surges) occur when the flow in a hydraulic conduit or piping system is changed suddenly. This condition typically occurs when opening or closing valves or the turbine wicket gates. The maximum pressure often occurs after a unit load rejection when the load is suddenly removed from a generator by tripping the unit circuit breaker. Chapter 12, Hydraulic Transients, discusses hydraulic design considerations in detail. While conduit or piping systems can be designed for the increased pressure, it is often more cost-effective to control pressure transients with pressure-regulator valves and design for lower pressures.

Pressure-regulator valves in the hydroelectric industry are valves used to minimize pressure rise in water conduits due to rapid closure of reaction turbine wicket gates or impulse turbine needles following load rejections. Howell-Bunger, angle-type, and sleeve-type valves can be used for this service. The valve is linked to the turbine wicket gates so that a proportional opening of the regulator valve accompanies the rapid gate closure. This permits the penstock flow to remain relatively constant during a load rejection and prevents excessive waterhammer pressure rise. Following the load rejection, the pressure-regulator valve is closed slowly to avoid significant pressure rise. Note that pressure-regulator valves are opened simultaneously with rapid wicket gate closure rather than by pressure changes as is the case with pressure-relief or pressure-regulating valves.

Air/Vacuum Valves

Air and vacuum valves are provided in piping systems to exhaust air from a pipeline system or spiral case, or to fill a vacuum to prevent conduit collapse. A typical air-vacuum valve has a valve housing that supports a movable float or piston. This float or piston lowers to exhaust air or fill a vacuum and floats to seal the piping system when the air is exhausted. When used on penstock or pipeline systems, particularly in remote locations, it is recommended to use extra valves in case some of the air admission valves do not open.

Bypass Valves

Valve applications where water is conveyed around a turbine, powerhouse, or dam often are referred to as bypass valves. Energy-dissipating and pressure regulator valves often are used in this application. Needle valves and other valve types also are used in bypass lines to balance the pressure across large butterfly or spherical valves before they are opened or closed.

Flow-Control Valves

Valve applications that control the flow of water often are referred to as flow-control valves. For large water conduits, energy-dissipating valves control the flow of water while simultaneously breaking the head in the releases. Fixed-cone dispersion and hollow-jet valves commonly are used to control releases from low-level outlets while sleeve valves are used for "in-line" piping applications. Flow-control valves commonly are used for pressure-regulator service.

Throttling or flow control with large diameter butterfly or spherical valves rarely is recommended because of potential cavitation and structural damage from the rough operating conditions. For smaller applications, a wide variety of flow-control valves have been developed for specific fluids and operating service. For water service, globe valves commonly are used to throttle flow without detrimental cavitation or high noise levels.

References

[1]ASME Boiler and Pressure Vessel Code, Section VIII, Division 1, American Society of Mechanical Engineers, New York, 1995.

[2]Ehrhart, R.W., "Auto Oscillation in the Hyatt Plant's Penstock System," *International Water Power & Dam Construction,* April 1979, pages 38-41.

Bibliography

AWWA Standard for Rubber Seated Butterfly Valves, ANSI/AWWA C504-87, American Water Works Association, Denver, Colorado, June 1988.

Control Valve Source Book, Fisher Controls International, Marshalltown, Iowa, 1988.

Zipparo, Vincent J., and Hans Hasen, Editors, *Davis' Handbook of Applied Hydraulics,* Fourth Edition, McGraw-Hill, Inc., New York, New York, 1993.

Chapter 6
Trashracks and Cleaning Devices

Trashracks, which usually consist of rows of parallel bars, protect equipment and waterways from objectionably large or damaging debris and act as a safety rack for unwary swimmers or boaters. Trashracks are located in the inlet of a waterway or conduit that leads to the equipment needing protection. The inlet or conduit "serves" the equipment. Additionally, trashracks may be installed at all depths in reservoirs for safety purposes since scuba divers periodically inspect the intakes. Trashracks also may be used to protect fish.[1]

Water flowing through clean trashracks has an insignificant head loss. However, debris stopped by the trashracks can sustain head losses that almost stop the flow of the water. If there is a need to intercept trash with a trashrack, then there is a need to remove the intercepted trash so that the flow of the water will not be hindered. Therefore, cleaning devices have to be provided in order that the trash can be removed from the trashracks. This cleaning device can be anything from a manually operated hand held rake to a sophisticated automatic hydraulic or electrically actuated machine.

Debris can be placed into two categories:

- Natural debris such as tree trunks, branches, grass, moss, and weeds; and
- Man-made debris such as tires, plastic cans, and bottles.

Natural debris is the most common debris that trashracks must catch. For reservoir and run-of-the-river power plants, natural debris is most prevalent during flood conditions. Steep-sided reservoirs can receive seasonal debris loads similar to the local vegetation found on those slopes. Submerged logs and other non-buoyant trash may drift into the trashracks below the water surface. Aquatic growth and other vegetation are serious problems in canals. Man-made debris, although not as common as natural debris, still can be a problem that requires protection at the turbine intake. To eliminate having damaging debris pass through the turbine or pump-turbine, trashracks are installed at the entrance of the penstock for the turbine and on the inlet to the suction tube of a pump-turbine.

TRASHRACKS

Types of Trashracks

The details and general construction of trashracks vary with the service required, configuration of the trashrack structure, depth of water, and accessibility for replacement. Trashracks usually consist of rectangular cross-sectional vertical bars held together with lateral bars or structural shapes. Although round or streamlined trash bars may reduce hydraulic head loss, they are not used commonly because of the added expense and their tendency for vibration. The Design of Vibrating Loads section later in this chapter discusses the vibration of the individual trash bars. Trashracks may be divided into three types according to construction and installation requirements: end bearing, side bearing, and integral trashracks.

FIGURE 6-1: ***Integral Trashrack on Penstock Intakes at the Grand Coulee Third Power Plant.*** *An integral trashrack combines several panels of trash bars with lateral support beams or members. The support members make a multi-sided rigid frame that carries the loading into the trashrack supporting structure.* *(Courtesy Bureau of Reclamation)*

End-bearing trashracks. End-bearing trashracks are the simplest and, for shallow trashracks, usually the least expensive of the three kinds of racks. The trash bars, which run from top to bottom, individually carry the loads into the trashrack structure. End-bearing trashracks are commonly used for canal headworks and where a single rack section can extend from top to bottom of the rack-protected area. Due to the end-bearing characteristics, these racks can be installed side-by-side covering the desired area. End-bearing trashracks are ideal where low-head conditions exist and large trashrack areas are necessary. Usually end-bearing trashracks are in an inclined position; however, they can be used in a vertical or near vertical position if proper care is taken to secure the racks with clips or expansion anchors. In installations where the trash bars are excessively long, it becomes economically advantageous to install one or more lateral support beams. These support beams go to the sides of the trashrack structure, providing intermediate support to the trash bars so that their loaded spans are shorter. This reduces the size of the trash bars.

Side-bearing trashracks. Side-bearing trashracks are supported by the trashrack structure on both sides of the rack. The trash bars run from top to bottom and are supported by lateral bars or beams that carry the loads into each side of the rack. The racks are supported or retained by guides or grooves provided in the trashrack structure. Usually side-bearing trashracks are in a vertical position, but they can be used effectively in an inclined position and occasionally are used in the prone position. Side-bearing trashracks are used for high, medium, or low head conditions. Sometimes side-bearing trashracks are substituted for end-bearing trashracks when the trashrack supporting structure cannot withstand high bearing loads at the top. The major limiting design concern of a side-bearing trashrack is the span. The longer the horizontal span, the larger the lateral beams have to be to carry the trash load to the sides of the trashrack structure. An economic decision has to be made between building a stronger trashrack or building an intermediate pier in the trashrack structure. Side bearing racks may be stacked in tiers to obtain any designed height or placed side by side with intermediate piers to obtain any desired width.

FIGURE 6-2: ***Integral Trashrack at the Jordanelle Dam Outlet Works.*** *Integral trashracks simplify the trashrack structure by eliminating most of the concrete supports usually required for large trashracks. These are used often in deeply submerged applications such as penstocks or multilevel selective withdrawal outlet systems.*

(Courtesy Bureau of Reclamation)

Integral trashracks. An integral trashrack combines several panels of trash bars with lateral support beams or members. The panels are connected by either welding or bolting the support members together. The support members make a multi-sided rigid frame that carries the loading into the trashrack supporting structure. Integral trashracks simplify the trashrack

Chapter 6

structure by eliminating most of the concrete supports usually required for large trashracks. Integral trashracks often are used in deeply submerged applications such as penstocks or multilevel selective withdrawal outlet systems. Integral trashracks are not intended to be replaced. Figures 6-1 and 6-2 show integral trashracks.

Design Considerations

A major design consideration for trashracks is the approach velocity. This value is the overall trashrack area divided by the maximum flow. The approach velocity for each structure has to be determined on a case-by-case basis. An approach velocity of 0.3 to 0.6 meter/second (1 to 2 feet/second) is desirable but not always practical. This slow approach velocity reduces the tendency to collect debris against the racks, minimizes the possibility of trashrack vibration, and provides a relatively safe condition for intruders such as boaters or swimmers. Minimizing the approach velocity can reduce the hydraulic head loss through trashracks. Trashracks designed with low velocities generally do not have destructive vibration problems and probably can be designed without taking into account the trash bar and panel vibration. To limit the physical size of the structure and thus decrease the capital cost, one can use larger approach velocities of up to 1.5 to 2 meters/second (5 to 7 feet/second).

Tests on trashracks have shown that velocities are not uniform through the trashrack area. Localized velocities can be as much as twice the average velocity depending on the geometry of the trashrack structure. In addition, at intakes with more than one opening per unit, the flow of water is not distributed uniformly between the openings. With two openings per unit, the trashrack design should allow for two-thirds of the water flowing through one of the two openings. With three openings per unit, trashrack design should allow for 45 percent of the water to flow through one of the three openings.

Many factors enter into the decision concerning how large an approach velocity to use when designing a trashrack. These factors include:

- Economics;
- Safety considerations;
- Preservation of fish [1];
- Location of the trashrack structure within the system;
- Amount of expected debris;
- Submerged conditions;
- Age of the structure (i.e., a new structure would probably have more debris than an old structure);
- Type of usage (i.e., a canal outlet versus a river intake structure);
- The amount of sediment that the water may be carrying; and
- Intake hydraulics and eddy currents.

If a trashrack is to be raked, special consideration must be made. Different raking methods require different slopes for the trashrack. Generally, hand-raked trashracks have as flat a slope as possible. Trashracks that are to be raked mechanically should have a slope angle between 5 to 30 degrees from the vertical. Side support trashracks that are to be raked and are stacked in tiers must be kept in alignment. Usually large dowel pins between the panels maintain the alignment.

Design for nonvibrating loads. Trashrack bar sections should be rectangular. The depth of a trashrack bar section should be at least 4, but less than 12, times its thickness. The lateral unsupported length should not exceed 70 times its thickness. Normally lateral supports are spaced between 450 and 760 millimeters (18 and 30 inches) from each other. To avoid damaging other parts of the structure, assume that the trashrack bars will fail when the stress in the bar reaches the failure stress value calculated in Equation 6-1.

EQUATION 6-1:

$$S_F=1.10F_Y$$

where:

S_F=failure stress; and

F_Y=minimum yield stress of the material, kilopascal (pounds/square inch), kPa (psi).

Equation 6-1 is used only for failure when the support structure is not designed for full differential head. Sometimes, trashrack slots are used for head gates. In this case, the structure is designed for full differential heads, and Equation 6-1 is not valid.

Similarly, the safe working stresses for all components of a trashrack used to support flashboards should not exceed the safe working stress value calculated with Equation 6-2.

EQUATION 6-2:

$$S_S=0.6F_Y$$

where:

S_S=safe working stress; and

F_Y=minimum yield stress of the material kPa (psi).

Design for vibrating loads. Trashracks used in pump-turbine intake/discharge structures usually experience velocities higher than 1.5 meters/second (5 feet per second). In past years, trashrack panels in the inlet and discharge of pump turbines have experienced failures. Trashracks used in pump-turbine intake/discharge structures usually experience localized velocities as large as 4.5 to 6 meters/second (15 to 20 feet/second). The load stresses are normally low, but the racks usually have to be analyzed for vibration. The trashrack's bar shape should be designed to avoid vibration frequencies induced by vortex shedding that will produce resonance in the trashrack. It has been found that round or elliptically shaped trashrack bars should not be used because they are subject to vortex shedding. This can lead to failure of the bar. It is recommended that rectangular bars with an aspect ratio of 1:4 (at least) be used. This provides a stable cross-sectional area in the flowing water.[2]

As shown in Figure 6-3, there are three modes of vibration for a trashrack:

- Vertical heave acts in the direction along the long axis of the bar;
- Horizontal heave acts in the horizontal direction perpendicular to the long axis of the bar; and
- Plunge acts at right angle to the angle of incline.

The natural frequency—Equation 6-3—gives the fundamental natural frequency of an individual rack bar in water.[3,4]

EQUATION 6-3:

$$f_n = \frac{\alpha}{2\pi} \left[\frac{EIg}{ML^3} \right]^{0.5}$$

where:

f_n = fundamental bar natural frequency, Hertz;
α = coefficient of bar end fixity (α=22 for clamped ends, α=10 for pinned ends), dimensionless;
E = modulus of elasticity of the bar material, kilograms/meter squared (pound/feet squared);
I = moment of inertia of the bar about the neutral axis parallel to flow, meter^4 (feet^4);
g = acceleration of gravity, meter/second squared (feet/second squared);
M = effective mass of the bar, kilogram (pound); and
L = unsupported length of the bar, meter (feet).

Equation 6-3 gives only a quick check for the fundamental natural frequency of an individual bar. It is not sufficient for design of the trashrack panels. The use of the natural frequency of a single trashrack bar can be justified if high localized flow velocities are assumed. No flow through a trashrack has completely uniform flow. The high localized velocity may tend to cause a single trashrack bar to vibrate independent of the trashrack panel. If this vibration is close to the natural frequency of the single bar, the bar may fail.

In recent research it has been shown that the natural frequency of the trashrack panel, as determined with a finite element analysis, can be as much as one-third the calculated

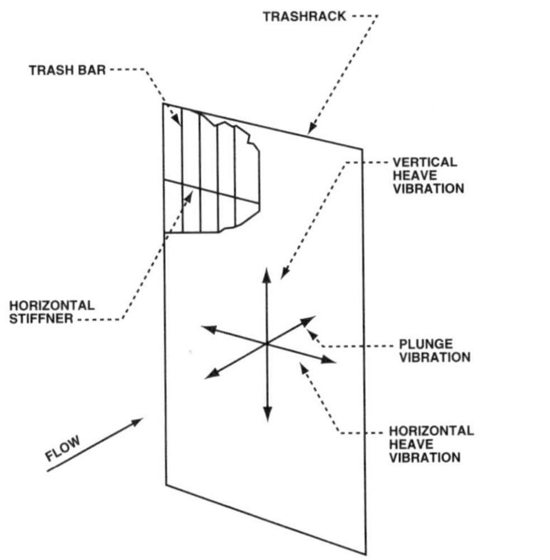

FIGURE 6-3: *Three Modes of Vibration for a Trashrack.* *There are three modes of vibration for a trashrack shown in this figure. Vertical (or axial) heave vibration acts along the long axis of the bar. Horizontal (or side) heave vibration acts perpendicular to the long axis of the bar. Plunge vibrations act at right angles to the angle of incline.*

natural frequency of the individual trashrack bar.[5] It should be noted that rigidly connecting individual trashrack panels can cause them to vibrate at higher harmonics and ultimately fail. A less rigid connection between individual panels may eliminate this vibration.

The natural frequency of the trashrack bar or trashrack panel has to be sufficiently different in magnitude than any forcing frequencies produced by the turbine or pump-turbine so that resonance of the forcing and natural frequency will not cause the trashrack to fail. Forcing frequencies can come from pressure pulsations in the turbine or pump-turbine or from vortex shedding of the individual bars of the rack. Some of the most important forcing frequencies follow.

Equations 6-4 and 6-5 give the forcing frequencies perpendicular and parallel to the flow respectively. [2,6]

EQUATION 6-4:

$$f_{fh} = \frac{N_{SN} V}{w}$$

where:

f_{fh}=forcing frequency perpendicular to the flow, Hertz;
N_{SN}=Strouhal number is dependent on the bar cross-section shape and the spacing of the bars, and has a large variation in magnitude, dimensionless;
V=fluid approach velocity, meters/second (feet/second); and
w=the bar width perpendicular to the flow, meters (feet).

EQUATION 6-5:

$$f_{fp} = \frac{2 N_{SN} V}{w}$$

where:

f_{fp}=forcing frequency parallel to flow, Hertz;
N_{SN}=Strouhal number, dimensionless;
V=fluid approach velocity, meters/second (feet/second); and
w=bar width perpendicular to flow, meters (feet).

Pressure pulsations from a rotating runner (blade passing frequency) are calculated using Equation 6-6.[7]

EQUATION 6-6:

$$f_f = \frac{N N_b}{60}$$

where:

f_f=forcing frequency to flow, Hertz;
N=unit rotational speed, rpm; and
N_b=number of runner blades, dimensionless.

Chapter 6

The draft tube forcing frequency is given by Equation 6-7 (commonly called the Rheingans formula).[8]

Equation 6-7:

$$f_f = \frac{N}{60k}$$

where:
f_f=forcing frequency to flow, Hz;
N=unit rotational speed, rpm; and
k=Rheingans number

The value of “k” as first noted by Rheingans, is usually from 2 to 4. On a given machine, “k” varies depending on the operating conditions. For some machines and plants, “k” may be as high as 8.

The natural frequency of the trashrack bars should be not less than 2.3 to 2.5 times larger than the forcing frequency found in the flowing water.[7,9] Higher harmonics of the fundamental forcing frequency can begin to induce resonance even when the fundamental forcing frequency is one-half the natural frequency of the trashrack bar.[2] Therefore, just because the natural frequency of the trashrack bar is larger than the fundamental forcing frequency, the trashrack bars may still vibrate from higher harmonics that may be present. For example, forcing frequencies from rotating electrical and mechanical equipment inside the powerhouse may be transmitted to the trashracks and cause them to vibrate.

Design Criteria

Bars of totally submerged trashracks that function under more than 6 meters (20 feet) of head should be designed to fail at approximately 6 meters (20 feet) of differential head, assuming that the trashrack structure is designed to withstand it. Usually submerged trashracks for waterway inlets are designed to meet the operating conditions of the trashrack supporting structure. If the trashracks are equipped with automatic trashrakes or have head loss sensors that inform operators when the racks require cleaning, they can be designed for a lower differential head. Usually, trashracks that are submerged 6 meters (20 feet) or less are designed for a differential head equal to two-thirds of the maximum depth of submergence. However, if these racks are intended to serve as supports for flashboards, they should be designed for a safe stress under maximum load conditions as calculated using Equation 6-2. In all installations, if the trashracks should ever become totally obstructed, they must be designed to yield before any damage is sustained to the support structure. In some cases instrumentation is used to shut down the unit at a predetermined head differential across the trashrack.

Care should be taken to avoid heavy loadings from trashracks near or on the exposed edges of concrete structures. The normal practice is to connect the horizontal beam members on the side bearing trashracks to an angle that will disperse and carry the concentrated load farther back into the guide slot of the concrete support structure.

In locations where freezing is a problem, trashracks must be designed for loads resulting from ice. Ice loadings can induce much greater loads than what would be expected from other trash loadings. In addition to designing for higher load stresses, the trashracks also should be properly secured to prevent being lifted from the supporting structure by floating ice.

For many structures, it is easier to prevent icing around the trashracks than to design for loadings resulting from ice accumulation. In some situations, ice is kept from forming in the trashrack area with air blast or bubbler systems, which bring up the warmer waters from below the surface to eliminate ice formation. Occasionally, heaters are used either inside of hollow trash bars or near the trashracks. In some cases, the correct design will provide good flow patterns that will prevent freezing around the trashrack. Periodically, vibrating trashracks with an exterior exciter have been used for controlling the formation of frazil ice on the trash bars. Also, certain trash bar materials or coatings on metal trash bars are able to prevent frazil ice from forming or allow operation at a lower water temperature before ice forms. Designing discontinuities into the metal trash bars to prevent cold air temperatures from penetrating deep into the water has been partially successful for controlling ice formation. The discontinuity is created either by having a physical gap of several centimeters in each trash bar or by adding a material along the trash bar which has poor heat transfer properties. Either of these methods stops the transfer of heat from the lower portions of the trash bars to the colder surface water.[10]

Corrosion Protection

Economic analysis has shown that it is cheaper to replace structurally weakened racks than it is to repaint them periodically. The trashrack's design should allow for some corrosive deterioration without sacrificing structural integrity. When it is practical, the recommended minimum trash bar size used in trashrack design is 12 by 50 millimeters (0.5 by 2 inches) to allow for corrosive deterioration.

Welds are very vulnerable to corrosion. When determining weld size, the allowance for corrosion of the weld is combined with the requirement for the strength of the weld. Welds that are highly stressed or in pure shear are usually oversized to allow for corrosion.

Typically, trashracks are painted with some highly protective coating compatible with currently acceptable environmental regulations. Galvanizing the trashrack is not an acceptable method of protection for normal water conditions. During shipping and installation, care should be taken not to damage the protective coating. If the coating does get scratched or chipped, it should be repaired properly. If a rack has to slide a considerable distance on a concrete surface during installation, a wearing bar should be used to protect the coating.

On some installations, cathodic protection systems are used to protect the racks. Most of the phenomena involving corrosion of metals submerged in water is electrochemical in nature. When two dissimilar metals in contact are immersed in water, the system becomes a galvanic cell. The metal that is above the other in the Electromotive Force Series of Metals becomes the anode in the galvanic cell and suffers corrosion. The other metal becomes the cathode and the corroding anode protects it. In the case of a trashrack, a mass of metal that is higher on the Electromotive Force Series of Metals (anode) would be connected to the iron trashrack (cathode), and both immersed in water. A small voltage is applied to the system ensuring that a galvanic cell is initiated. The corroding anode then protects the trashrack.

Trash Bar Spacing

Trash bar spacing is selected to suit the equipment being protected. The usual practice is to provide as large a clear opening as possible and still protect the downstream equipment. For a Francis, Kaplan, or propeller turbine, the spacing must be smaller than the minimum opening in the turbine runner or the maximum opening of the wicket gates (whichever is smaller). Closer spacing of trash bars than is needed will result in unnecessary hydraulic head loss and will cause premature clogging of the racks.

For a Francis turbine, the trash bar spacing may be estimated from the following equations. Equations 6-9 and 6-10 are only appropriate for Francis units with specific speeds calculated from Equation 6-8 between 76 and 380 metric (20 and 100 English).

Calculate the Francis turbine specific speed.

EQUATION 6-8:

$$N_S = \frac{N\ (\text{Power})^{0.5}}{(\text{Design Head})^{1.25}}$$

where:

N=the turbine rotational speed, rpm;
Power=power at full gate, kW (horsepower); and
Design Head=head in meters (feet).

Calculate the dimensionless factor F.

EQUATION 6-9:

$$F = 0.80 - \frac{N_{sm}}{762} \qquad \text{(metric)}$$

$$F = 0.80 - \frac{N_{se}}{200} \qquad \text{(English)}$$

where:

N_{sm}=metric specific speed; and
N_{se}=English specific speed.

Calculate the trashbar spacing.

EQUATION 6-10:

$$S = \frac{F D_3}{n}$$

where:

S=the clear space between trashrack bars for Francis turbines, meters (feet);
F=dimensionless number that is calculated using Equation 6-9;
D_3=the runner discharge diameter, meters (feet); and
n=the number of runner blades (use 19 if not known).

For a Kaplan turbine, the trash bar spacing may be estimated from the following equation. Equation 6-11 is based on a turbine with 24 wicket gates.

EQUATION 6-11:

$$S = \frac{D_3}{12}$$

where:

S=the clear space between trashrack bars, millimeters (inches); and
D_3=the runner discharge diameter, millimeters (inches).

Equation 6-11 is appropriate for Kaplan turbines having a discharge diameter of less than 1.5 meters (5 feet). Larger turbines require a maximum trashrack bar spacing of 150 millimeters (6 inches).

Equation 6-12 is based on a turbine with 20 wicket gates.

EQUATION 6-12:

$$S = 0.097D_3 + K_1$$

where:

S=the clear space between trashrack bars, millimeters (inches);
D_3=the runner discharge diameter, millimeters (inches); and
K_1=4 millimeters (0.2 inch).

Equation 6-12 is appropriate only for Kaplan turbines having a discharge diameter less than 1.5 meters (5 feet). Larger turbines require a maximum trashrack bar spacing of 150 millimeters (6 inches).

For an S-Type turbine, the trash bar spacing may be estimated from the following equation.

EQUATION 6-13:

$$S = 0.167D_3 - K_2$$

where:

S=the clear space between trashrack bars, millimeters (inches);
D_3=the runner discharge diameter, millimeters (inches); and
K_2=100 millimeters (4 inches).

Equation 6-13 is appropriate only for S-Type turbines having a discharge diameter less than 1.5 meters (5 feet). Larger turbines require a maximum trashrack bar spacing of 150 millimeters (6 inches).

These calculated trashrack bar spacings should be confirmed by actual dimensions from manufacturer's data after the turbine has been designed.

Cleaning Devices

If there is a debris or icing problem that could cause an overload of the trashrack or associated structure, some type of cleaning device is required for the trashrack. Also, a cleaning device is needed if the debris or icing problem could cause undesirable head losses or reduce the water delivery capability. The type of cleaning device needed will vary with the service required, configuration of the trashrack and structure, depth of water, availability of power, and economics.

When the maximum flow is traveling down a waterway, the trash load will be the largest, and thus the need for cleaning the trashracks will be the most important. The cleaning device should have the capability to operate and clean the trashracks when maximum flow is moving through the intake structure. This will ensure that the power plant can generate maximum energy during high flow conditions.

Submerged logs, "dead heads," cannot be removed by raking. Some rack cleaning equipment includes grapples to handle logs. Even with grapples, it may be necessary to have a diver help remove logs.

Types of Debris-Cleaning Devices

There are two general categories for debris-cleaning devices for trashracks: manual and mechanical.

Manual cleaning devices. The most common type of manual cleaning device for trashracks is the hand-held rake. This consists of a rake head (similar to a garden rake) connected to a long pole. A hand-held rake requires a person to physically lower and position the rake against the trashrack and then drag it and the debris to the top of and over the rack. Variations of the hand held rake include a squeegee-type rake head in lieu of teeth or a portable pool-type vacuum cleaner. The squeegee-type rake has been used successfully at canal turnouts where debris can be pushed off the trashrack to continue on downstream in the canal.

Divers sometimes clean trashracks that mechanical devices are unable to clean. A removable section of trashrack should be provided to allow divers inside the trashrack structure to clean downstream of the trashrack.

Mechanical cleaning devices. Some of the common types of mechanical cleaning devices are: hydraulic trashrakes, catenary trashrakes, hoist and carriage trashrakes, backhoe-type rakes, and sluicing gates.

The hydraulic trashrake mimics the physical motions used in manual raking operations. It consists of a telescoping rake head that is lifted clear of the rack face during the downward travel of the rake head. Once fully extended, the rake head is swung into position to bear on the rack face. Then the rake head is retracted along the rack face with the hydraulic system maintaining contact between the rake head and the rack face. Trash is dragged to the top of the rack. At the top of the rack, the rake head continues to retract, pulling debris completely off the rack face, dumping the debris onto the deck or into a conveyor hopper. These types of rakes can be provided with or without teeth. The rakes are completely out of the water when not operating, thus allowing easier maintenance and also reducing corrosion problems. The hydraulic trashrake requires electrical power. It can be operated manually with push buttons or automatically by means of timers or a differential water-level control device that measures the water levels across the trashrack. Usually, automatic limit switches and overload devices are provided to assist in operation.

This type of trashrake is successful with long stringy weeds and mosses, with bulky weeds, with tree branches, and with logs. With an indexing feature, which permits the rack to move laterally along the intake deck on a rail structure, one trashrake can be used to rake multiple intake bays.

The catenary trashrake consists of raking beams, spaced approximately 3 meters (10 feet) apart, that are attached to continuous chains driven by an electrically operated motor. The raking beams travel up the rack face, then return along a guided path upstream of the rack. At the top of the rake's travel path, a scraper mechanism removes the collected debris from the raking beams, dumping the debris onto the deck or into a conveyor. The raking beams contain a plate with teeth cut to match the trashrack bars. When raking, the teeth are at a right angle to the trashrack. The catenary trashrake can be operated manually or automatically.

The catenary trashrake performance is generally poor for stringy mosses, stringy weeds, and bulky tumbleweeds. It is difficult to remove the stringy mosses and weeds from the raking beams. The catenary trashrake has problems in picking up tumbleweeds. Other disadvantages are that a part of the rake is always in the water, the capital cost is very high, and multiple rakes are required on trashrack intakes that have multiple bays or where a wide intake precludes the use of a single rake.

Each trashrack bay has to have a catenary trashrake since it is not mobile. If there are multiple bays, one trashrake breakdown will not hinder operating trashrakes on other bays. With other trashrake systems, a trashrake breakdown will stop trash cleanup of all bays.

The hoist and carriage trashrake has a large rake head that is raised and lowered by an electrically operated hoisting system. The rake head runs in guides mounted to the sides of each trashrack structure bay, parallel to the trashrack. The rake head folds away from the trashrack face during the downward travel, then at the bottom of the rack, the head swings into position and the teeth extend into the spaces between the rack bars. The rake head then is raised along the trashrack. At the top of the rack, it dumps debris onto the deck or into a conveyor. The hoist and carriage trashrake can be operated manually or automatically.

The hoist and carriage trashrake performance is generally poor for stringy mosses, stringy weeds, and sediment areas. With stringy mosses and weeds, it is difficult initially to get the debris off the trashrack bars because of the gap between the rake teeth and the rack bars. Sediment accumulation can cause problems. When the rake head reaches the bottom of its travel, a cam mechanism is supposed to swing the rake teeth between the trashrack bars. If the cam mechanism is on the guides rather than on the rake, submerged logs, sediment, or other debris can prevent the rake head from traveling down far enough to hit this cam. If this occurs, the rake head will not swing into position and, when the hoist raises the rake head, the rake teeth are still parallel to the trashrack.

The backhoe-type rake is basically a backhoe machine that has a rake or cleaning type head installed on it. It can be used to clear weeds, trash, and even large, bulky debris from trashracks. The rake can dump the debris on either side of it, can push the debris away from the trashrack (at canal or river turnouts) to be conveyed on downstream, or can be carried along the top of the structure and dumped at a specific site. The backhoe-type rake can be designed to operate using electric motors or internal combustion engines. The rake can be installed on rails mounted on the trashrack deck or can be provided with rubber tires. In either case, some type of counterweight or support rail is required for stability when lifting. The backhoe-type rake requires a person to operate it. One problem is that the operator, when sitting in the cab, has a hard time seeing what the rake head is doing, especially below the water surface and on steeply sloped (near vertical) trashracks. This could cause excessive loading on the trashracks from operator error. Advantages of using this type of cleaning device are that a single unit can be used for cleaning multiple intakes, and, when not in operation, all components are out of the water, minimizing corrosion and making maintenance easier.

Sluicing gates that open to allow collected debris to pass can be provided in structures. This method of cleaning debris from a structure should be used only when it is appropriate and lawful to pass debris downstream.

References

[1]Committee on Hydropower Intakes of the Energy Division of the American Society of Civil Engineers, *Guidelines for Design of Intakes for Hydroelectric Plants,* Chapter 7, American Society of Civil Engineers, New York, New York, 1995.

[2]Crandall, S.H., S. Vigander, and P.A. March, *Destructive Vibration of Trashracks due to Fluid-Structure Interaction,* American Society of Mechanical Engineers Paper No. 75-DET-63, presented at the Design Engineering and Technical Conference, Washington, D.C., September 17-19, 1975.

[3]Levin, L., "Etude Hydraulique des Grilles de Prise d'Eau," *Proceedings from the International Association for Hydraulic Research,* 7th General Meeting, Volume I, Bertrand, Ida., Lisbon, Portugal, 1957, pages C11-1 to C11-11.

[4]Levin, L., "Problemes de Perte de Charge est de Stailite des Grilles de Prise D'Eau," *La Houille Blanche,* Volume 22, No. 3, 1967, pages 271-278.

[5]Behring, A.G., and C.H. Yeh, "Flow-Induced Trashrack Vibration: Pump Turbine Schemes, Planning, Design, and Operation," Paper Presented at the Joint American Society of Mechanical Engineers-Canadian Society of Mechanical Engineers Applied Mechanics Fluids Engineering and Bioengineering Conference, Niagara Falls, New York, June 18-20, 1979, pages 125-134.

[6]Strouhal, V., "Über eine Besondere art der Tonerzeugung," *Ann. der Physik und Chemie,* Volume V, 1878, page 216.

[7]Sell, L.E., "Hydroelectric Power Plant Trashrack Design," *Journal of Power Division, American Society of Civil Engineers,* January, 1971, pages 115-121. Discussion by E.E. James and F.T. Katakure, ASCE P04, 1971. Author's Closure, American Society of Civil Engineers, P01, 1972.

[8]Rheingans, W.J., "Power Swings in Hydroelectric Power Plants," *ASME Transactions,* April, 1940.

[9]Todd, R.V., "Design Considerations for Trashracks At Pumped-Storage Facilities," *Hydro Review,* October, 1992, pages 46-52.

[10]Haynes, Donald F., "Seeking Solutions for Icing at Dams and Hydro Plants," *Hydro Review,* Volume XII, Number 7, December 1993, pages 58-64.

Bibliography

Donaldson, R.M., "Hydraulic Turbine Runner Vibration," *Transactions of ASME,* July 1956.

"Koski, James A., "Getting Rid of the Trash at Cheatham Dam," *Hydro Review,* Volume XI, No. 1, March 1992, pages 52-54.

Schneeberger, Heinz, "Using Automated Equipment for Cleaning Trashracks," *HRW-Hydro Review Worldwide,* Volume 2, No. 2, Summer 1994, pages 33-38.

Stutsman, Richard D., Gary D. Mackey, and James J. Strong, "Taking the Muscle out of Trashraking," *Hydro Review,* Volume X, No. 5, August 1991, pages 22-28.

Zowski, T., "Trashracks and Raking Equipment: Part I – Trashracks," *International Water Power & Dam Construction,* September 1960.

Checklist for Chapter 6

Trashracks

1. Has the trash load been determined?
2. What ice loading is appropriate?
3. For what differential head should the trashrack be designed?
4. Will the trashrack support structure carry the loaded trashrack?
5. What trash bar spacing is required?
6. For what corrosion factor should the trashrack be designed?
7. Will trash bar vibration occur at the average velocity or local higher velocities through the trashrack?
8. What is the maximum velocity at any place on the trashrack?
9. What is the maximum flow through the trashrack?
10. What forcing vibrations from equipment in the powerhouse can be transmitted through the support structure to the trashrack?
11. What forcing vibrations from the rotating machinery can be transmitted through the water to the trashrack?
12. Has the capital cost been balanced properly with the cost of the head loss through the rack?
13. Will ice prevention be required?

Cleaning Devices

1. What type of trashrack cleaning devices are needed?
2. Should the cleaning device be automatic or manually operated?
3. Does the trash have to be conveyed automatically away from the area or will it be stockpiled and then carried away manually?
4. What maximum weight or size of debris needs to be handled by the cleaning device?
5. Has the trash load been determined?
6. Is the cleaning device capital cost and operating cost properly balanced with the trash load impacts? (Manual raking may be all that can be justified.)
7. Are provisions needed, such as rails on the intake deck, to allow for cleaning devices?

Chapter 7
Gates

Some gates and valves may be used interchangeably in similar applications. Therefore, Chapter 5, Major Valves, should be consulted for valves that may be used in place of some of the gates described in this chapter. Chapter 8, Lifting Equipment, describes hoists for gates.

Major gate applications at hydropower facilities can be categorized by the type of service: flow control, closure, and emergency closure. Flow-control gates throttle flow to vary the flow rate passing the gate structure. Closure gates provide a positive conduit closure under no flow conditions. Emergency closure gates provide a positive conduit closure under flow conditions.

A large number of unique and complex gate designs have been developed to satisfy these service needs. Many of these designs have been discontinued because of lack of dependability, high maintenance costs, or high initial costs. However, many of these obsolete gates are still in service. Consequently, this chapter discusses the obsolete gates, for information only, along with the newer equipment. See Table 7-1 for a general use guide.

Flow Control Gates

Radial Gates (Tainter) and Top-Seal Radial Gates

Radial gates, also called Tainter gates, have been used successfully to regulate flow from spillways and outlets in dams and for checks and turnouts in canals. Radial gates are used extensively because their design is simple. They cost less to build and are less expensive to install than other types of gates used for the same service.

Radial gates, as shown in Figure 7-1, have the skinplate in the shape of a portion of a cylinder with an axis horizontal. A gate's skinplate is the water-bearing surface on the gate's frame. The water pressure is applied against the convex side and is resisted by beams and girders on the concave side. Structural arms, extending radially from the skinplate to stationary trunnions in a structure, resist the

TABLE 7-1: ***General Use for Gates****

Gate Name	Type of Service: Closure or Emergency Closure	Type of Service: Flow Control	Maximum Design Head+ meters (feet)	Currently Considered in Designs
PRIMARY GATES++				
Radial (surface)		X	20 (70)	Yes
Radial (top-seal)		X	75 (250)	Yes
Jet-flow		X	150 (500)	Yes
Clamshell		X	60 (200)	Yes
Drum		X	10 (30)	No
Ring		X	5 (12)	No
Roller		X	10 (26)	No
Cylinder	X		100 (325)	No
SECONDARY GATES+++				
Roller-mounted	X		150 (500)	Yes
Wheel-mounted	X	X	150 (500)	Yes
Slide	X	X	50 (150)	Yes
High-pressure	X	X	75 (250)	No
Outlet	X	X	150 (500)	Yes
Ring-follower	X		150 (500)	Yes
MISCELLANEOUS GATES				
Flap		X	6 (20)	Yes
Inflatable		X	6 (20)	Yes

*This table is based primarily on existing U.S. Bureau of Reclamation installations.

+The heads given in the Maximum Design Head column are not absolute. They may be exceeded if appropriate consideration is given to seal arrangement, energy dissipation in the fluidway, and gate and structure vibration induced by flow.

++Primary gates are required to control the flow of water through a dam for flood control or reservoir evacuation and to shut off the water supply or power outlets in case the outlet ruptures.

+++Secondary gates are equipment often used for emergency closure in the event that the primary gate, valve, or other flow control device fails to close. Secondary gates also are used to dewater or facilitate maintenance of the primary equipment or other equipment whose function is auxiliary to the primary equipment.

pressure force on the skinplate. Pedestals anchored to either side of the water passage house the trunnion bearings located on the horizontal axis of the cylinder. The size of a radial gate is limited because designing for deflection and rigidity makes large gates very heavy. Radial gates have been designed as large as 15 meters wide and 20 meters high (50 by 70 feet) and 33 meters wide by 9 meters high (110 by 30 feet).

Surface radial gates are radial gates that are installed so that the maximum water surface is at the top of the gate. They are used to regulate releases under relatively low head. Wire rope hoists, chain-type hoists, or hydraulic hoists operate the gates. Electric motors or internal combustion engines can drive the wire

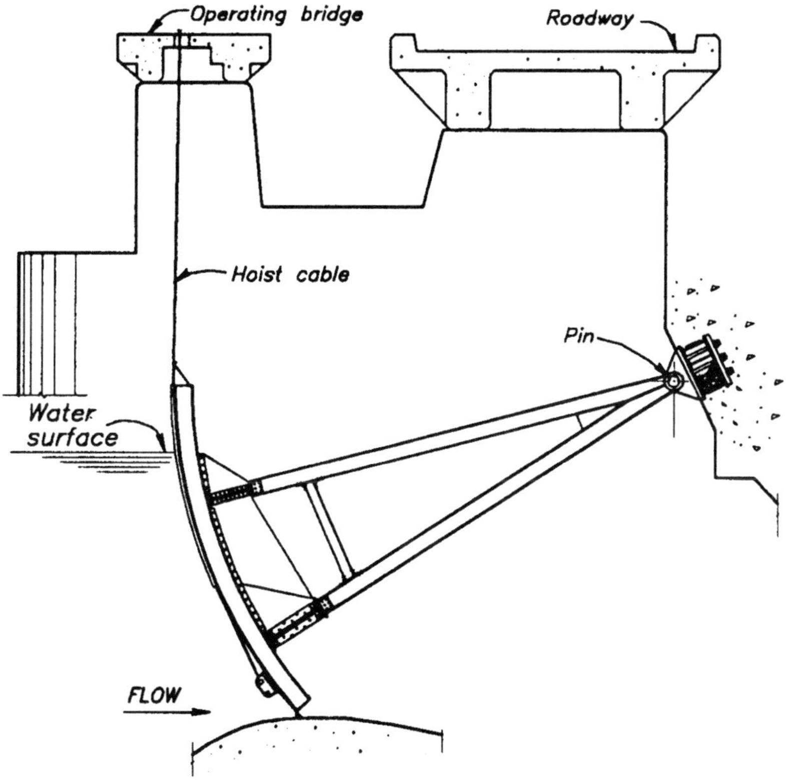

FIGURE 7-1: ***Radial Gate.*** *Radial gates have the skinplate in the shape of a portion of a cylinder with an axis horizontal. The size of a radial gate is limited because designing for deflection and rigidity makes large gates very heavy. Radial gates have been built as large as 15 meters wide by 20 meters high (50 by 70 feet) and 33 meters wide by 9 meters high (110 by 30 feet).*

---

rope or chain hoists.

A few radial gates have been installed in openings below the water surface to regulate flow in high-head installations. These are top-seal radial gates. In these installations, the gate must seal across the top as well as along the sides and bottom. Top-seal radial gates are difficult to seal across the top and at the corners. Top-seal radial gates have been designed for heads up to 135 meters (450 feet).

Theoretically, surface gates have more than enough weight to use gravity to close against the force of flowing water. Usually, surface gates have an electric brake to control the rate at which a chain or wire rope hoist closes the gate. These brakes require an outside power source. Depending upon the head and gate design, some top-seal radial gates may require assistance from the hoist to push them closed.

Additional weights may partially counterbalance the weight of a radial gate and arms. This would reduce the size of the required hoist. Some spillway radial gates have been counterbalanced with floats, on the upstream face, that open the gate automatically when the reservoir water surface exceeds a preset elevation.

The surface-type radial gates open below the water surface like roller- and wheel-mounted gates. They do not pass floating debris or ice without a considerable waste of water, although they can be designed with a hinged flap on the top edge that rotates down and allows debris and ice to be passed downstream. In cold weather applications, ice prevention systems should protect the radial gates from ice loads and from freezing closed. Ice forming on the hoist chain links or roller links on the hoist sometimes is a problem.

A certain amount of torsional rigidity must be designed into a radial gate so that the gate will not twist and bind within the gate structure. The face plate of a radial gate offers little torsional resistance to the deflections associated with unequal movements of the support arms. When building radial gates, designers should consider various factors as ways to make up for the lack of rigidity, such as adding structural members to increase torsional rigidity, synchronizing the gate hoists, and using guide shoes.

A box section, used as a gate beam, will provide more torsional rigidity to the gate structure than the more common beams and girders used to stiffen the face plate. With a box section the torsional rigidity is about its longitudinal axis, which is horizontal and normal to the direction of flow. The box beam also adds rigidity across the width of the gate, but to a much lesser degree than on the longitudinal axis.

There are examples of gates designed with generously sized box beams to stiffen the face plate. The added torsional resistance of a horizontal box section would vary considerably depending on the gate proportions. It is possible to design a small gate to be sufficiently rigid so that a single hoist operating on only one side of the gate can raise it. For a narrow (tall) gate, added torsional resistance may be sufficient (with guide rollers or guide shoes) to allow raising or lowering the gate from only one attachment point on the side. For a square gate, such one-sided operation may or may not be possible. For a wide (short) gate, the added torsional resistance of a box beam would be negligible, and it is very unlikely that this gate could be raised from one side without binding the gate in the gate bay.

Unequal lifting forces can cause a non-level raising of a radial gate and induce twisting that might cause permanent distortion of gate members. Unequal frictions or other resisting forces at either side of the gate could cause unequal lifting forces. However, if the gate is raised in a level manner from equal travel of the hoisting elements, gate twisting and distortion are not induced.

It is necessary to synchronize the gate hoists to eliminate the twisting of the gate in the gate structure. Wire rope hoists provide for a level gate lift, even though operating forces from one side of the gate to the other may vary significantly. A common load split is to consider 40 percent of the load at one side of the gate and 60 percent at the other side. Hydraulic hoist systems are not inherently synchronized. They could develop significantly different operating forces unless the rod travels are controlled or the volumes of oil pumped to each hoist are equalized.

The hoist design can influence how the gate responds to pulsating hydraulic forces that may occur when the gate is discharging at low flows. High stiffness in the hoisting system can help reduce a tendency to chattering. (Chattering is an audible or visual low-frequency vibration of a gate.)

Large radial gates usually have guide shoes at either side of the gate to:

- Restrict lateral motion between gate bay walls;
- Prevent metal-to-metal contact between wall plates and gate leaf;
- Limit seal deflection and ensure seal contact with wall plates; and
- Minimize vertical friction forces due to lateral operating thrusts from sway or non-vertical lifting.

In general, a radial gate should have guide shoes at either side to ensure that lateral gate movement in one direction does not open a gap between the side seal and the wall plate at the other side of the gate. On some small radial gates, the stiffness of the side seals is sufficient to allow the side seals to act as guide shoes.

In actuality, guide rollers are not practical since the bushings freeze because of corrosion, icing, or accumulating debris. If this occurs, the roller slides on the wall plate like a fixed shoe.

Vibration has been a problem on gates with improperly designed bottom leaf seals and beams. The vibration in a radial gate most likely comes from submerged flow conditions (water surface on both sides of the gate) rather than free discharge conditions (water pressure on only one side). Turbulent flow through a partially open gate can produce eddy currents that may cause differential pressures on surfaces of gate leaf components such as beams and seals. These differential pressures can cause the gate to vibrate. A well-designed radial gate should have the following features:

- The lowest edge of the gate leaf should provide a definite point for flow control rather than a curved surface along which the control point could shift. A continuous surface such as the bulb of a music note seal may cause problems because of the lack of a fixed control point. The square edge of a rectangular seal provides a distinctive control point;
- If transverse beams are used on the downstream side of the gate leaf, the lowest one should not be a wide flange (WF) or I beam. The lower downstream beam flange forms a pocket similar to a gate slot, induces eddy currents that may cause pressure differentials on the beam, and results in gate vibration;
- The lowest transverse beams, on the downstream side of the gate leaf, should have holes drilled through the web to allow for equalizing pressure pulsations across the beam; and
- The minimum gate opening should be large enough to avoid the effects of short tube flow. Short tube flow is when the ratio of the gate opening to the flow

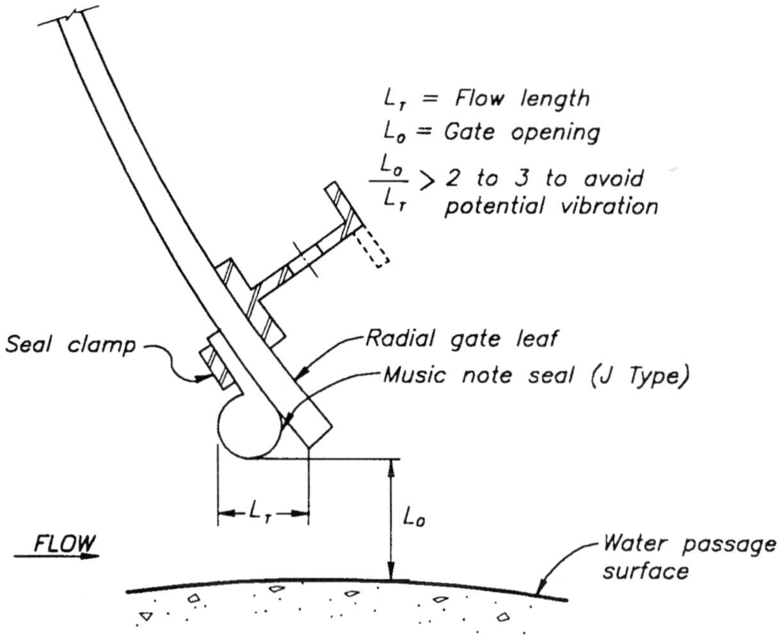

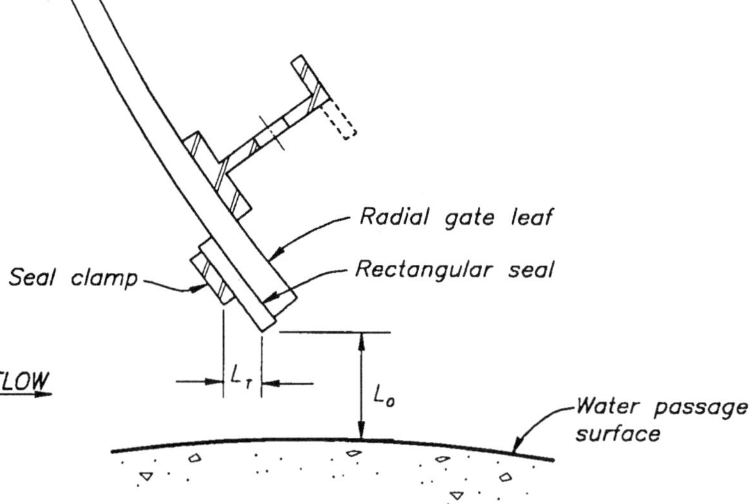

FIGURE 7-2: ***Radial Gate Arrangement for Submerged Flow.*** *Short tube flow, which can cause gates to vibrate, occurs when the ratio of the gate opening (L_O) to the flow length (L_T) is less than 2 or 3. The ratio of the opening of the gate to the thickness of the seal and skinplate arrangement, in the direction of flow, should be greater than 3. This will eliminate short tube flow effects and ensure that the gate opening acts like an orifice. Both a music note (J type) seal (top) and rectangular seal (bottom) are shown to differentiate where L_O and L_T are measured.*

length is less than 2 or 3, as shown in Figure 7-2. The ratio of the opening of the gate to the thickness of the seal and skinplate arrangement, in the direction of flow, should be greater than 3. This will eliminate short tube flow effects and ensure that the gate opening acts like an orifice. Short tube operation can cause pressure pulsations that may induce gate vibrations. However, operating at smaller openings is acceptable if gate vibration does not occur.

Cavitation damage to the bottom of the leaf can result when operated at small openings.

A properly designed radial gate requires only minor maintenance such as inspecting and replacing the seals, wire ropes or chains and lubricating roller links on the hoist. However, ice forming on the links is sometimes a problem during operation in a cold environment.

Jet-Flow Gates

The jet-flow gate acts as a regulating gate under either free or submerged discharge conditions. The gates can be operated under heads up to 150 meters (500 feet) for free discharges, and have been used for submerged discharge under 75 meters (250 feet) of head. The gates can be operated at any opening without sustaining cavitation damage.

The jet-flow gate is a good choice for flow regulation because it is easier to design for the forces it encounters than the forces a fixed-cone valve encounters. Historically, the jet-flow gate has had less vibration-related problems than a fixed-cone valve. Because of the simplicity and excellent flow regulation characteristics, use of these gates has expanded, and gates ranging in size from 150 to 2,400 millimeters (6 to 96 inches) in diameter. Table 7-2 provides a comparison of the jet-flow gate and the fixed-cone valve.

The jet-flow gate, as seen in Figure 7-3, consists of a flat, rectangular-shaped leaf, encased in a steel body. It has a circular, cross-sectioned fluidway opening in

TABLE 7-2: ***Comparison of Jet-Flow Gate and Fixed Cone Valves***

Jet-Flow Gate	Fixed Cone Valve
Not widely commercially manufactured	Commercially manufactured
	Less expensive
Excellent flow regulation	Excellent flow regulation
Will regulate to smaller openings and will not cavitate	Will not regulate to openings of less than 5 % open without the possibility of vibration and cavitation problems
Requires less maintenance	
Will operate in the submerged condition with less vibration	
Less ice buildup in cold weather	

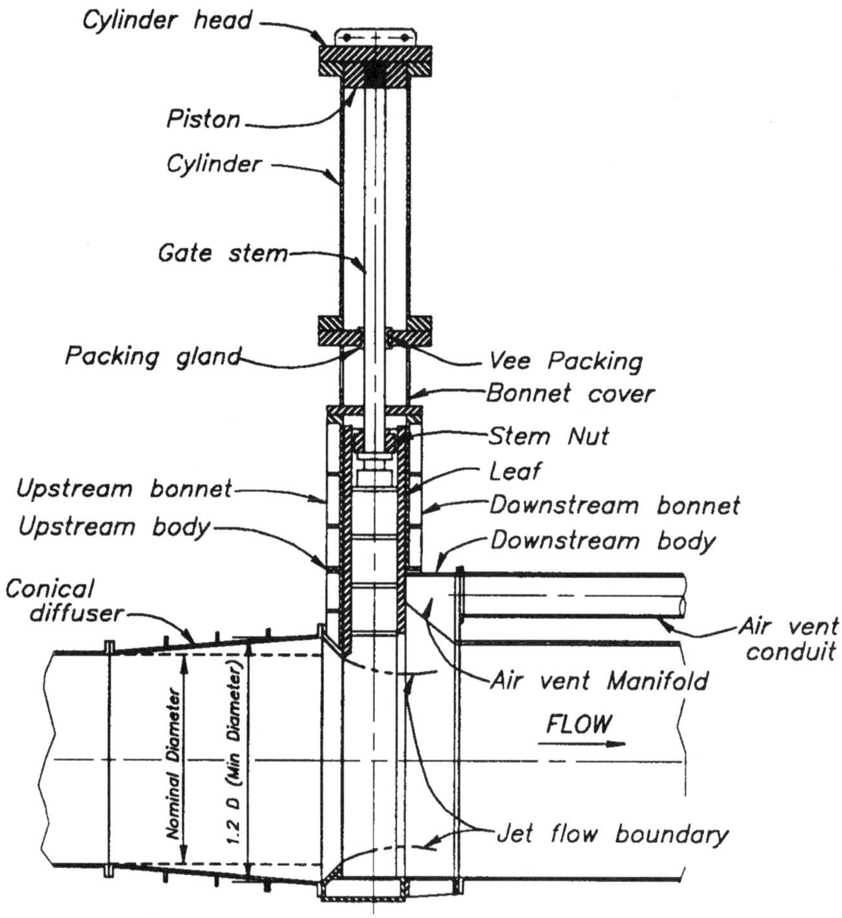

FIGURE 7-3: *Jet-Flow Gate.* *This figure shows a jet-flow gate that consists of a flat, rectangular-shaped leaf encased in a steel body. The fluidway is one of the unique features of the jet-flow gate. Starting at the upstream circular opening in the body, the fluidway forms a nozzle in the shape of a truncated 45-degree cone. The nozzle in the fluidway causes the discharging jet to contract and spring free of the slots in the body (referred to here as the jet flow boundary).*

the upstream body and a modified circular, cross-sectioned opening in the downstream body. The top of the body connects to a bonnet with a bonnet cover upon which the operating hoist is mounted.

The fluidway is one of the unique features of the jet-flow gate. Starting at the upstream circular opening in the body, the fluidway forms a nozzle in the shape of a truncated 45-degree cone. The inlet diameter of the nozzle is at least 20 percent greater than the outlet diameter of the nozzle. The nozzle in the fluidway causes the discharging jet to contract and spring free of the slots in the body.

The shape of the opening in the downstream body varies; its cross-section is either circular, horseshoe-shaped, or egg-shaped. When the gate is used under free discharge conditions, the downstream body is approximately 20 percent larger in cross-sectional area than the nozzle outlet. When used under submerged discharge conditions, the diameter of the downstream body should be approximately three times the orifice's diameter to prevent gate vibrations and surging.

The leaf is contained in slots on either side of the body. Moving the leaf down to intercept the discharging jet regulates the flow. The seal arrangement is a unique feature of the jet-flow gate. A floating seal, actuated by water pressure, is mounted at the downstream end of the nozzle and rides against the flat upstream face of the leaf. The jet-flow gate design eliminates the installation problem associated with keeping a nearly perfect plane on the seal/seat arrangement for the high-pressure and outlet-type of regulating gates.

The gate must be sized so that the diameter of the upstream conduit or pipe is equal to or greater than the orifice's diameter. When the conduit diameter is smaller than the large (entrance) diameter of the nozzle, it must have a conical diffuser with an included angle (an angle enclosed between two lines or surfaces) of about 7 degrees.

Early versions of the jet-flow gate used wheels in the leaf that rode on tracks in the body to support the hydraulic load acting on the leaf. The wheels had electric-motor-driven screw hoists to operate the leaf. The body and bonnet were embedded in concrete. Later versions eliminated the wheels and tracks. Metal sliding on metal supports the hydraulic load, and a hydraulic hoist moves the leaf. Under free discharge conditions, it is optional in most installations whether the body and bonnet are embedded in concrete. This is because the seal is on the upstream side of the leaf, and the bonnet and downstream body are not subjected to water pressure.

There is no good way of predicting the vibration magnitude and forcing frequency from the submerged release of high pressure water. Pressure pulsations from submerged water releases may or may not be destructive to equipment or structures. One installation uses twin, side-by-side, 1,200-millimeter (48-inch) jet-flow gates discharging 3.4 to 14 meters (11 to 46 feet) below the surface of the tailwater pool surface. Vibrations and cavitation noises were reported when these gates operated. After extensive examinations and vibration measurements were made, it was concluded that no damage was occurring to the gates or their associated structures. The tailwater pool absorbed most of the total energy of the submerged jets. However, this energy creates a great deal of turbulence in the tailwater pool and causes some vibration of the power plant and retaining walls because of the fluctuating water pressure on them. The designer should understand that equipment and structures used to regulate submerged flow may vibrate during flow discharge. Because there is no good inexpensive way of predicting vibration magnitudes of submerged flow during design, there is no substitute for robust design of equipment and structures that may be affected by that flow.

Large quantities of air will be drawn into enclosed conduits downstream of a jet-flow gate. Air-vent sizing is discussed in the "Air Vents for Gates in Conduits" section.

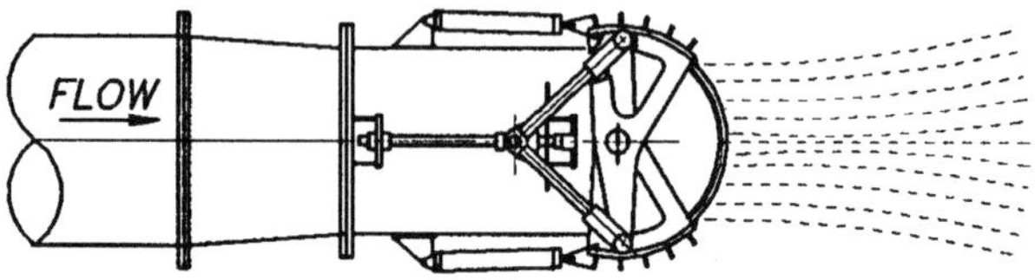

GATE IN OPEN POSITION

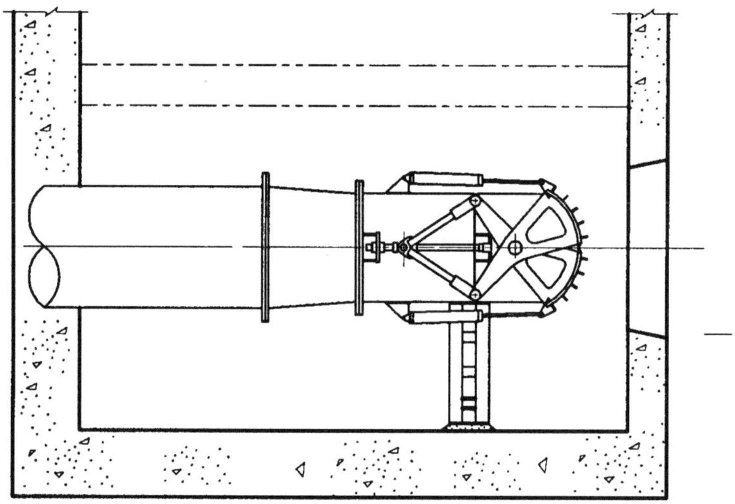

GATE IN CLOSED POSITION

FIGURE 7-4: ***Clamshell Gate.*** *The clamshell gate, shown in this figure, has a cylindrical steel section with a single radius (in the direction of the flow) machined from the centerline to the top and bottom of the cylindrical section on the downstream end with two radial segments that meet along the transverse centerline of the body. Moving the two radial segments away from each other opens the gate. The gate operates over its range of openings without vibration. This makes it ideal for regulating flows. It can be used for river outlet works and other large conduits where flow regulation is required.*

Clamshell Gates

In 1976, Thomas J. Isbester patented the clamshell gate while working in the Research Laboratory of the Bureau of Reclamation. The clamshell gate was designed to be used as a flow regulating device either under free or submerged discharge.

The body of the gate is a cylindrical steel section with a single radius (in the direction of the flow) machined from the centerline to the top and bottom of the cylindrical section on the downstream end, as seen in Figure 7-4. The operating members are two radial segments that meet along the transverse centerline of the body.

Moving the two radial segments away from each other opens the gate. The gate opening is symmetrical about the transverse centerline of the valve body. Trunnion pins attach the radial segments to the cylindrical body. The radial segments rotate around the trunnion pins. The pins are fixed to the outer wall of the body at the centerline, 180 degrees apart, and extend perpendicular to the longitudinal centerline of the valve body. The gate operates over its range of openings without vibration. This makes it ideal for regulating flows. It can be used for river outlet works and other large conduits where flow regulation is required.

The seal for the gate is mounted in the valve body. The radial segments slide over the seals as they open and close. Any of a number of seal shapes can be used, including wedge and "music note" shapes. Leakage may occur between the mating surfaces of the radial segments, especially if the gate operating mechanism does not continually have hydraulic pressure on it. This will occur in remote areas where the hydraulic power unit is not operated continually or where accumulator tanks are not used.

The operating mechanism can be designed in many different ways and configurations. Hydraulic cylinders are the most convenient because two radial segments have to be moved simultaneously. It is possible to use mechanical links, rotating shafts, and gears to open and close the radial segments. The force required to move the radial segments decreases as the gate opening increases.

The clamshell gate is suited for both free and submerged releases, but is not suited for in-line service unless there is an expansion chamber downstream from the gate.

Free releases. During free releases to the atmosphere, the general appearance of the jet issuing from the gate varies considerably from small to large openings. Figure 7-5 shows photographs of the flow pattern on the gate at various openings when discharging to the atmosphere. The main stream of the jet is well-guided in the downstream direction, and only thin fins, associated with the intersection of the sides of the gate body and the radial segments, are responsible for the change in appearance. The fins contain very little energy and may be confined easily with deflector walls.

Submerged releases. Submerged gate releases are possible. The thin jet from small gate openings is slowed more quickly than the thick or circular jet produced from large gate openings. Therefore, small clamshell gate openings should produce less turbulence in the stilling pool.

Tests on a model and prototype that are 760 millimeters (30 inches) in diameter at a head of 27 meters (90 feet) show no indication of cavitation damage or other problems associated with high velocity flow.

The coefficient of discharge of the clamshell gate approaches 1 in the fully open position. Table 7-3 shows the comparison of flow coefficient and physical size of the clamshell gate to other flow regulating gates and valves for the same

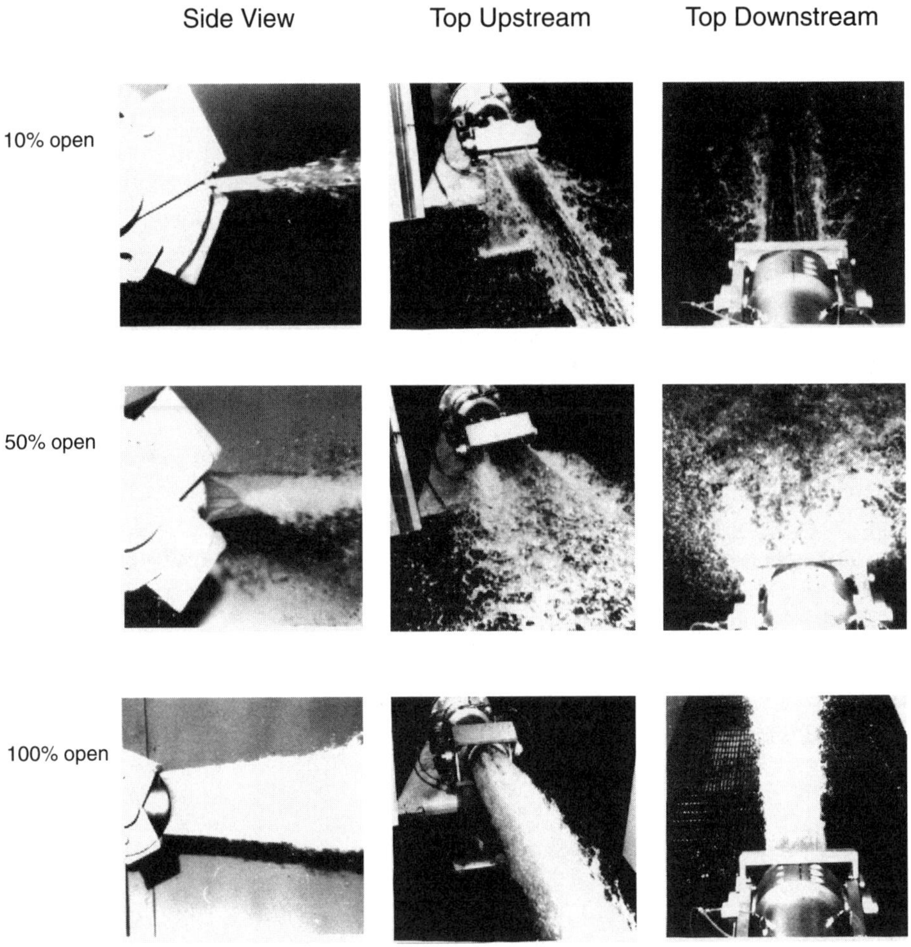

FIGURE 7-5: ***Clamshell Gate Flow Patterns.*** *These photographs show the side view, top upstream view, and top downstream view of the flow pattern from a clamshell gate at 10, 50, and 100 percent openings. These photographs, of an 8-inch model gate being tested at the Bureau of Reclamation Research Laboratory in Denver, Colorado, were taken when the gate was discharging to the atmosphere. (Photo courtesy Bureau of Reclamation)*

discharge. The reduced size of the clamshell gate is the result of the high coefficient of discharge as compared to that of other flow regulating gates and valves.

CLOSURE GATES

Roller-Mounted Gates (Stoney, Caterpillar, Tractor, and Coaster)

The roller-mounted gate is basically a flat, structural-steel gate leaf that has a skinplate, beams, girders, and a roller train system. The roller train system carries the

TABLE 7-3: ***Comparison of Flow Coefficient and Physical Size of the Clamshell Gate to Other Regulating Gates and Valves for the Same Discharge***

Gate or Valve Type	Clamshell mm (inch)	Jet Flow mm (inch)	Hollow-jet mm (inch)	Fixed cone mm (inch)	Needle valve mm (inch)
Coefficient of Discharge	0.964	0.8-0.84	0.7	0.85	0.45-0.6
Diameter	2,500 (100)	2,800-2,700 (110-107)	3,000 (117)	2,700 (105.5)	3,700-3,200 (146-127)

hydraulic load from the gate leaf to tracks embedded in concrete on either side of the water passage. This concept is applicable to a number of gates, each having a different name, but each gate is just a modification of the basic concept.

The roller-mounted gate, formerly known as a coaster gate, has one or more sets of rollers contained in continuous chains that are mounted on fixed carriages on each side of the gate leaf. The gate may be installed vertically or on a slope. It is raised to open and lowered to close. It may be installed either on the upstream face of a dam, in a gate shaft located in a tunnel, or in a tunnel plug having a gate chamber that incorporates a bonnet and a bonnet cover. Hydraulic cylinders using pressurized oil from electric-motor-driven oil pumps commonly operate the gates.

All roller-mounted gates contain a number of small moving parts in the roller trains that are vulnerable to seizing from corrosion or debris. Once a roller seizes, rolling friction becomes sliding friction, which is many times higher. If enough rollers seize, the fiction force acting on the gate will increase enough so that the hoist will not be able to raise the gate, nor will the gate weight be sufficient to lower the gate.

Later designs of the roller-mounted gate incorporate double-stem seals that are pressure-actuated from the reservoir head. The clamps and water passage ports for these seals must be kept free of corrosion, debris, and ice so the seals can extend properly once the gate is closed.

The Stoney gate was named after its inventor, F.G.M. Stoney. It resembles the roller-mounted gate except the rollers are not in the form of a continuous chain. They are mounted between rigid vertical bars at each side of the gate, but are not attached to either the gate or the guide. The rollers transmit the hydraulic gate load from the leaf to the track and move vertically between the two as the gate moves. The roller moves only one-half the distance of the gate movement. The caterpillar gate is like a roller-mounted or Stoney gate that is installed on a sloping surface—except that it is lowered to open and raised to close. The hoist stem crosses through the flow path. The gate is used exclusively in spillway applications. Screw-type hoists driven by electric motors operate the gates. The caterpillar gate, particularly the gate stem, is vulnerable to damage from floating debris and ice when used on spillways.

The tractor gate involves a more complex design. It is also like the roller-mounted gate except that the roller carriages themselves move relative to the gate leaf when the gate seats. This allows the water load on the leaf to transfer from the rollers to the seating surfaces. The tractor gate has been used as a high-head emergency gate installed on the upstream face of the dam. The hoist consists of a double-drum, electric-motor-driven unit, wire ropes, and multiple rope sheaves that are housed in cross-head members.

In the past, roller-mounted gates in sizes up to 15 by 15 meters (50 by 50 feet) have been installed in spillways and used to regulate releases from reservoirs under a relatively low head. However, they are no longer used in this way because of their initial expense and high maintenance cost. Wire rope or chain hoists operate the low head, and spillway gates are often counterweighted. These hoists required an external power source to raise or lower the gate because of the gearing in the hoist. Later applications used the roller-mounted gate as an emergency gate with a hydraulic hoist to take advantage of the closing features of the hydraulic hoist. The roller-mounted gate can be closed without an external power supply if the gate weight is greater than the sum of the friction and hydraulic forces tending to resist the gate's closure. The resulting downward hydraulic force on a gate (downpull) is the difference between the upward and downward hydraulic forces acting on the gate surfaces.

Some designs also have seals that are actuated by the reservoir water pressure. Seals may be mounted on either the upstream or the downstream side of the gate leaf.

Ring-Follower Gates (Paradox, Ring-Seal)

The ring-follower gate was designed exclusively for use as a guard and emergency gate in circular conduits or pipes at heads over 150 meters (500 feet). It was not designed to be used as a flow control valve and will sustain cavitation damage if used for this. The follower becomes a circular part of the conduit when it is raised into the open position. This creates an almost resistant-free water passage suitable for high velocity flows. The ring-follower gate is a modification of the more basic high-pressure gate. The high-pressure gate, which is rectangular in shape, requires a round-to-square inlet transition and square-back-to-round outlet transition when it is installed in a circular conduit. The ring-follower eliminates the need for transitions.

The ring-follower gate, seen in Figure 7-6, has a flat leaf with a follower attached to the bottom of the leaf. The leaf and follower are encased in a steel body having a circular water passage opening on both sides. Attached to the top of the body is an upper bonnet with a bonnet cover. Attached to the bottom of the body is a lower bonnet with a cover containing one or more drains. The body, both bonnets, and the lower cover are embedded in concrete. A hydraulic hoist is mounted on top of the upper bonnet cover with the piston stem connected directly to the leaf. The leaf and follower are moved up to open the gate and down to close it. The follower and the leaf are the same thickness and move up and down in continuous slots of equal width in the body and both bonnets. The follower has a circular opening through it that matches the opening in the upstream and downstream body halves when the gate is open. The follower opening completely covers the gate slots and, as a result, only a negligible hydraulic head loss occurs in flows through the open gate. The metal seal and seat all around the opening in the downstream body half supports the hydraulic load on the leaf. An air inlet manifold, incorporated in the top of the downstream body half, admits air downstream from the leaf during emergency gate closures.

The paradox and ring-seal gates are modifications of the basic ring-follower gate.

Paradox gate. The paradox gate is similar to the ring-follower gate in that it has a circular opening, ring-shaped follower on the bottom of the leaf, lower bonnet, and air inlets. When the leaf of the paradox gate is in any position other than seated, continuous chains of rollers in carriages carry the hydraulic water load acting on the leaf. The carriages are mounted along the sides of the leaf and follower and roll on tracks in the downstream halves of the body and lower bonnet. During closure, lugs in the lower bonnet stop the leaf movement, but the roller carriage continues to descend. The withdrawal of the inclined roller trains moves the leaf downstream to seat the metal seal ring on the leaf against the metal seat ring in the downstream body half. The action is reversed when the leaf is raised. Thus, the vertical movement of the leaf and follower is accomplished without sliding contact. The forces required for raising and lowering the leaf are relatively small. Twin, electric-motor-driven screw hoists mounted on the upper bonnet cover can supply these forces.

Ring-seal gate. The ring-seal gate also is similar to the ring-follower gate, but instead of carrying the hydraulic leaf load by sliding contact on seal and seat surfaces, it uses anti-friction, continuous roller trains (or wheels) on the leaf and wheels on the follower, both of which ride on tracks in the downstream body and bonnet halves. The most unique feature of the ring-seal gate, from which the name is derived, is the action of the seal when the gate is closed. Water taken from the conduit hydraulically propels a moveable ring laterally forming a watertight seal between the leaf and the body. On some gates, the ring is in an annular recess in the downstream gate body and is concentric with, and around, the opening in the body. The seal moves to press against the leaf. On other gates, the ring is in an annular recess in the leaf that is concentric with the opening in the body when the leaf is closed. The seal moves to press against the seat in the downstream body. Thus, the hydraulic leaf load always remains on the roller and track system when the leaf is closed. However, like the paradox gate, the force required to move the leaf is relatively small, and twin, electric-motor-driven screw hoists can be used to operate the leaf. It, too, has an air inlet manifold to admit air to the downstream side of the leaf during emergency closures.

The ring-seal gate is not suitable for operation at partial openings and was

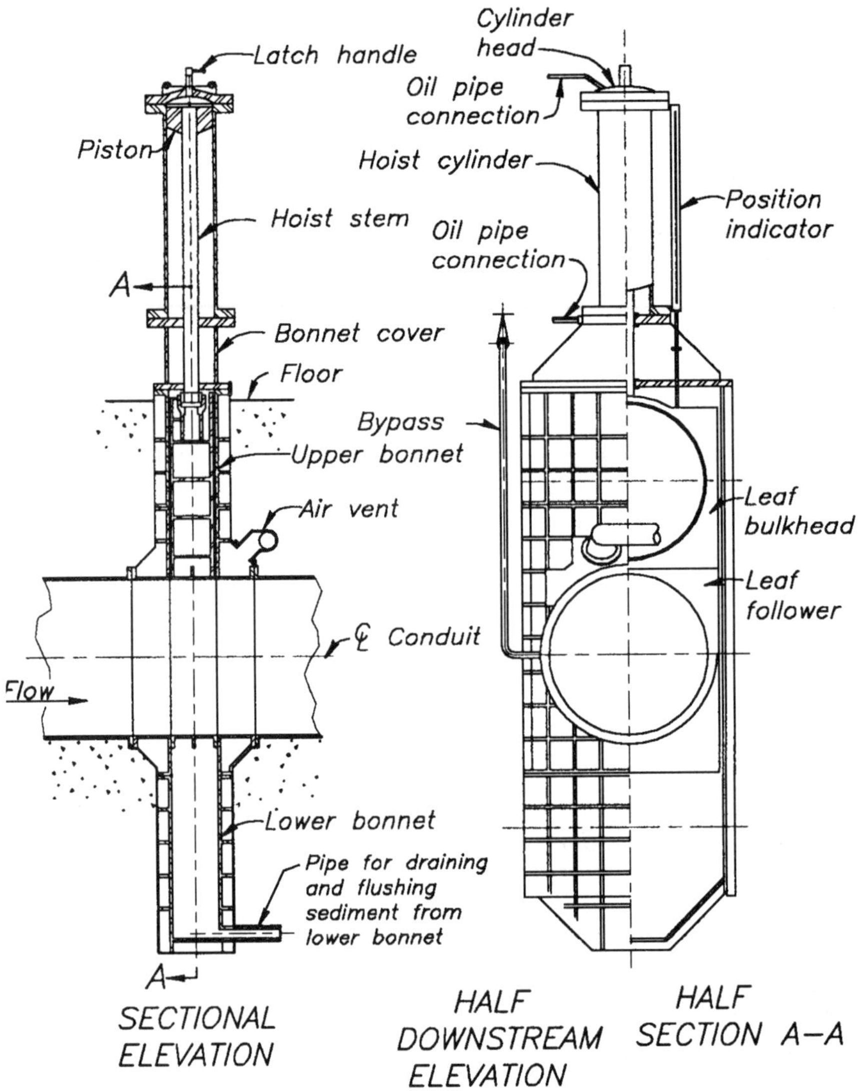

FIGURE 7-6: *Ring-Follower Gate.* *The ring-follower gate, seen in this figure, has a flat leaf with a follower attached to the bottom of the leaf. The leaf and follower are encased in a steel body having a circular water passage opening on both sides. The downstream side of the leaf of a ring-follower gate must be adequately vented during emergency closure to prevent excessive vibration or possible damage downstream from the gate. The ring-follower gate, with its sliding seal/seat arrangement, is relatively simple and rugged. It still is used mainly in outlets having circular conduits (pipes) because no transition is required and flow disturbances are minimal.*

designed to be operated only as an emergency gate. However, ring-seal gates can be used as service gates in a manner similar to that used for the paradox gates. The gates have been used by the U.S. Bureau of Reclamation under heads of 90 meters (300 feet).

Both the paradox and the ring-seal gates contain many small moving parts such as roller trains, carriages, wheels, seals, and guides that are vulnerable to seizing caused by corrosion, mineral deposits, and silt. Once the rollers start to slide instead of roll, the required hoisting effort increases radically and the screw hoists may no longer move the leaf.

The ring-seal gate relies on seal movement produced by water pressure applied to the back side of the seals, but the water system is vulnerable to clogging by corrosion, mineral deposits, and silt.

The downstream side of the leaf of a ring-follower gate must be adequately vented during emergency closure to prevent excessive vibration or possible damage downstream from the gate. The paradox gate also was designed exclusively for full open or closed operation, like the ring-follower gate. It has been used under heads in excess of 180 meters (600 feet). The paradox gate also must have adequate venting. However, paradox gates have been used as service gates that can be opened or closed unbalanced when the quantities of water handled are so large that reservoir releases can be regulated by completely opening one or more of these gates. Thus, the gates are only operated either fully open or closed.

When used within the stated limitations, all three types of gates perform satisfactorily. However, because of the high manufacturing and maintenance costs, new paradox and ring-seal gates have not been used in recent years and ring-follower gates are used almost exclusively.

The ring-follower gate, with its sliding seal arrangement, is relatively simple and rugged. It is still used mainly in outlets having circular conduits (pipes) because no transition is required and flow disturbances are minimal; however, the gate does have some design deficiencies. Because the gate has metal-to-metal seal and seat, care must be exercised during the manufacturing process and while embedding the gate in concrete to ensure that the seal and seat are in a near-perfect plane so the gate will not leak excessively.

One problem is common to all three types of gates: ring-follower, paradox, and ring-seal. When the gate is in the open position, the circular opening in the follower must align very closely with the circular openings in the upstream and downstream body halves. Otherwise, an offset in the water passage is produced that can cause cavitation damage.

Gates Used for Both Flow Control and Closure

Wheel-Mounted Gates (Fixed-Wheel Gates)

Wheel-mounted gates can be used in spillways to regulate releases from the reservoir under a relatively low head or act as emergency gates in high-head applications. They can be installed either on the upstream face of the dam on the inlet of a penstock or in a gate shaft or a tunnel plug having a gate chamber that has a bonnet and bonnet cover.

The wheel-mounted gate (also known as fixed-wheel gate), illustrated in Figure 7-7, consists of a flat structural-steel gate leaf with steel wheels. The wheel-mounted gate is similar to a roller-mounted gate. The gate leaf is made up of a skinplate, beams, and girders, but instead of a roller-train system, the leaf has steel wheels to carry the hydraulic load from the gate leaf to tracks embedded in the concrete on either side of the water passage.

Wheel-mounted gates can be designed and built as large as 15 meters wide by 15 meters high (50 by 50 feet). In the past, wire rope or chain hoists operated wheel-mounted gates that were often counterweighted. The hoists required an external power source to raise or lower the gate because of the gearing in the hoist. Recent applications use the wheel-mounted gate as an emergency gate with a hydraulic hoist to take advantage of its self-closing features. The hydraulic hoist-operated, wheel-mounted gate can be closed without an external power supply if it is designed so the gate weight is large enough to overcome the friction and uplift forces acting on the gate. Recent designs also have gate seals that are propelled by reservoir water pressure and may be mounted on either the upstream or the downstream side of the gate leaf. The clamps and water passage ports for the pressure-actuated seals also must be kept free of corrosion and debris so the seals can extend properly when the gate is closed.

Both radial gates or wheel-mounted gates are used in spillway applications. However, using a wheel-mounted gate as a spillway gate is limited because of the aesthetics and costs of the large superstructure required for the overhead hoist. Economic considerations favor using a radial gate for spillway applications. Also, when used on spillways, wheel-mounted gates are not particularly well-suited to passing floating debris without a considerable water waste. This occurs because the water being released flows beneath the leaf rather than over the leaf as it does on a drum gate or a radial gate with a top flap. Gate types may be mixed to accommodate this concern.

The wheel-mounted gate should not be used to regulate high-head flows. Surfaces immediately downstream from gate slots are subject to cavitation damage under high-velocity flows.

Wheel-mounted gates contain a limited number of moving parts (wheels and pins) and require an exact alignment of the wheels and tracks so each wheel will carry its share of the hydraulic load on the gate leaf. The wheels are vulnerable to seizing from corrosion and debris, and once they no longer turn the gate may not close or open in an emergency because of the increased friction force.

Early designs used greased wheel bearings. However, because they were relatively inaccessible, these wheels were difficult to grease and maintain. The greased wheels had the potential to contaminate the water. Using self-lubricating or anti-friction bearings have reduced these problems.

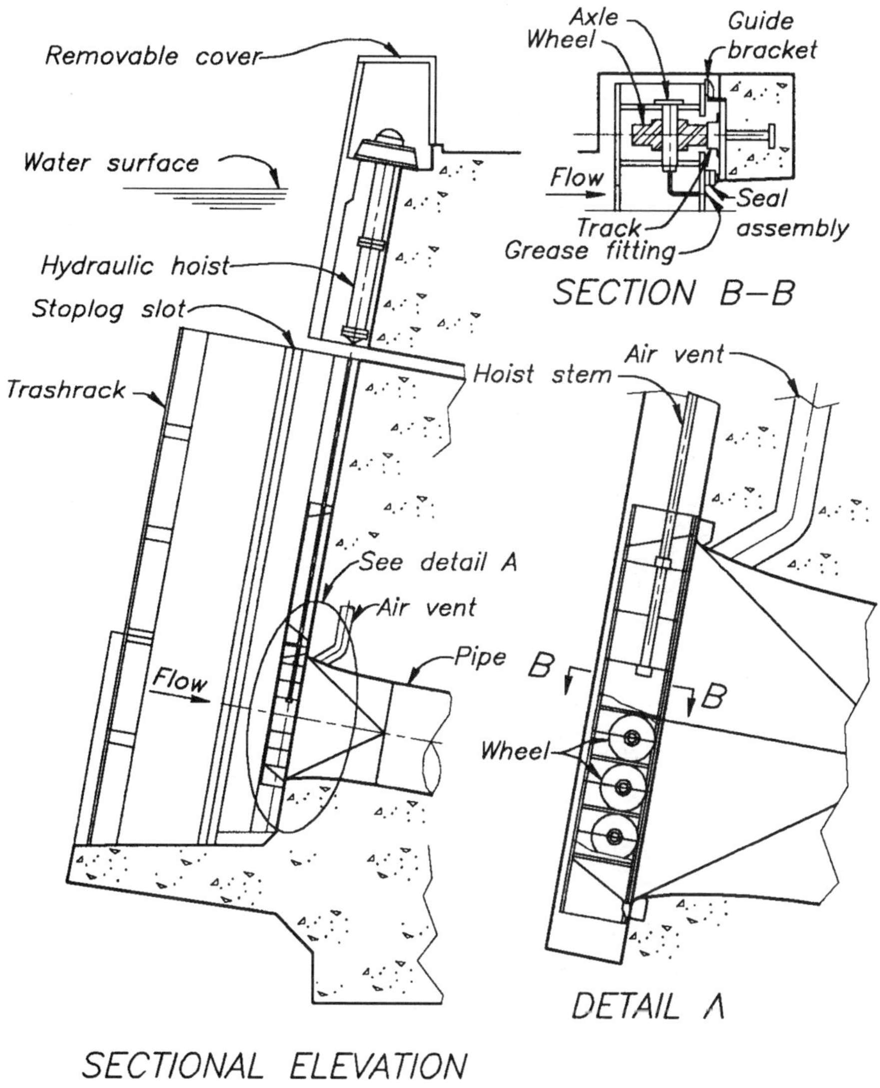

FIGURE 7-7: Wheel-Mounted Gate. *The wheel-mounted gate, shown in this figure, consists of a flat structural-steel gate leaf with steel wheels. It is similar to a roller-mounted gate. Wheel-mounted gates can be designed and built as large as 15 meters wide by 15 meters high (50 by 50 feet). Wheel-mounted gates are used in some spillway applications. However, using a wheel-mounted gate as a spillway gate is limited because of the aesthetics and costs of the large superstructure required for the overhead hoist. Detail A shows the wheels mounted on the leaf of the gate. The gate leaf is covering the entrance to the water passage. Section B-B shows the relationship of the gate leaf, wheel, seal, and track. Note the need for a grease pipe from inside the dewatered water passage.*

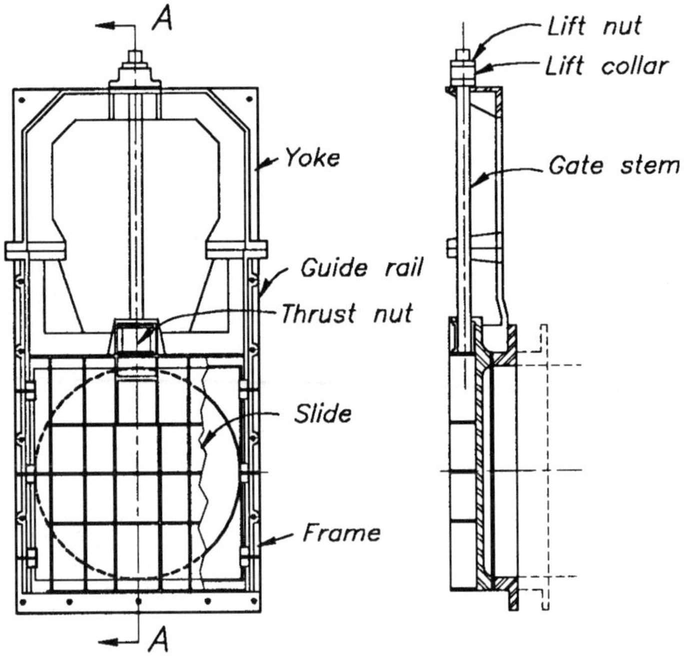

FIGURE 7-8: ***Slide Gate.*** *The slide gate, seen in this figure, is one of the simplest types of gates used to control water flow. It consists of a flat, either rectangular or square, slide (leaf) that moves within the side guides over a circular or rectangular opening that is the entrance to an outlet pipe or conduit. Slide gates give satisfactory service. However, if these gates are operated continuously, they do not perform well because of excessive seat wear.*

Slide Gates

The slide gate is designed to be installed on the upstream face of a dam or vertical shaft at the entrance to an outlet pipe or conduit. It can act as either an emergency/guard or a regulating gate under heads as high as 45 meters (150 feet). Slide gates are commonly used for sluicing applications and may be called sluice gates.

The slide gate, seen in Figure 7-8, is one of the simplest types of gates used to control water flow. It consists of a flat, either rectangular or square, leaf that moves within the side guides over a circular or rectangular opening that is the entrance to an outlet pipe or conduit. Compressive loads acting on seal seats on the sides of the opening support the hydraulic load on the leaf. Seal seats across the top and bottom of the opening also support the leaf when it is closed. Most of these gates have cast iron leaves and frames, but they can be fabricated out of steel plate. Generally, the frames are anchored to the upstream face of the dam or the upstream face of a vertical shaft. Manufacturers offer a complete range of sizes, heads, materials, and adapters for a wide range of installations. The slide gates made from cast iron are used extensively for low-head installations because they have good corrosion-resistant properties.

Either a manually operated screw hoist, a motor-operated screw hoist, or a hydraulic hoist moves the leaf, usually vertically. Stems connect the hoist (often called lift) to the leaf. Guides attached to the concrete above the opening support the stems laterally.

Slide gates give satisfactory service. However, if these gates are operated continuously, they do not perform well because of excessive seat wear.

These gates have deficiencies because of their low-cost design. One type of gate has a frame that merely bolts to the concrete face of a dam or shaft, and the joint between the two is grouted or caulked. Frequently, this joint leaks. Another type uses a thimble that is embedded in the concrete and has a machined face to which the frame is bolted. This design was developed to overcome the shortcomings of the bolt-on design. However, it often leaks, particularly if poor workmanship is involved in placing the concrete. The frames have adjustable guides, wedges, and stops that also are vulnerable to poor workmanship. If they are not carefully and precisely adjusted, the gate will leak excessively.

Slide bearing surfaces can be cast iron on cast iron, steel on steel, stainless steel on bronze, stainless steel on Teflon, or stainless steel on ultra high molecular weight polyethylene (UHMWPE). Coefficients of sliding friction can vary from 0.4 (steel on steel) to less than 0.1 (UHMWPE).

High-pressure gates and outlet gates are custom-built slide gates in which the rectangular or square leaf is encased in frames that have a rectangular or square cross-section for water passage. The tops of the gate frames connect to a bonnet. The bonnet has a cover upon which is mounted a hydraulic hoist for moving the leaf up and down. The frames and bonnet are embedded in concrete. Like the cast iron slide gate, compressive loads acting on the metal seats at the sides of a rectangular or square opening support the hydraulic load on the leaf. Seats across the top of the opening also support the leaf when it is closed. The bottom of the fluidway and the gate sill is flush with the water passage. Because the gate requires venting for unbalanced operation, the top of the downstream gate frame has an air inlet manifold to admit air downstream from the leaf.

The high-pressure gate, seen in Figure 7-9, can act as an emergency or guard gate in a tunnel, conduit, or pipe and as a regulating gate at the end of a conduit or pipe at heads of up to 75 meters (250 feet). The early designs of high-pressure gates with flat-bottom leaves are not suitable for use as regulating gates. The flat-bottom leaves are subject to damage from cavitation. If they are used as regulating gates, the bottoms of the leaves must be modified by attaching stainless steel plates to protect the existing metal.

Most problems with cavitation occur in the gates that are used for regulation. High-pressure and outlet gates, used for emergency or guard service, are generally trouble free if the leaf is held full open and does not protrude into the flow path.

The outlet gate is a high-pressure gate modified so it can be used for regulating flows up to heads of 150 meters (500 feet) as compared to the maximum of 75 meters (250 feet) allowed for the high-pressure gate.[1] Some of the modifications are as follows:

- The body, bonnet, bonnet cover, and leaf are steel only, either cast steel or steel plates and bars;
- The flatness and waviness of the steel plates used in the water passage surfaces for the bodies and liners are controlled to close tolerances in the finished and installed product;
- The offsets in the water passage surfaces at the flanged joints between the body halves and the liners also are controlled to close tolerances. This prevents the onset of cavitation and subsequent damage downstream from the offset;
- The embedment process is controlled more carefully to minimize the chances of distorting the body halves. The outlet gates must be made and embedded with extreme care to comply with close tolerances. These tolerances are difficult to obtain and, generally, problems with cavitation or leakage can be traced to noncompliant tolerances;
- Many of the areas susceptible to cavitation damage in the water passage surfaces of the body halves or on the leaf's control surfaces are either overlaid or clad with stainless steel;
- Typically, the bottom of the leaf is sloped 45 degrees relative to the flow and terminates with a sharp edge. The bottom seating surface of the leaf is kept as narrow as possible to ensure adequate aeration of the seating surface to prevent damage from cavitation; and
- The gate slots are kept as narrow as possible. The width of the water passage surface in the body half upstream from the slot has no contraction. There is a slight outward offset at the sides of the slots (3 to 13 millimeters (0.125 to 0.5 inch)) and at the top of the bonnet opening—all on the downstream sides.

MISCELLANEOUS GATES

Flap Gates

Flap gates are installed on the crest of a spillway to regulate the reservoir water elevation by controlling the discharges over the spillway. They may increase maximum reservoir storage capacity or sluice trash and ice past the dam crest. In all cases, the spillway must be capable of passing the design flood when fully lowered.

A flap gate is a bottom-hinged gate located on the crest of a spillway. In the lowered position, the upstream skinplate of the gate blends with the fixed concrete crest structure to form the spillway crest profile.

There are two types of flap gates: the torque tube and the shell. The torque-tube type, also called a Bascule gate, consists of a torque tube with a leaf extension. These are practical up to 3 meters (10 feet) high. Saddle-type bearings anchored at intervals along the length of the spillway support the torque tube. A lever arm extends from either end of the torque tube or from just one end.

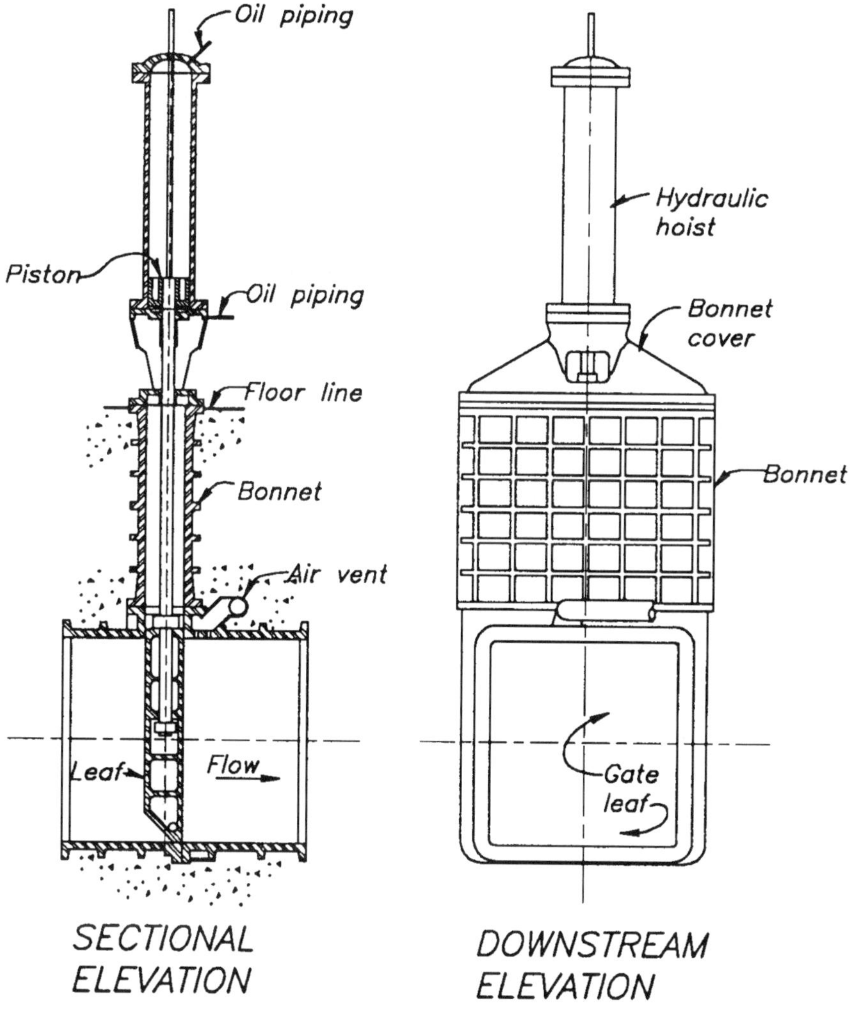

FIGURE 7-9: *High-Pressure Gate.* *The high-pressure gate, shown in this figure, is basically a slide gate. The high-pressure gate can act as an emergency or guard gate in a tunnel, conduit, or pipe and as a regulating gate at the end of a conduit or pipe at heads up to 75 meters (250 feet).*

An hydraulic cylinder attached to the lever arm operates the gate. Figure 7-10 shows an example of this type of flap gate.

Figure 7-11 illustrates the shell-type of flap gate. It consists of a closed shell formed from the upstream and downstream convex skinplates. The cavity between the skinplates has bracing and bulkhead ribs. The ribs extend through the bottom of the gate and form supports for the gate hinge pins. Hydraulic cylinder operators are located either on piers at the ends of the gate or on the

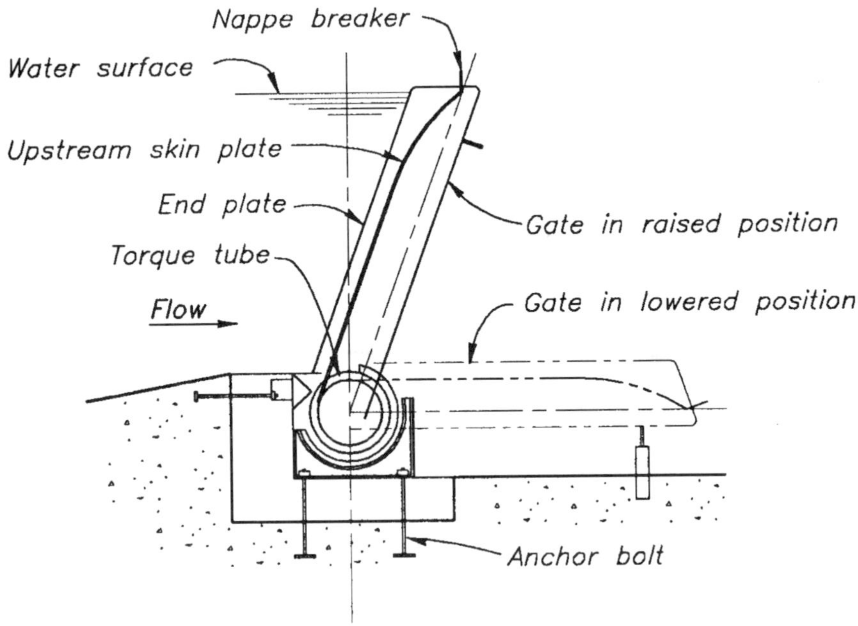

FIGURE 7-10: ***Crest Torque Tube Gate.*** *There are two types of flap gates: the torque tube, seen in this figure, and the shell (seen in Figure 7-11). The torque-tube type consists of a torque tube with a leaf extension. These are practical up to 3 meters (10 feet) high. The torque-tube type of gate reaches a practical size limitation when it becomes too costly to manufacture.*

downstream side of the gate. This type of gate has been used for heights up to 6 meters (20 feet).

The shell-type gate can be made in sections if hydraulic equipment for operating the sections can be located in the spillway structure below the crest. Then, only the spillway geometry would limit the size. The torque-tube type of gate reaches a practical size limitation when the torque tube becomes too costly to manufacture.

Although the gates are designed with nappe breakers and are ventilated at the ends of the gate, it may be possible for negative pressures to occur on the downstream side of the gate. There must be sufficient hydraulic capacity to overcome any anticipated additional gate loads due to negative pressure. The downstream portion of the spillway must be designed so the tailwater does not block off any vent pipe openings that are furnished.

The normal type of nappe breakers are rectangular plates attached at equal intervals along the top edge of the upstream skinplate to effectively break the nappe and allow air underneath it. Under certain conditions, the nappe breakers may not function and the gate may vibrate due to the change of the air pressure

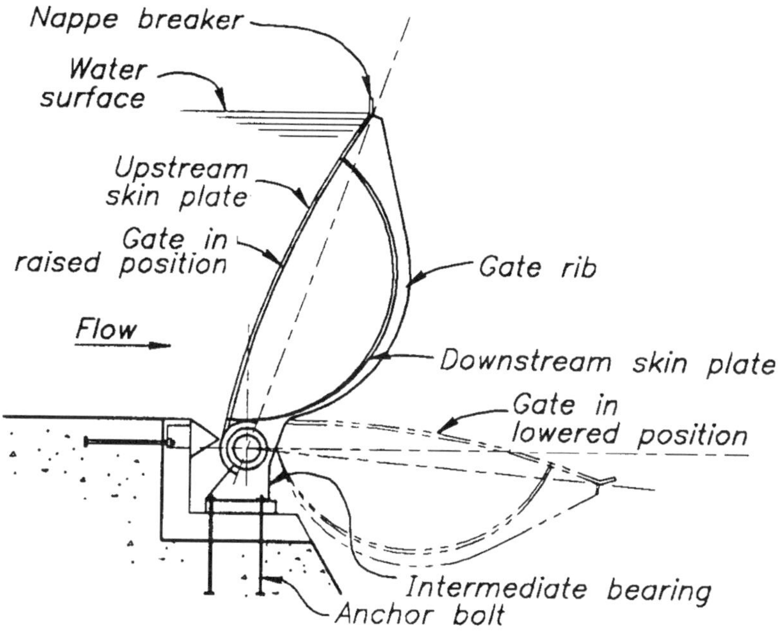

FIGURE 7-11: ***Shell (Crest) Gate.*** *This figure shows the shell-type of flap gate, which is one of the two types of flap gates. It consists of a closed shell formed from the upstream and downstream convex skinplates. This type of gate has been used for heights up to 6 meters (20 feet).*

under the nappe or the change in the velocity and direction of the flow of the nappe. If the flap gate vibrates, flow splitters may have to be added. Flow splitters are a long, deep, wedge-shaped structure mounted on the flap gate skinplate. The splitter starts upstream of the downstream edge of the gate and extends a short distance downstream, off of the downstream edge of the gate. Flow splitters are designed to break a much thicker nappe than the nappe breakers.

Even though the hydraulic control system will have a relief valve to prevent overloading the hydraulic cylinder, the site may have large pieces of ice or debris that come to rest on top of the gate. Under these conditions, it would be advisable to have a boom upstream of the gate to keep ice and large debris away from the gate to prevent overloading the gate operating system.

Inflatable gates. Inflatable gates (inflatable weir) are installed on the crest of a spillway, on the crest of a dam, bottom of a canal, or another similar water passage to increase the hydraulic height of the structure it is installed on. The function of the gate can be simple: to raise the crest height. Or, it can be more complicated: to regulate the upstream water surface elevation. Within limits, changing the pressure in the inflatable gate controls the upstream water surface level. The gate can be deflated to allow flood flows carrying large debris to pass over it unobstructed. The inflatable gate can act as a sluice gate to move unwanted deposits from upstream of the gate.

Inflatable gates are made from a long sheet of rubber or synthetic material that is formed into a cylinder by folding the long sides and ends together. A series of anchor bolts and plates fasten the folded ends and sides to the water passage, which makes a tight, inflatable cylinder that can act as a dam or gate. The gate can be readily adapted for a new structure or an old structure. Inflatable gates have been made as long as 120 meters (400 feet) and as high as 5 to 6 meters (16 to 20 feet).

Either water or air can inflate the gate. Air has the following advantages over water:

- Air requires smaller pipes for the inflation process;
- Air system may be easier to install than the water system;
- Water quality inside a water inflated gate probably would deteriorate and when released to the environment could pollute the water supply;
- Air would not freeze in a cold climate; and
- Depending on the relative size of water pumps and compressors, the air system would probably inflate more quickly than a water system.

Maintenance is low since there is no painting required to protect the synthetic rubber material. Painting the rubber dam with a white latex paint may extend its life expectancy. However, the life expectancy of an inflatable gate is less than that of a steel gate. This should be considered in evaluating the alternatives.

Flashboards are difficult and time-consuming devices to install and remove in flooding conditions and are, therefore, prime candidates to be replaced with inflatable gates. The inflatable gate can be automatically inflated or deflated relative to the elevation of the upstream water surface. Thus, if the inflatable gate needs to be deflated to pass a flood, no one needs to be present to do it.

An air-inflated gate can oscillate while water is passing over the top if the top is not designed properly. Since water is a denser fluid than air, a water-inflated gate will have a tendency to be more stable than an air-inflated gate. Some designs have protuberances built on the surface of the gate to break up the flow so that the inflated gate will not oscillate. Early designs, when deflated and laying flat, sustained damage when large flows carrying heavy debris passed over it. This damage occurred at the far downstream fold that did not lay absolutely flat when deflated. Some manufacturers have solved these problems with recent designs. Inflatable gates tend to pass more water over the center of the gate. This may not be desirable.

Air Vents for Gates in Conduits

Generally, air vents are provided downstream of emergency closure gates and regulating gates when they are discharging into a tunnel, conduit, or confined energy dissipater. Air vents permit air to be introduced inside a closed conduit to prevent sub-atmospheric pressures that can collapse the pipe or allow cavitation. The vent also permits air to escape during filling operations. The designer can use two types of vents: a valveless standpipe that extends from the conduit being protected to an elevation above the maximum water surface and a short vent pipe equipped with an air and vacuum valve on the end. Either type provides protection for the conduit during transient flow conditions or during steady-state gate operation at partial gate openings. The valveless standpipe is the most reliable because it has no moving valve parts.

There are three basic flow conditions that have to be considered while designing a vent system in a conduit:

1. Flow in a partially filled conduit;
2. Flow in a conduit where a hydraulic jump exists before the end of the conduit; and
3. Flow in a completely filled conduit.

Hydraulic jump is an abrupt rise in water surface that may occur in an open channel when water flowing at supercritical velocity changes to subcritical velocity.[2]

Flow in Partially Filled Conduits

Flow in a partially filled conduit is treated as open channel flow. The required air flow can be determined from Equation 7-1.

EQUATION 7-1:

$$\frac{Q_a}{Q_w} = \frac{A_d}{A} - 1$$

where:

Q_a=Volume rate of flow of air, cubic meters/second (cubic feet/second);
Q_w=Volume rate of flow of water, cubic meters/second (cubic feet/second);
A_d=Cross-sectional area of conduit, square meters (square feet); and
A=Cross-sectional area of water prism, square meters (square feet).

Equation 7-1 results in excessively large values for the airflow rates and can be reduced by using a more complicated sizing technique. Using this equation will result in a larger than necessary air vent, but it will operate satisfactorily.

Flow Having a Hydraulic Jump That Fills the Conduit

If the conduit is horizontal or sloping upward in the direction of flow and a hydraulic jump fills the conduit, Equation 7-2 can be used to calculate the amount of air entrained in the flowing water. Thus, this gives the airflow rate needed to size the vent. Prototype tests have confirmed Equation 7-2.[3]

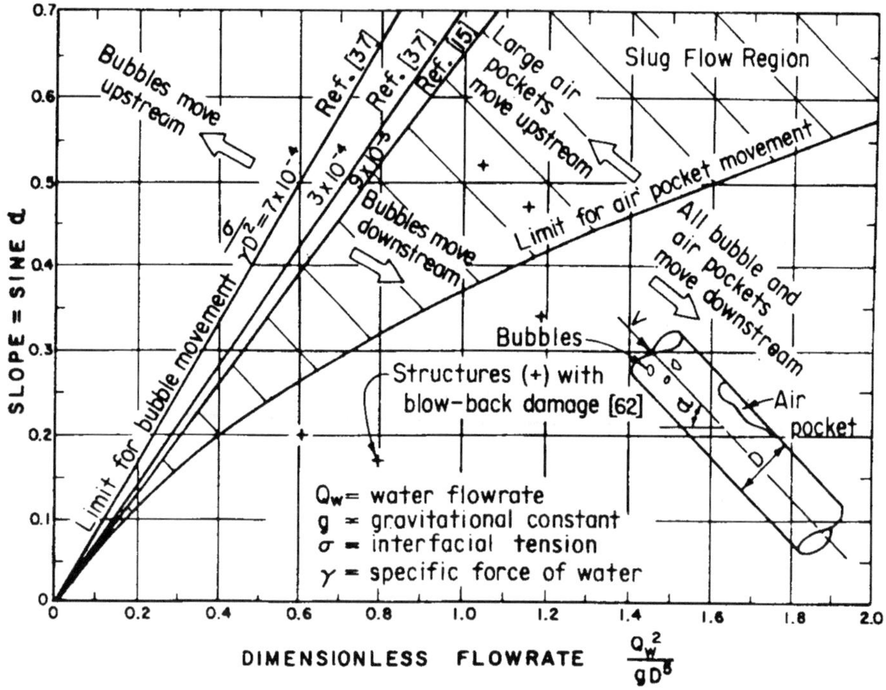

FIGURE 7-12: ***Bubble Motion in Closed Conduits-Full Flow.*** *It is important that, when designing a conduit, all bubbles and air pockets must move downstream (see lower right side of graph). Values from this graph can be used to determine the diameter of the conduit so that the entrained air will flow downstream. When the conditions in this graph are met, Equation 7-2 may be used to determine the air vent requirements.[3] (Courtesy Bureau of Reclamation)*

EQUATION 7-2:

$$\frac{Q_a}{Q_w} = 0.0066\,(F\text{-}1)^{1.4}$$

where:

Q_a=Volume rate of flow of air, cubic meters/second (cubic feet/second);
Q_w=Volume rate of flow of water, cubic meters/second (cubic feet/second); and
F=Froude number upstream of the hydraulic jump, dimensionless.

Equation 7-2 is valid only if all the entrained air is passed downstream. Use Equation 7-3 to calculate the Froude number in a circular pipe.

EQUATION 7-3:

$$F=V/(gy_e)^{0.5}$$

where:

V=Mean flow velocity, meters/second (feet/second);
g=Gravitational constant at the specific location, meters/second squared (feet/second squared); and
y_e=Effective depth of water before the hydraulic jump, meters (feet).

Use Equation 7-4 to calculate y_e.

EQUATION 7-4:

$$y_e = A/T$$

where:
A=Cross-sectional area of water prism, square meters (square feet); and
T=Top width of flow passage, meters (feet).

Use Equation 7-5 to calculate T.

EQUATION 7-5:

$$T = 2[y(D-y)]^{0.5}$$

where:
y=Flow depth, meters (feet); and
D=Conduit diameter, meters (feet).

If the conduit slopes downward in the direction of the flow, air bubbles can either move downstream or upstream relative to the pipe wall. In this case, values from the graph in Figure 7-12 must be used to size the conduit so that the entrained air will flow downstream. Then, when the conditions outlined in the graph are met, Equation 7-2 may be used to determine the air vent requirements.

Emergency Penstock Intake Gate Closure

An air vent should be provided in the penstock downstream of the intake gate. This vent will prevent a vacuum in the penstock during an emergency closure of the intake gate. The vent should be sized for an air flow equal to or greater than:

- The maximum flow of water possible that can discharge through the outlet works pipe or a Pelton turbine before the intake gate begins to close;
- The flow of water discharging through the turbine at 100 percent wicket gate opening at rated generator-capacity head for a Francis runner of less than a specific speed of 60 (English power units); or
- The flow of water at 100 percent wicket gate opening and runaway speed at generator capacity rated head (i.e., when flow through the turbine is at maximum) for a Francis and Kaplan turbine runners with specific speeds larger than 60. This flow can be as much as twice the rated capacity.

The equations and rule of thumb given in this section treat the subject of conduit venting in a simplistic manner. For a more complete discussion of air vent requirements in conduits, the reader should seek more detailed information.[3]

HOIST CAPACITY REQUIREMENTS

There are two obscure forces on a gate of which the designer has to be aware: breakaway and downpull. When determining the hoist capacity, the designer of the gate hoist must take into account the weight of the gate leaf, the buoyant force, the frictional forces, and the breakaway and downpull forces. The maximum required hoist effort to move a gate occurs in the opening direction just as the leaf begins to move. The breakaway force occurs just as the gate begins to open and is equal to the weight of the gate plus the friction resistance between the leaf and its seat or surroundings. If a gate leaf were allowed to be silted in, the breakaway force could be large enough to resist the maximum force that the hoist could generate.

The reduction of pressures resulting from the flow of water under the leaf of a gate produces downpull. This pressure can be positive or negative. Its magnitude may be many times greater than the weight of the gate leaf. It is taken as positive in the direction of a closing gate. Breakaway and downpull forces are of concern to the person who designs or furnishes the gate hoist.[4]

Stoplogs and Bulkhead Gates

Modern stoplogs and bulkhead gates look very much alike. Stoplogs generally come in more than two sections and extend above the water surface of the reservoir. Bulkhead gates generally come in one piece and totally cover the entrance to the water passage. If they are not too heavy to install and store conveniently, bulkhead gates normally are used instead of stoplogs to isolate a water passage. Stoplogs can isolate a water passage when a bulkhead gate is too large or too heavy to install or store conveniently.

Stoplogs and bulkhead gates seal the water passage from the reservoir so that it may be dewatered for inspection or maintenance. They are usually placed at the farthest extreme of the water passage so that the maximum length can be dewatered. Stoplogs and bulkhead gates usually are placed across water passage entrances or outlets under balanced pressure. They may be used in an emergency.

The term "stoplogs" comes from their original form of rough hewn timbers. The only similarity of modern stoplogs to the original is there are a number of similarly sized parts placed in slots at the entrance or exit to a water passage. Modern stoplogs are either framed with structural steel shapes and covered with a skin plate, or they are made of concrete. The edges of the ends and bottom normally have elastomer seals. The elastomer seals on the ends mate with the sealing surface on the embedded stoplog guides. The elastomer seals on the bottom mate with the sealing surface on the embedded bottom seat or the top of another stoplog. These seals have to be maintained so that they will effectively do their job. Since it is continually underwater, the embedded sealing surface in the water passage can be made of 18-8, or similar, stainless steel to eliminate corrosion.

For large openings, the upper stoplog may have a significantly less water load than the lower stoplog and may not have to be designed with as large a cross section.

Bulkhead gates usually are constructed of structural steel and skinplate. They have elastomer seals along the edges of the gate leaf. In some instances, such as in diversion tunnel closing, the bulkhead gate can be built heavily enough to be closed with flow into the water passage. Closing a water passage opening with water flowing is not a normal bulkhead gate use.

Installing and removing stoplogs and bulkhead gates requires a crane (portable, gantry, mobile, or monorail). This can be a significant capital cost when building a power plant. Careful consideration should be given to the need for a permanent crane to handle the stoplogs and bulkhead gates. Recent trends have been to rent a mobile crane when stoplogs or bulkhead gates need to be installed and removed, thus saving the capital cost of the crane. However, renting a mobile crane can be a bothersome and time-consuming maintenance event. It may be very difficult to obtain in an emergency. The size of the crane required to install and remove bulkhead gates can be a factor in deciding whether to use a bulkhead gate or stoplogs.

Normally, a lifting frame is provided to install and remove the stoplogs and bulkhead gates from the entrance of a water passage and to place them into their storage position. The lifting frame facilitates handling the stoplogs, and hoist ropes do not have to remain attached to the crane or stoplog when using a lifting frame. A lifting frame provides added height to a stoplog during installation to prevent its tilting in the stoplog slots and getting stuck so that neither the lifting frame nor the stoplog will move. The height-to-width ratio should be at least 0.60 to prevent them from getting stuck.

Needles

Needles are a vertical variation of stoplogs where members are placed side by side, footed against a sill secured to the bottom of the opening, and supported on the top by a horizontal strain member to the work platform. Typically, needles are placed by extending them horizontally over the water and allowing them to tip into the current until the bottom end swings down in contact with the bottom of the opening. They are then drawn upward until the lower end rests on the sill and are rolled sideways into place against those needles already installed. They are removed by lifting each individual member from the seat and hauling it out. It is possible to substitute steel members such as sheet piling for typical wood timbers.

Gate Seals

No matter how well a gate is constructed, there will be leakage around the gate unless appropriate seals and sealing surfaces are provided. The best gate seal will be a long-lasting, resilient, flexible body forced against a noncorrosive rigid surface without transmitting large frictional forces to the gate.

The embedded sealing surfaces can be made of 18-8, or similar, stainless steel to retard corrosion.

If the gates have to be operated in freezing environments, ice prevention at sealing surfaces is essential. Seals can be heated with heating cable or by electrical resistance methods. Heating cables can be embedded in the concrete behind the sealing surfaces. They either can be attached to the embedded side of the sealing surfaces or enclosed in a pipe that is attached to the embedded side of the sealing surfaces. If the cable is attached just to the sealing surface, when it fails, the cable cannot be replaced, and the heating effect is lost. If the cable is installed in a pipe, it can be replaced if it fails. Also, resistance heating of gate members that need protection from ice buildup is possible.

GATES NOT CONSIDERED FOR NEW DESIGNS

A large number of unique and complex designs of gates have emerged. Many of these designs have been discontinued because of lack of dependability, high maintenance costs, and high initial costs. However, many of these obsolete gates are still being used. Consequently, this chapter includes a discussion of these gates for information only. The following gates are not recommended for new designs.

Drum Gates

Drum gates are designed to be installed near the top of a dam to regulate the reservoir water elevation by controlling the discharges over the spillway. They have also been designed strictly for increasing maximum reservoir storage capacity. In both cases, the spillway must be capable of passing the design flood when fully lowered.

A drum gate, shown in Figure 7-13, is a long, buoyant vessel made of structural steel that is hinged to one side of a gate chamber, which is recessed in a broad-crested spillway structure. Drum gates 40 meters (135 feet) long and 8.5 meters (28 feet) high have been built and installed in the United States. The cross-section of the gate is shaped like an acute circular sector. Filling the chamber beneath the gate with water from the reservoir raises the gate and draining the chamber lowers it. Valves, either manually set or controlled automatically for a preset gate position, control the flows into and out of the gate chamber. The gates are noteworthy because they allow an unobstructed flow path for floating debris during floods as they are lowered to allow releases to overtop the drum.

Using drum gates results in a more costly spillway gate design. Thus, radial gates are used more commonly than drum gates.

Drum gates are buoyant vessels, and flexible hoses attached to the gate remove any internal leakage to an external drain. Failure of the skinplates or any of these hoses will allow the interior of the gate to fill with water and will cause the gate to inadvertently sink to the fully open position. This could be catastrophic if there are flow restrictions for the downstream channel.

Seals are provided all around the gate to keep the reservoir water from leaking into the chamber beneath the gate and to keep the water in the chamber from leaking out so that the gate position can be controlled.

Needle valves often are used in the chamber drains when automatic control is provided. These valves are susceptible to malfunction, which will negate the automatic features.

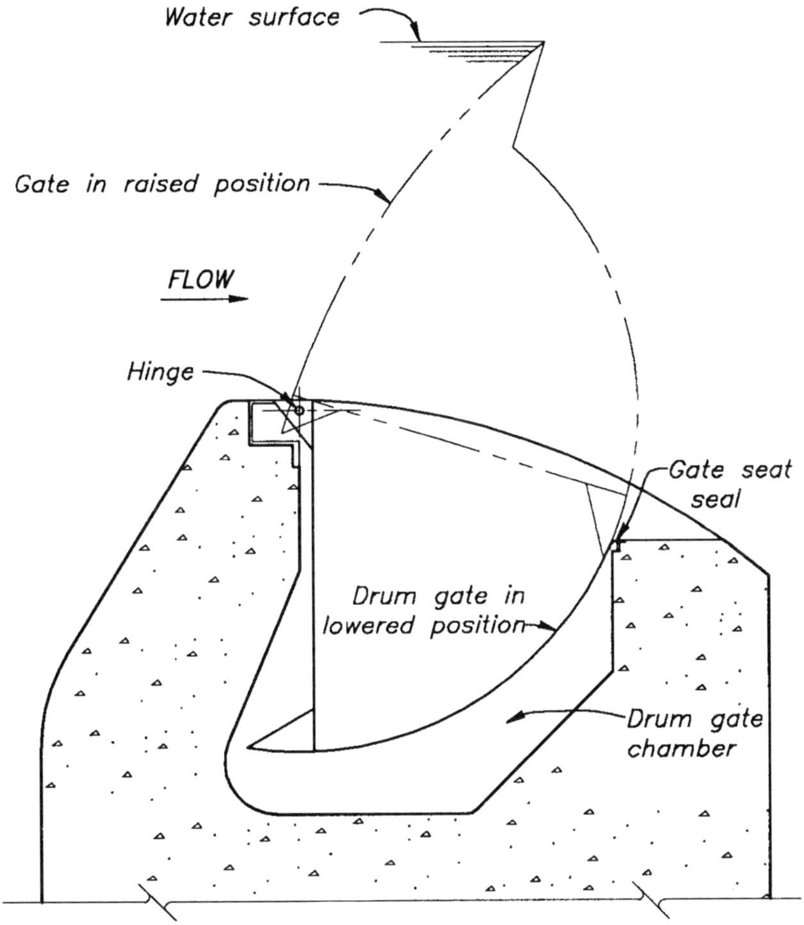

FIGURE 7-13: ***Drum Gate.*** *A drum gate, shown in this figure, is a long, buoyant vessel made of structural steel that is hinged to one side of a gate chamber, which is recessed in a broad-crested spillway structure. Drum gates 40 meters (135 feet) long and 8.5 meters (28 feet) high have been built and installed in the United States. These gates are noteworthy because they allow an unobstructed flow path for floating debris during floods as the gates are lowered to allow releases to overtop the drum.*

The gate should have flow splitters so the area below the nappe will aerate. Under some flow conditions, usually with a thin nappe, drum gates will vibrate. This occurs if there is no flow splitter to allow air behind the nappe of the water that is flowing over the gate.

Freezing temperatures and ice formation also cause problems in drum gate installations. Ice can damage the concrete or the hinge anchorage, or the gate seals can freeze the gate in one position. If the gate freezes in a raised position and the chamber beneath is drained, the gate will fall when the ice thaws. This can damage the gate or the concrete in the gate chamber.

Heavy silt loads in the water used to operate the gates can settle out, cake, harden, and prevent the gates from operating freely.

Ring Gates

A ring gate regulates flows into a morning-glory type of spillway. The gate is a buoyant vessel that floats up and down in a chamber beneath the gate. It has a ring shape in the plan view, while in profile, looks like a large oil storage tank. When the gate is lowered to its fully open position, the top surface of the gate and the top of the concrete structure form a smooth bellmouth-shaped entrance for the spillway tunnel. Filling the chamber with water from the reservoir raises the gate and draining the chamber lowers it. Water-operated needle valves can control the flows from the gate chamber. A leveling device (racks, pinions, and shafts) keeps the gate in a horizontal plane at all times. The spillway crest has seals that rub against the inner and outer diameters of the gate to prevent water from leaking out of, or into, the chamber beneath the gate. Like the drum gates, the ring gate allows an unobstructed flow path for floating debris.

Ring gates are no longer recommended to use in intake structures because:

- The gate is a buoyant vessel like the drum gate, and internal leakage in excess of the capacity of the flexible drains, or failure of the flexible drains, will cause the gate to sink. When the gate sinks with a full reservoir, it goes to the fully open position, which could be catastrophic if there are flow restrictions for the downstream channel. Maintenance on the gate is only possible when the reservoir is lowered below the spillway crest;
- The seals on the gate prevent water from leaking into or out of the chamber. Cocking of the gate can bind the seals causing them to leak. Excessive leakage past these seals will adversely affect gate control; and
- Freezing temperatures and ice formation can cause problems with ring gates if the reservoir is high enough to be up on the gate. If ice is allowed to freeze against the gate, it could crush the gate. Also, the water flowing from the chamber drains can cause icing problems in the tunnel, which, in turn, can cause seal damage or immobilize the gate.

Roller Gates

Roller gates are designed to be installed on a broad-crested spillway structure, much like a drum gate. They can regulate the reservoir water surface elevation through the spillway discharges.

The roller gate consists of a large-diameter steel roller with internal trusses. The roller is placed with the axis horizontal and perpendicular to the flow in the openings between piers on a broad-crested spillway structure. Encircling each end of the roller is a heavy annular rim with massive gear teeth and a bearing surface.

It transfers the hydraulic load from the gate to a rack with similar teeth and bearing surfaces. The rack is anchored to a sloping ledge on the upstream side of the piers. The gate roller also has an extended lip along the lower edge that seats against the spillway crest. Heavy chains are wound under and over the roller, at either one or both ends, and are connected to a chain hoist supported overhead by a hoist deck or the piers. Lifting on the hoist chains raises the roller to discharge water beneath it. This causes the roller to climb up the fixed racks. Reversing the process lowers the roller.

The racks and chains are susceptible to clogging from floating debris. The gates must be kept free of any heavy ice accumulations; otherwise, they become inoperable. Also, these gates are not particularly suited to passing floating debris without a considerable waste of water. Some roller gates are designed so they can be lowered small distances to allow ice or floating debris to pass over the top of the roller.

Roller gates have given satisfactory service; however, roller gate use is rare because of the high cost. Other gates such as radial or wheel-mounted are now used more extensively.

Cylinder Gates

Figure 7-14 shows the schematic for a cylinder gate. These gates are primarily used in intake towers upstream from dams. The use of cylinder gates dates back to the late 1800s. The moveable portion of the gate resembles a large barrel, with no top or bottom, reinforced to withstand external pressure.

The cylinder gate has a stationary casing containing sills and guides that are embedded in the concrete of a circular intake tower. The moveable barrel, called a cylinder gate, is only slightly smaller than the internal diameter of the tower shaft so it can move up or down across the rectangular, steel-lined openings embedded radially in the walls of the tower. If the cylinder gate is raised, the water from the reservoir flows through the lined entrances and into the vertical shaft of the tower that connects to the outlet tunnel. At Hoover Dam, two cylinder gates are used in the same tower with the water from the upper cylinder gate flowing through the center of the lower gate. The cylinder gates are operated by three symmetrically placed stems that are raised and lowered by individual center-drive electric motor-driven screw hoists.

The cylinder gate is satisfactory for operation when the gate is opened or closed under no-flow conditions, but it is not recommended for emergency or regulating purposes. The long stems can act as elastic members and accentuate the variable forces acting on the bottom of the gate that are induced by flow. Therefore, the gate has a tendency to vibrate when it is used to regulate flow or is closed under emergency conditions.

Since the gate barrel has to slide into a very close tolerance casing, it can become misaligned and catch in the casing. The clearances between the outside of the gate and the casing containing the opening ports have to be small to prevent

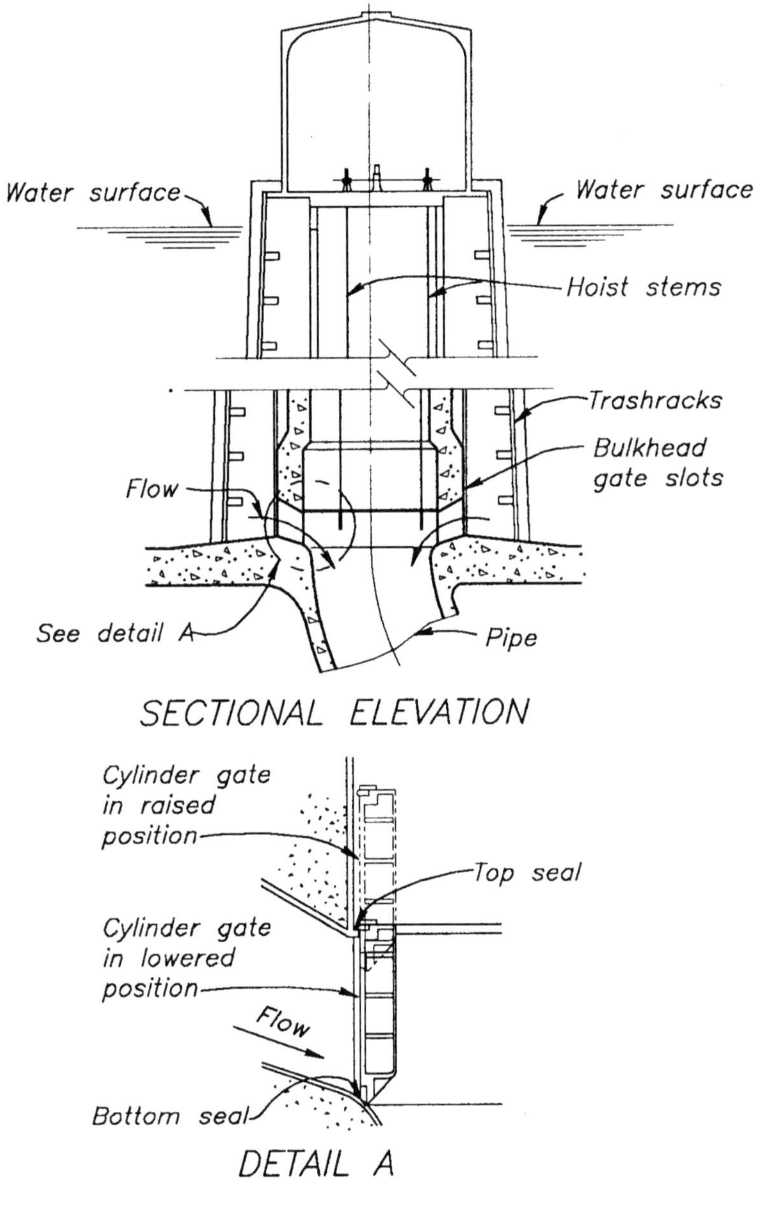

FIGURE 7-14: *Cylinder Gate.* *This figure shows the schematic for a cylinder gate. These gates are primarily used in intake towers upstream from dams. Their use dates back to the late 1800s. The moveable portion of the gate resembles a large barrel, with no top or bottom, reinforced to withstand external pressure. The cylinder gate is satisfactory for operation when the gate is opened or closed under no-flow conditions, but it is not recommended for emergency or regulating purposes.*

the gate from cocking. The multiple long stems must be raised or lowered at exactly the same rate to prevent the gate from cocking. With use, the guide shoes can work loose, increasing the clearance between them and the guides built into the tower. The increased clearance results in less restraint on the gate causing damage to the gate when it moves radially and increasing the possibility of the gate catching in the casing.

The gates were designed to have both upper and lower seating rings seat simultaneously. As there are a number of dimensions involved in this seating arrangement, it is very difficult to ensure uniform contact at the seating areas.

Some variations of the cylinder gate are still being used in place of turbine shut-off valves.

References

[1] *Guidelines for Safety Evaluation of Mechanical Equipment,* U.S. Department of the Interior, U.S. Bureau of Reclamation, Denver, Colorado, 1992.

[2] *Design of Small Dams,* Third Edition, U.S. Department of the Interior, U.S. Bureau of Reclamation, Denver, Colorado, 1987.

[3] Falvey, H.T., *Air-Water Flow in Hydraulic Structures,* Engineering Monograph No. 41, U.S. Department of the Interior, U.S. Bureau of Reclamation, Denver, Colorado, 1980.

[4] Murry, R.I., and W.P. Simmons, *Hydraulic Downpull Forces on Large Gates,* Research Report No. 4, U.S. Department of the Interior, U.S. Bureau of Reclamation, Denver, Colorado, 1966.

Bibliography

AWWA Standard for Cast Iron Sluice Gates, ANSI/AWWA C501, American Water Works Association, Denver, Colorado.

Construction papers, Uprating and Refurbishing Hydro Powerplants II, *International Water Power and Dam Construction* Conference, Zurich, Switzerland, October 1989.

Kohler, Warren H., *Selection of Outlet-Works Gates and Valves,* Reprint 1057, Presented at American Society of Civil Engineering Meeting, Chicago, Illinois, October 13-17, 1969.

Manual of Steel Construction, Allowable Stress Design, Ninth Edition, American Institute of Steel Construction, Chicago, Illinois.

Sagar, B.T.A., "How Safe Are Your Gates?" *Hydro Review,* Volume X, Number 5, August 1991, pages 56-63.

Submersible-Type Tainter Gate For Spillway, Marseilles Lock and Dam; Hydraulic Model Investigation, T.R. HL-89-1, Department of the Army, Waterways Experiment Station, Corps of Engineers; Vicksburg, Mississippi, January 1989.

Zipparro, Vincent J., and Hans Hasen, Editors, *Davis' Handbook of Applied Hydraulics,* Fourth Edition, McGraw-Hill, Inc., 1993.

Checklist for Chapter 7

1. Will the gate have to be operated under balanced pressure or unbalanced pressure?
2. Will the gate be required to regulate flow?
3. Gate hoists should be designed for breakaway loads. These loads can be two to three times the lifting load.
4. Consider corrosion between gate leaf and guides. Stray currents between the gate leaf and guide should be eliminated.
5. Is the protective coating selected for the gate appropriate for the water quality and the surrounding region?
6. Alkali-aggregate reaction in the concrete can cause swelling and gate clearances can be lost.
7. If freezing can be a problem, heaters or a deicing system, which agitates the water to keep it from freezing, must be provided.
8. Have stoplogs been designed with the smooth surface toward the water?
9. Is the gate seal design appropriate for the application?
10. If silt concentration is heavy, gate opening or stoplog removal may be difficult. Provisions for sluicing silt from the downstream side of a gate or stoplog may have to be provided.
11. Have the air venting requirements for the gate been considered?
12. Has a gate emergency closure system been provided in case of a catastrophic failure of the penstock, spiral case, headcover, wicket gates, or wicket gate operating system?

Chapter 8
Lifting Equipment

This chapter will provide an engineer, unfamiliar with cranes and hoists, the basic background to start a discussion with a manufacturer on the type of lifting equipment required to meet a particular condition.

A hydropower plant uses a wide variety of lifting equipment. Equipment can range from a simple block-and-tackle arrangement, through fixed and traveling hoists, to massive bridge and gantry cranes with capacities of hundreds of tons. Each type of lifting equipment is used in numerous configurations that, like much other equipment in a hydropower plant, is customized to some degree for its purpose.

Simple lifting devices of low capacity most often raise hatch covers and handle small items of equipment in areas out of the reach of the main powerhouse crane. Small, traveling overhead, underhung hoists are used for the same purpose and for transporting loads to areas within the service area of larger lifting equipment. Fixed and traveling hoists handle intake gates, draft tube gates, and stoplogs, and operate spillway control gates. A traveling overhead bridge crane commonly handles turbine and generator parts in all but the smallest of indoor powerhouses. Gantry cranes find a similar use at outdoor-type powerhouses. In addition, gantry cranes often handle hydraulic gates where conditions provide advantages for this type of equipment over fixed and traveling hoists. Portable and mobile lifting equipment is sometimes a good alternative for smaller loads.

The scope of this chapter does not allow a description of every type of lifting device in all its configurations. It, therefore, discusses common permanently installed devices, one or more of which is likely to be found in a visit to any hydroelectric plant. Special types of lifting equipment used when constructing hydroelectric plants, such as tower cranes and cableways, are purposely omitted. Such information will be found in references listed in the bibliography at the end of this chapter. For the same reasons, the chapter does not cover many interesting and ingenious devices designed for special purposes, such as removing turbine runners through the draft tubes, operating in confined areas, and handling equipment in powerhouses of unconventional shapes. A literature search and reference material from appropriate manufacturers and other authoritative sources can give information for lifting equipment for unusual conditions.

Crane engineers will note the omission of equipment such as DC motors, dynamic and regenerative brakes, and air-cushion devices. This is not intended to infer that the omitted items should not be used. Indeed, they can and have been used in hydroelectric plants where special conditions indicate using them, but the applications are not common. They should not be overlooked.

The ANSI/ASME B30 series of standards are an important reference for crane specification and design.[1] They cover all major types of cranes and are concerned with general construction and installation, inspection, testing, maintenance, and operating requirements. They are particularly concerned with safety, and, for this reason, their directions should be carefully followed. Also, the Crane Manufacturers' Association of America, Inc. (CMAA) Specification #70 is an important reference, but following its recommendations does not ensure compliance with applicable federal Occupational Safety and Health Administration (OSHA), state, and local regulations and codes.[2] Hydro powerhouse cranes must be designed, tested, and meet mandatory OSHA requirements for the maximum load to which they will be subjected. Additionally, the latest issues of the regulations and codes must be researched, particularly about testing requirements for installed equipment.

Various devices are available for installation on the lifting equipment to assist the operator in making a "safe" lift. However, a lack of calibration of instruments, inadequate maintenance of equipment or operator disregard can be disastrous.

Bridge Cranes

Bridge cranes are used chiefly in indoor-type hydro powerhouses. They are used for two main purposes. One is for construction—when the cranes are used to install the main hydro generating units and other items of heavy equipment. In underground powerhouses, they have even greater use in placing concrete and meeting other civil engineering requirements. The second purpose is for the maintenance and repair of installed equipment after construction is complete. Both purposes require handling a variety of loads with a broad range of hoisting and travel speeds. The construction phase constitutes a major use during the operating life of a bridge crane in large powerhouses. The crane may need to be overhauled before transfer to maintenance and repair duties.

Service requirements for powerhouse bridge cranes limit the designs to a few basic types, generally of girder or box-type bridge construction with end trucks traveling on rails and top-running trolleys. Figures 8-1 through 8-3 show commonly used configurations. Each type shown in the figures may be designed for almost any required load capacity, but individual service conditions tend to associate the types with certain load capacity ranges.

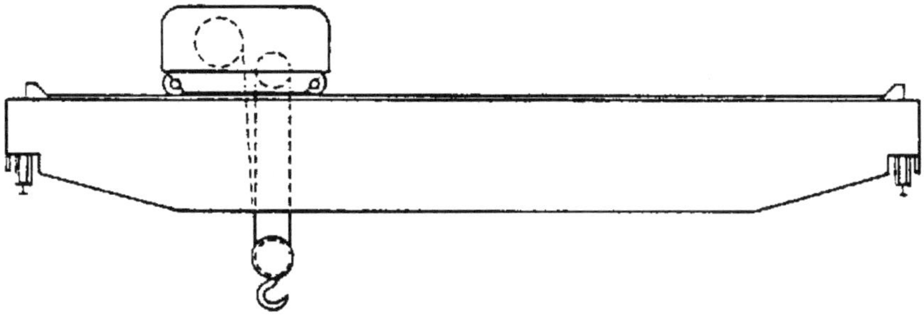

FIGURE 8-1: ***Single Trolley, Single Hoist Crane.*** *Service requirements for powerhouse bridge cranes limit the designs to a few basic types such as this traveling overhead bridge crane. The single trolley, single hoist design shown here is associated with capacities in the 4,500- to 13,600-kilogram (5- to 15-ton) range. In this case, the single hook and moderate hoist and travel speeds are adaptable for construction purposes and subsequent maintenance and repair duties in small powerhouses.*

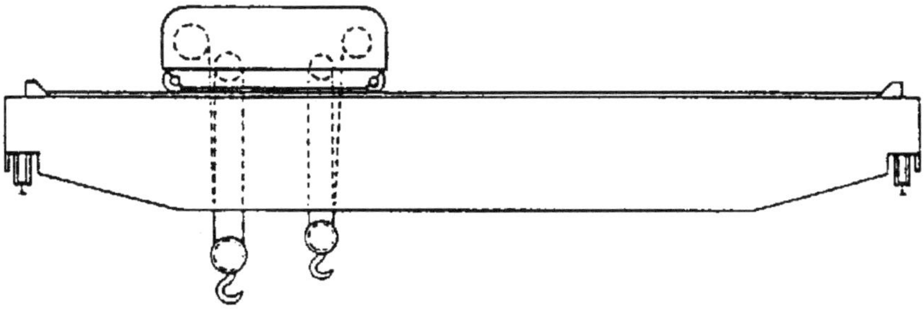

FIGURE 8-2: ***Traveling Overhead Bridge Crane Having a Single Trolley with Main and Auxiliary Hoists.*** *This traveling overhead bridge crane has a single trolley with main and auxiliary hoists. The two-hoist trolley shown is most useful when load capacity requirements are in the 13,600- to 136,000-kilogram (15- to 150-ton) load range and when low speeds are necessary for precise handling of items such as runner and shaft assemblies and generator rotors. The lower speed main hook handles the heavier loads; and the higher speed auxiliary hook, with a load capacity in the 2,700- to 23,000-kilogram (3- to 25-ton) range, is used for lighter loads.*

The single trolley, single hoist design shown in Figure 8-1 is more associated with capacities in the 4,500- to 13,600-kilogram (5- to 15-ton) range. In this case, the single hook and moderate hoist and travel speeds are adaptable for construction purposes and subsequent maintenance and repair duties in small powerhouses.

When load capacity requirements are in the 13,600- to 136,000-kilogram (15- to 150-ton) load range where low speeds are necessary for precision handling of items such as runner and shaft assemblies and generator rotors, the two-hoist trolley shown in Figure 8-2 is more typical. The low speed main hook handles heavier loads and the higher speed auxiliary hook, with a load capacity in

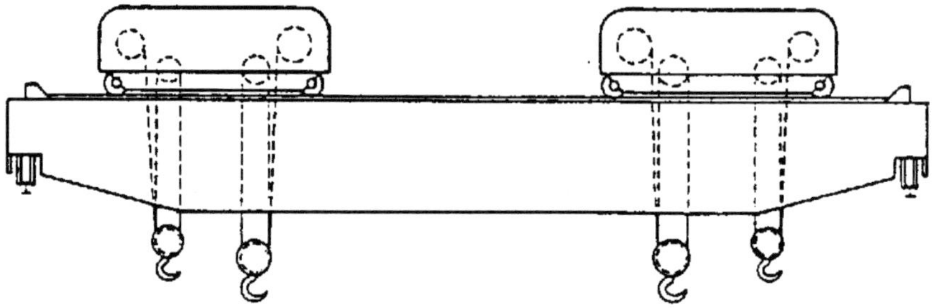

FIGURE 8-3: ***Traveling Overhead Bridge Crane Having Two Trolleys with Main and Auxiliary Hoists.*** *For load capacities exceeding 136,000 kilograms (150 tons), using two trolleys can keep the trolley size down and attain a better load distribution on the bridge. This figure shows a variation of the double-trolley design in which the main and auxiliary hoists on the trolleys are arranged in such a way that the smaller hooks can approach closer to both building walls. This arrangement also minimizes the length of a lifting beam connecting the main hooks of the two trolleys.*

the 2,700- to 23,000-kilogram (3- to 25-ton) range, is used for lighter loads. For load capacities exceeding 136,000 kilograms (150 tons), using two trolleys can keep the trolley size down and attain a better load distribution on the bridge.

The largest load, usually an assembled generator rotor, can be attached to a lifting beam connected to the hook blocks of the two hoists. Figure 8-3 shows a variation of the double trolley design in which the main and auxiliary hoists on the trolleys are arranged in such a way that the smaller hooks have a close approach to both building walls. This arrangement also minimizes the length of a lifting beam connecting the main hooks of the two trolleys.

A lifting beam connecting the main hook blocks on two separate bridge cranes can best handle loads exceeding about 450,000 kilograms (500 tons). Two double-trolley bridge cranes can suspend very heavy loads of about 900,000 kilograms (1,000 tons) and more. This configuration, in which a lifting beam connects the main hook blocks on each crane and a third lifting beam attached to the load is connected at its ends to the first two beams, forms a four-point hoist suspension. Multiple suspension points for heavier loads reduce crane size and spread the loads over greater lengths of bridge rail. This helps the design of the rail-supporting structure and the amount of excavation in underground powerhouses. There is also an advantage during the construction stage in having two cranes available for assembly work in powerhouses containing several hydro generating units.

Load ranges described above indicate only a general trend. The important point is that the crane should have the best combination of characteristics to meet the requirements of the construction and subsequent maintenance and repair. ANSI/ASME B30.2 and other references listed at the end of this chapter give much guidance in crane selection. Particular attention should be given to CMAA Specification #70 for Electric Overhead Traveling Cranes.[2] The CMAA, an organization of leading electric overhead traveling crane manufacturers in the U.S.,

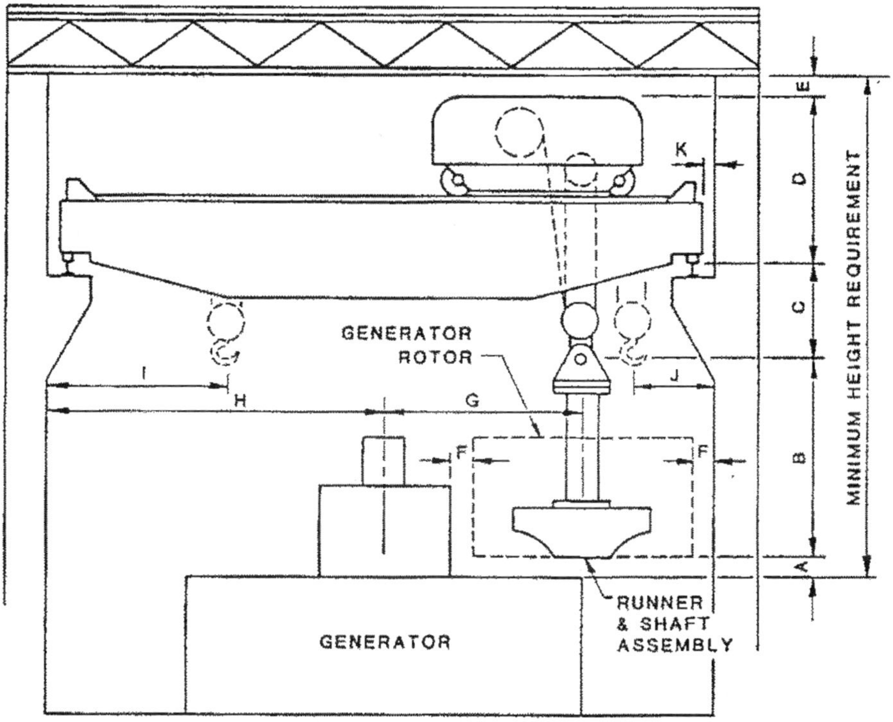

FIGURE 8-4: ***Crane Clearance Diagram.*** *This diagram can determine important dimensions for the crane itself and the powerhouse in which it operates. It is particularly important to make a diagram for a conventional-type powerhouse with an erection bay at one end and two or more hydro units requiring unobstructed passage of assemblies between the erection bay and installed positions. This figure shows a typical crane clearance diagram for such a condition. The diagram can be preliminary at first with some dimensions being finalized after the contract for the crane is awarded. A, B, C, D, and E mark the dimensions that affect the height of the powerhouse. The other dimensions, marked F, G, and H, influence the span of the crane bridge and width of the powerhouse. The hook minimum approaches, I and J, to the walls, and bridge end clearance, K, with the wall also affect the bridge span and powerhouse width.*

developed this specification for promoting standardization and providing a basis for equipment selection. Although intended by the CMAA only for providing "guidelines for technical procedure," it is unusually detailed for this type of document and is recommended as a basis for specification of hydro powerhouse cranes.

A crane clearance diagram can determine important dimensions for the crane itself and the powerhouse in which it operates. It is particularly important to make a diagram for a conventional-type powerhouse with an erection bay at one end and two or more hydro units requiring unobstructed passage of assemblies between the erection bay and installed positions. The diagram can be preliminary at first with some dimensions being finalized after the contract for crane supply is placed. Figure 8-4 shows a typical crane clearance diagram for a conventional powerhouse. In large powerhouses, the biggest items transported on the crane are usually the generator rotor and the runner and shaft assembly. Generating unit transformers and switchgear may require consideration if they are installed indoors. Mini-powerhouses may require handling the turbine and the generator as complete individual assemblies. Often, hydro units for micro-powerhouses are delivered with the turbine and generator mounted on a single base frame or contained within a unified housing. Whatever the configuration, the major consideration is for handling the load in the construction phase. If this is done, clearance requirements for maintenance and repair usually will be satisfactory.

Figure 8-4 shows the most important dimensions. A, B, C, D, and E are the dimensions that affect the height of the powerhouse. The vertical clearance A with the underside of the load should be enough to allow for transport over cover plates, bolt heads, and other low protrusions. Normally, a minimum of 150 millimeters (6 inches) is adequate. The distance B from the lowest point on the load to the saddle of the hook must take into account lifting beams, rope slings, and other devices used to attach the load. The hook is indicated in its highest position and is shown below the top of the crane rail. In some crane designs the maximum height of the hook may be above the crane rail, making distance C a negative amount in summation of the dimensions for total height requirements. Certain cranes are designed to minimize the distance D from the top of the bridge rail to the highest point on the trolley. This could have economic advantages in limiting the height of the powerhouse, particularly for an underground installation. Although, a dimension E of approximately 75 millimeters (3 inches) is required for the top of the trolley to clear the lowest obstruction below the roof structure, consideration should be given to the availability of adequate room above the trolley for convenient repair and maintenance work at an appropriate location in the crane's operating area.

The other dimensions, marked F, G, and H, influence the span of the crane bridge and width of the powerhouse. Side clearances, F, between the load and obstructions should be at least 300 millimeters (12 inches), and possibly 600 millimeters (24 inches) for larger equipment, to allow for some slight swinging of the load. Clearance F and size of the load determine the minimum distance, G, required from the centerline of the hydro units to the crane hook and the powerhouse wall. These dimensions, in combination with distance, H, from the centerline of the hydro units to the opposite wall, are principal considerations in determining the crane bridge span. Other dimensions with an effect on the bridge span and powerhouse width, depending on the installed equipment and its configuration, are hook minimum approaches, I and J, to the walls, and bridge end clearance, K, with the wall.

Approaches I and J for the main and auxiliary hooks, with their corresponding approaches to the powerhouse end walls, establish the horizontal working area for the crane. The powerhouse and crane should be designed to locate as many items of heavy equipment as possible for convenient lifting within the working area. Besides turbine and generator parts, large powerhouses may include such items as governing equipment, pumps, compressors, tanks, transformers, and switchgear. Convenient lifting can have a significant affect on the original construction time and subsequent downtime for removal of equipment for repairs, modification, or replacement.

Each crane manufacturer has developed several bridge crane designs and a standard range of sizes for each design. Manufacturers' literature contains many of the dimensions indicated in the crane clearance diagram. The crane manufacturer will recommend a rail size for the crane runway, but supply of the rail and end stops, unless otherwise specified, and design of the rail supports are for other suppliers. Design and installation of rails should meet the requirements stated in CMAA Specification #70.

CMAA Specification #70 classifies electric overhead traveling bridge cranes according to their type of service (A to F), loads (L1 to L4), and load cycles (N1 to N4).

When the crane will experience the maximum load only a few times when installing equipment during the construction phase and rarely, if ever, afterward for repair purposes, this type of service places the crane in Class A for standby or infrequent service and Load Class L1 for seldom hoisting of rated load and normal hoisting of very light loads. Load Class L2 may be required under some conditions for infrequent hoisting of the rated load and infrequent hoisting of normal loads of about one-third of rated load. The lowest Load Cycle N1 for 20,000 to 200,000 cycles is ample for this service.

For two-hoist bridge cranes, the lift (vertical hook travel) required for the main hook is usually that required to install the lowest heavy assembly of the hydro generating unit. The auxiliary hook, on the other hand, may be required to lower lighter parts and inspection and maintenance equipment through hatchways to lower levels of the powerhouse, such as the spiral case, draft tube, and draft tube access gallery.

Gantry Cranes

The gantry crane has an advantage over the traveling bridge crane in locations where no convenient structure, such as a powerhouse wall, can support an elevated rail or when providing a support solely for an elevated rail may be uneconomical, obstructive, aesthetically unpleasant, or undesirable for other reasons. As a result, gantry cranes in hydropower plants are found mostly on spillway decks and powerhouse intake and draft tube decks where they are used mainly for handling gates, stoplogs, bulkheads, and trashracks. A fixed hoist can be provided on the gantry in simple, single-purpose applications, such as raising and lowering intake gates. Usually, however, the plant has a traveling hoist moving in a direction perpendicular to the gantry tracks. The greater operating flexibility provided by a traveling hoist equipped with main and auxiliary hooks, if required, allows the crane to be used for a variety of equipment handling and powerhouse service purposes.

Outdoor- and semi-outdoor-type powerhouses with multiple generating units use the gantry crane where it can be positioned to remove deck covers and service individual units. An innovation with the gantry crane is having vertical roller doors between the legs of powerhouse gantries. This allows the generating unit, over which it is positioned, to be protected from adverse outdoor conditions during maintenance and repair periods. Bridge overhangs on powerhouse gantries can be used for gate handling and other purposes already described.

Figures 8-5 and 8-6 show the usual gantry crane design configurations encountered at hydropower plants. Figure 8-5 shows a full gantry with equal length legs at both ends of the bridge. Figure 8-6 shows a semi-gantry with travel trucks mounted on rails at different elevations. As indicated in the diagrams, both types may have one or both ends of the bridge extended to provide an overhang for additional travel of the trolley.

Except for special structural considerations in the design of the frame, most requirements for a gantry crane are similar to those for a bridge crane, and much of CMAA Specification #70 applies.[2] The gantry may have a housing on top to provide protection for the trolley and hoisting equipment in climates where snow and icing conditions occur. Outdoor-type mechanical and electrical equipment is required when the housing is not provided.

Gate Hoists

Gate hoists are commonly found on spillway decks and intake structures for raising and lowering permanently installed water flow control gates. They are found in sluiceways, fish handling systems, low level outlets in dams, and similar installations. Operation may be as simple as opening and closing a gate in which the gate is completely lowered to stop flow or is raised clear of the water passageway for unrestricted flow. Also, the hoist may be required to regulate flow by accurately positioning and holding the gate at an intermediate position. This section considers only hoists of the type normally custom designed for operating large gates, often under high hydraulic heads. Standard, commercial-type, underhung hoists or manual and motor-driven operators similar to those used for valves are the likely choice to operate smaller gates. Standard types of smaller slide gates and sluice gates can be purchased as packaged units with their hoists.

Permanent gate hoists typically are used for large spillway and intake gates. A common configuration contains a drive motor at the center that is connected to a worm gear reduction with a double-ended output shaft. Two drive shafts extend from the worm gear to each end of the gate slot where they drive two rope drums through, if necessary, additional gear reductions. Wire ropes connecting drums to the upper ends of the gate can have a variety of arrangements. Ropes reeved (passed through) in a form similar to that found on bridge cranes can connect vertical gates in slots with ample room for multiple sheave blocks. Gate slots that do not allow room for a conventional sheave block may require a special rope reeving arrangement to avoid obstructions. Multiple separate parallel ropes can be

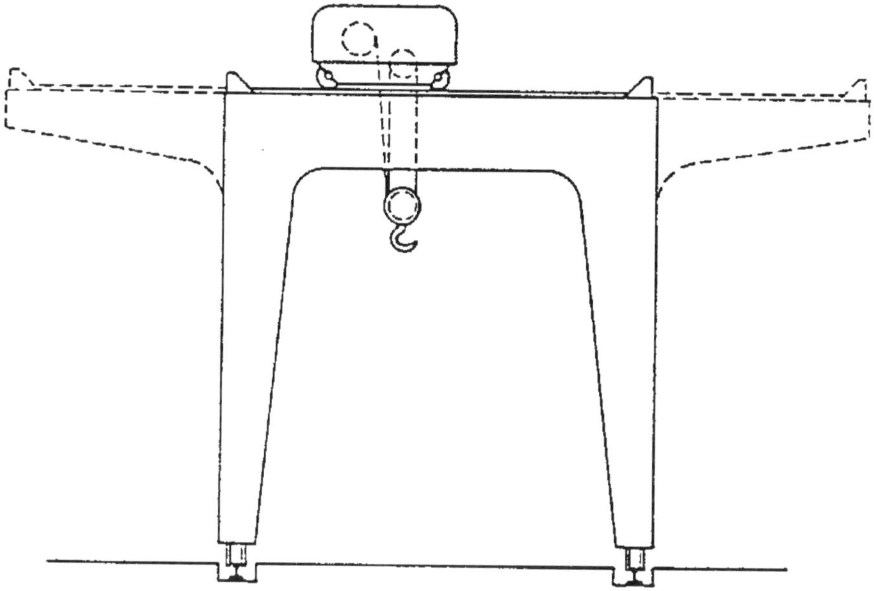

FIGURE 8-5: *Full Gantry Crane.* *This figure shows a full gantry crane with equal length legs at both ends of the bridge. As indicated here and in Figure 8-6, this crane may have one or both ends of the bridge extended to provide an overhang for additional travel of the trolley.*

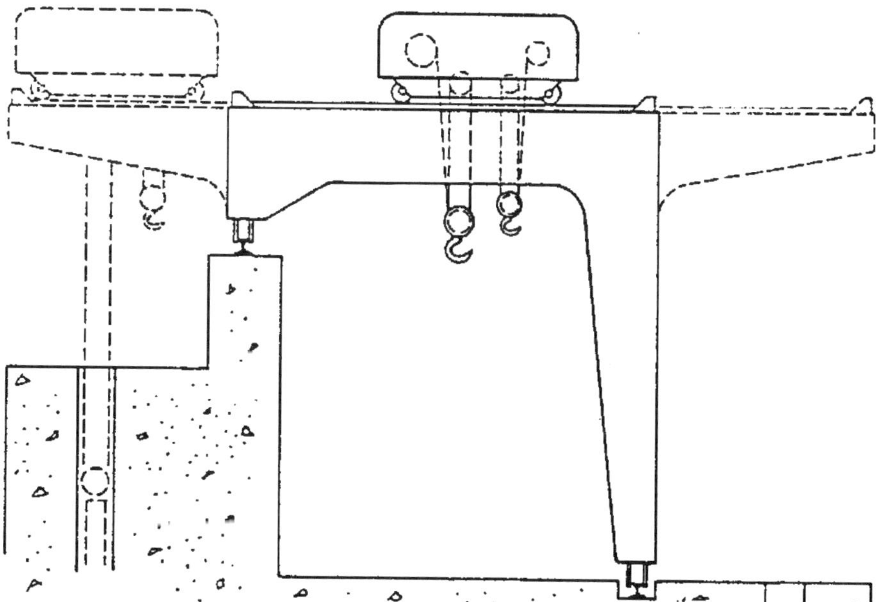

FIGURE 8-6: *Semi-Gantry Crane.* *This figure shows a semi-gantry crane with travel trucks mounted on rails at different elevations. As indicated here and in Figure 8-5, this crane may have one or both ends of the bridge extended to provide an overhang for additional travel of the trolley.*

used on radial arm spillway gates with the bottom ends anchored near the lower edge of the upstream face of the gate. This type of rope connection requires spiral winding of the rope between flanges on the drum. The ropes must be located in a position that allows submerged lengths of the rope to be held tightly against the face of the gate; otherwise, water flow patterns can cause damaging vibrations of the ropes. Where allowed by the length of the gate travel, chains can be used instead of ropes for connecting the gate to the hoist. Chains can be of the familiar loop-link or roller-chain type. Loop-link chains are wound on specially serrated drums designed to engage and cradle the links. Toothed sprocket wheels engage the roller chains, and the end of the chain not connected to the gate is allowed to hang in a loop or be folded and stored in a container. The robustness of chains allows them to resist wear and last longer than wire ropes, but they do not have the design advantages gained from various forms of rope reeving.

A traveling hoist is an economical solution for operating many gates in a multiple bay spillway. The hoist's design follows the general form of that for the trolley on a bridge crane, but it travels on rails set in the deck in a manner similar to a gantry crane. The rails straddle gate slots in the deck and allow the hoist to be positioned above any gate requiring operation. At least two traveling hoists are required at each installation for reliability of gate operation and safety. To conveniently connect and disconnect the hoist ropes or chains to and from the gate requires special devices. Where the operation is a simple open or close function, the gate remains resting on its bottom sill or is dogged at the top of the slot when not connected to the hoist. Gates that must remain partially raised to regulate flow should not remain suspended on a traveling hoist. The gate should be equipped with serrations or a series of short ribs along the vertical edges for engaging with a latching device to hold the gate in the required position. This frees the hoist for operating other gates and prevents blocking the passage of other hoists traveling on the same rails.

Large heavy gates commonly use double-drum hoists connected near the gate's upper corners. Gates of low weight and with long travels can use other configurations. A single drum positioned above the center of the gate and attached to the upper ends of the gate by the two arms of a saddle sling to equalize the pull between the sides can handle gates of about 4,500 kilograms (5 tons) or less. Where the amount of rope required for double-drum hoists becomes a design problem, gates with long travels, such as penstock intake gates near the base of high dams, can use a similar arrangement.

As an alternative to rope or chain hoists, hydraulic hoists can be used for gates of all sizes and weights. The basic component of the hydraulic hoist is the hydraulic cylinder. The design can use a single cylinder attached to the center of the upper edge of the gate, or, if there is a tendency for the gate to tilt and become wedged in its guides, two cylinders attached near the two upper corners can be used. Operating pressures of 7,000 kPa (1,000 pounds/square inch) or more are used, permitting small diameter cylinders of low visibility at deck level.

The cylinder is supported in a trunnion mounted at or just below deck level, and the end of the piston rod is attached to the gate. Where the cylinder is an obstruction on the deck, building the cylinder into the gate structure and attaching the end of the piston rod to the deck mount inverts the arrangement, thereby containing most or all of the hydraulic mechanism below deck level.

Hydraulic power raises the gate, but the gate's own weight allows it to descend with the speed controlled by throttling hydraulic fluid in the return circuit. Where required, the gate can be allowed to descend at a high rate of speed for emergency closure—even under conditions of power failure to the hydraulic controls. Ports in the cylinder choke hydraulic fluid flow during the last 5 to 10 percent of gate descent to cushion gate impact on the bottom seal and avoid undesirable hydraulic effects in the water flow. To raise the gate when a power failure occurs in the hydraulic system, there is often a hand-operated pump available.

A small amount of fluid leakage occurs in hydraulic control systems under pressure. The leakage is of no consequence with a gate resting on its bottom seal or held in a dogged or latched position because the hydraulic pressure system can be shut down. It will cause position drift, however, to a gate suspended on the cylinder in an undogged intermediate position or a fully open position ready for instant emergency closure. Providing a gate position sensor, to supply a feedback signal to the hydraulic system controls, remedies this condition. The controls can then compensate for piston drift and allow the gate position to be maintained within a small deadband.

Mounting the hydraulic cylinder above maximum headwater level and extending the piston rod with one or more links can accommodate long gate travels for deeply submerged intake gates. The links may be in the range of 9 to 12 meters (30 to 40 feet) long with the ends flexibly pinned to each other and the end of the piston rod. The cylinder is sized to lift the gate clear of the intake opening. Dogging the second link from the top, disconnecting and removing the top link, and then extending and attaching the piston rod to the second link for the next lift permits additional lifting. The process is repeated until the gate is in a position where it can be worked on or removed from the slot.

Power plant design must include adequate provision for inspecting, maintaining, and, possibly, repairing gates. If such work cannot be performed under locally dewatered conditions by inserting stoplogs on the upstream side of the gate, then the gate must be raised at least clear of the water or above deck level where it can be worked on or transported to an area provided for the work. This presents no particular problems when a gantry crane is used to operate and service the gates. The operations, however, may entail special consideration when hoists are used. Space considerations may require the hoist to be mounted on a structurally supported platform above the deck to raise the gate to the required height. A crane may still be required for removal of fixed hoists where total extraction of the gate from its slot is required. For these cases, mobile cranes are a possibility if a power plant crane is not available.

TABLE 8-1: *Operating Speeds for Floor-Controlled Cranes (in feet per minute)*

	Hoist			Trolley			Bridge		
Capacity (tons)	Slow	Medium	Fast	Slow	Medium	Fast	Slow	Medium	Fast
3	14	35	45	50	80	125	50	115	175
5	14	27	40	50	80	125	50	115	175
7.5	13	27	38	50	80	125	50	115	175
10	13	21	35	50	80	125	50	115	175
15	13	19	31	50	80	125	50	115	175
20	10	17	30	50	80	125	50	115	175
25	8	14	29	50	80	125	50	115	175
30	7	14	28	50	80	125	50	115	150
35	7	12	25	50	80	125	50	115	150
40	7	12	25	40	70	100	40	100	150
50	5	11	20	40	70	100	40	100	150
60	5	9	18	40	70	100	40	75	125
75	4	9	15	40	70	100	30	75	125
100	4	8	13	30	60	80	25	50	100
150	3	6	11	25	60	80	25	50	100

MISCELLANEOUS LIFTING EQUIPMENT

The above sections have dealt mainly with cranes and hoists for handling large heavy items. Many types of standard commercial lifting equipment for lighter service are useful for applications out of the reach of the main overhead bridge and gantry cranes in larger powerhouses. Such uses include handling pumps, air compressors, water strainers, motors, transformers, circuit breakers, and similar items located in mechanical and electrical equipment galleries. Standard commercial cranes and hoists are particularly applicable at mini- and micro-hydropower plants where the maximum load often is less than 4,500 kilograms (5 tons). This section discusses several types of lifting equipment suitable for these applications.

ANSI B30.17 deals with small overhead and gantry cranes that have a single steel beam for the bridge with an underhung hoist traveling on the lower flange of the beam.[1] Cranes of this type are useful in maintenance and machine shops sometimes associated with hydropower plants. Small gantries or A-frames are available on legs fitted with flat rim wheels or casters that are suitable for moving over level surfaces. These are worth consideration as economic alternatives to more conventional cranes in mini- and micro-hydropower plants.

TABLE 8-2: ***Suggested Operating Speeds for Cab-Controlled Cranes (in feet per minute)***

	Hoist			Trolley			Bridge		
Capacity (tons)	Slow	Medium	Fast	Slow	Medium	Fast	Slow	Medium	Fast
3	14	35	45	125	150	200	200	300	400
5	14	27	40	125	150	200	200	300	400
7.5	13	27	38	125	150	200	200	300	400
10	13	21	35	125	150	200	200	300	400
15	13	19	31	125	150	200	200	300	400
20	10	17	30	125	150	200	200	300	400
25	8	14	29	100	150	175	200	300	400
30	7	14	28	100	125	175	150	250	350
35	7	12	25	100	125	150	150	250	350
40	7	12	25	100	125	150	150	250	350
50	5	11	20	75	125	150	100	200	300
60	5	9	18	75	100	150	100	200	300
75	4	9	15	50	100	125	75	150	200
100	4	8	13	50	100	125	50	100	150
150	3	6	11	30	75	100	50	75	100

A variety of cranes of the 180-degree, rotating wall-mounted jib type and 360-degree, rotating pillar-mounted type can be useful for handling small loads within localized areas such as maintenance and repair locations for pumps, motors, and similar items of limited size. Jib cranes also can be used for handling gates and stoplogs, particularly in circular- and semicircular-shaped intake structures. Derricks may be used for similar purposes. They are not as convenient to use as jib and pillar cranes with traveling hoists, but could have an advantage in some cases where minimum cost is of prime importance.

The type of underhung hoists used on the above jib and pillar cranes can be used on monorails for handling and transporting small loads. Hooks embedded in concrete ceilings above items of equipment are often used for attaching such a hoist when required for removing equipment or lifting the item sufficiently to insert rollers or a cart underneath.

Small mobile cranes have a use in some power plants for handling manhole covers, hatch covers, sections of floor plates and grating, items of maintenance equipment, and other light work in areas where other types of lifting equipment are not available or cannot be used conveniently. The common fork lift truck can perform many of the same operations and can be used for handling power plant supplies.

Design Considerations

In addition to the type and class of lifting equipment, type and size of major equipment to be handled, size of working area to be serviced, upper and lower hook elevations, and clearance requirements, the specifying engineer for custom lifting equipment should state any special design requirements discussed in the sections below and referenced to CMAA Specification #70.

Hoist and Travel Speeds

The heavy loads and fine positioning (spotting) characteristics generally required for hydropower plant cranes and hoists result in the lowest speed ranges found for hook-and-travel movements on this type of equipment. Tables 8-1 and 8-2 contain suggested hoist, trolley, and travel speeds for floor-controlled cranes and cab-controlled cranes of various capacities.[2] These values are a good starting point when determining operating speeds, but CMAA #70 notes that consideration must be given to length of runway for bridge speed, span of trolley for trolley speed, distance of average travel, and required spotting characteristics.

Given the speed, it becomes the responsibility of the crane designer to meet other requirements by selecting the proper combination of motor speeds and torques, motor controls, brakes, gear reductions, wheel diameters for travel movements, and drum diameters and rope reeving arrangements for hoists.

Structural Design

CMAA Specification #70 describes structural design for crane bridges, gantries, and trolleys in considerable detail, and other listed references contain supplementary information. CMAA Specification #70 shows typical truck and wheel arrangements for supporting traveling bridges, gantries, and trolleys.[2] These arrangements include 8-, 12-, and 16-wheel combinations in equalizing and compensating configurations for each group of wheels. Not shown is the simple end truck arrangement for the 4-wheel combination used on light cranes. As might be expected, crane capacity, allowable wheel loads, and distribution of wheel loads on the runway rail determines the choice of design. Unless there are structural limitations concerned with the runway rail support, the crane designer should select the end-truck design and wheel assemblies.

Mechanical Equipment

Crane bridge drive arrangements. CMAA Specification #70 shows and describes crane bridge drive arrangements designated Al through A6.[2] All arrangements drive two wheels, one on each runway rail. In five of the arrangements, a system of gear reductions and drive shafts connects the two wheels to one or two common motors near the center of the system. Similar drives may be used for operating hoists. One arrangement, commonly found on gantry crane legs, uses separate motors and gear reductions at each wheel. Crane, trolley, and traveling hoist drives are usually confined to the simpler arrangements of Al, A2, and A3. Engineering advantages of one drive over another are not large, and the crane designer should be allowed to select a drive that best fits other features of the equipment.

Load brakes. Load brakes are of particular importance for safely using cranes. OSHA requires each hoist to have at least two brakes: one a holding brake at the motor (as described elsewhere in this section) and the other a load brake to assist the motor in maintaining control over the load in the lowering direction. Safety requires the load brake to hold the load, preventing a runaway condition, should the motor holding brake fail when the motor is de-energized. Control considerations call for the load brake to have a restraining action that causes the motor to require approximately the same amount of power to lower the load as to raise it.

The most commonly used load brake on larger capacity hoisting equipment is the mechanical type. It can be mounted on the motor shaft along with the holding brake, on a gear reduction shaft extension, or, more usually, incorporated inside the gear reduction casing alongside the gears. The two principal items in a mechanical load brake are a circular ratchet and pawl mechanism, and a drum or disc brake. The ratchet and brake drum or disc rotate freely in the raising direction, causing no braking action. When the motor stops or moves in the lowering direction, the load tries to rotate the ratchet in the reverse direction where it engages the pawl and is stopped, causing the brake band to contract on the drum, or the brake disc to be compressed between its flanges. Since the load applies the brake, the amount of braking action is approximately proportional to the load.

A gate hoist incorporating a worm gear reduction with adequate reverse friction to hold the load does not necessarily require a load brake. However, a designer needs experience in the design because of the variables involved. The amount of worm gear reduction is not a reliable guide. A ratio of 50:1, thought to be adequate for a radial gate hoist, was found in one case to allow the gate to drift slowly in the close direction despite a holding brake on the outboard end of the motor shaft. Because of the critical requirements for controlling river flow, the hoist was modified in the field to include a load brake between the motor and worm gear reduction.

Bridge brakes. OSHA requires bridge brakes in cab-operated cranes to have a capacity capable of stopping bridge motion within a distance equal to 10 percent of the rated load speed when traveling at rated speed with rated load. Usually a foot pedal in the cab operates the provided hydraulic brakes. The pedal should not need a force greater than 320 kN (70 pounds) to develop the required braking action. Power-assisted hydraulic brakes can be provided where required. Non-cab-operated cranes without a non-coasting mechanical drive can have variable frequency controls that gradually slow the bridge down until the spring-applied, electrically released brakes engage. These brakes are similar to holding brakes described elsewhere in this section.

Rope hoists. Rope hoists described in references for this chapter indicate the wide variety of rope drums, load blocks, rope types, reeving arrangements, and hooks that can be used to design the hoist. The designer must have some familiarity with this equipment when specifying rope hoists, but the actual combination to achieve the required hoist performance characteristics is best made the responsibility of the experienced hoist designer.

Electrical Equipment

Hoist drive systems. Two types of hoist drive systems currently are specified for powerhouse bridge cranes. One is the static, stepless, variable speed dc drive. The other is adjustable frequency flux vector ac drive. Both systems are solid state and provide precise control at slow and fast speeds. Both drive systems are programmable and can be set up to provide a no load (empty hook) speed that is faster than the full load speed. As the initial cost of either system is comparable, sometimes it is left to the crane manufacturer to determine which system to provide. One advantage of the ac flux vector drive is that it uses a standard squirrel cage motor which requires less maintenance than the dc motor. Because both systems inherently provide positive control of a lowering load, a mechanical load brake cannot be used.

Older cranes may have wound-rotor or squirrel-cage type motors which provide satisfactory discrete speed operation.

Holding brakes. Hydropower plant lifting equipment commonly uses holding brakes of the AC, solenoid-operated, shoe type. Generally they are mounted on the motor shaft and intended to prevent travel and hoist movement when the motor is not energized. There are some variations in the design, but, in principle, the brake consists of a drum on the motor shaft against which two brake shoes are applied by a spring acting through a lever mechanism. When power is applied to the motor, an AC solenoid operates through the mechanism to compress the spring and, thereby, release the brake shoes from the drum. Some designs include a hydraulic damper to cushion the application of the brake shoes to the drum, preventing a sudden grabbing action.

The AC, solenoid-operated brake is a well-proven design with the advantages of simplicity, reliability, ease of adjustment for wear, and certainty of application by the spring at all times when the motor is not operating.

Current conductors. Current conductors refers to the system of conductors along the crane runway for supplying electrical power to the crane from the structure in which it operates, and along the trolley runway for supplying electrical power to the trolley from the bridge. Two types of conductors are used: rigid (or semi-rigid) bare conductors supported on insulators and flexible insulated conductors (cable) supported in loops on a track or wound on a spring-loaded drum.

Rigid conductors are best for large capacity bridge and gantry cranes, particularly those with long runways, because there are often design difficulties in handling the size and weight of a flexible conductor necessary for the heavy currents required. Sliding-shoe or wheel-type collectors mounted on an arm on the crane contact rigid copper or steel conductors. Safety requires that bare conductors are guarded from accidental human contact at all times when they are not de-energized and under repair. The guard may be a simple shield of metal or non-conducting material totally enclosing the conductors, except for a narrow opening for the collector arm to reach through. Also, the guard may be individual flexible shields, enclosing each conductor, that the collector deflects at its point of travel. Both types of guards provide an additional advantage in protecting the conductor from dirt and moisture. In outdoor locations, these individual conductor shields are particularly effective.

Cable conductors may be used for cranes and hoists where the cable size allows adequate flexibility for the cable to be stored on a cable reel or to hang from slides on a track without bends harmful to the conductor. Cable on reels may have the outer end permanently connected to the power source or be fitted with a plug for transfer between outlets spaced along the runway. Flexible conductors offer some economy compared with the rigid type, but their disadvantages should be studied before making a final choice.

References

[1]ANSI/ASME B30 Series, published by the American National Standards Institute, consists of the following:

B30.1 *Jacks*

B30.2 *Overhead and Gantry Cranes (Top Running Bridge, Single or Multiple Girder, Top Running Trolley Hoist)*

B30.3 *Hammerhead Tower Cranes*

B30.4 *Portal, Tower, and Pillar Cranes*

B30.5 *Mobile and Locomotive Cranes*

B30.6 *Derricks*

B30.7 *Base Mounted Drum Hoists*

B30.8 *Floating Cranes and Floating Derricks*

B30.9 *Slings*

B30.10 *Hooks*

B30.11 *Monorails and Underhung Cranes*

B30.12 *Handling Loads Suspended from Rotorcraft*

B30.13 *Storage/Retrieval (S/R) Machines and Associated Equipment*

B30.14 *Side Boom Tractors*

B30.16 *Overhead Hoists (Underhung)*

B30.17 *Overhead and Gantry Cranes (Top Running Bridge, Single Girder, Underhung Hoist)*

B30.18 *Stacker Cranes (Top Running Bridge, Multiple Girder with Top or Under Running Trolley Hoist)*

B30.19 *Cableways*

B30.20 *Below-the-Hook Lifting Devices*

B30.21 *Manually Lever Operated Hoists*

B30.22 *Articulating Boom Cranes*

[2]*Specification for Electric Overhead Traveling Cranes,* CMAA Specification #70, Crane Manufacturers' Association of America, Revised 1983, Pittsburgh, Pennsylvania.

Bibliography

Shapiro, Howard I., *Cranes and Derricks,* McGraw-Hill, Inc., New York, New York.

Weaver, Wm. M., *Crane Handbook,* Whiting Corporation, Harvey, Illinois, 1991.

CHECKLIST FOR CHAPTER 8

1. What type of cranes are required at the powerhouse?
2. Are hoists needed for the gates?
3. What is the maximum load to be handled during construction?
4. Will a lifting beam be required to lift the maximum load?
5. Do the building walls allow adequate room for moving equipment with the cranes?
6. Can the crane reach all areas of the powerhouse to set down parts during overhaul?
7. How much vertical hook travel is required?
8. Is the structure strength adequate to support the crane and its load?
9. Are the hook and crane travel speeds slow enough to allow precise positioning of large heavy parts?
10. Are there any local regulations or codes that will require special equipment?
11. What will be used for a load to test the crane?
12. Who will operate the cranes during construction?
13. Who will maintain the cranes during construction?
14. What is the minimum acceptable condition of the cranes when they are turned over from construction to operation and maintenance personnel?

Chapter 9
Auxiliary Systems

The auxiliary systems in a power plant are essential to the power plant's successful operation. Auxiliary systems should include the following functions: cooling water, draining water from the equipment and structure, detecting and suppressing fire, compressing air, providing potable and service water, handling transformer and lubricating oil, treating sewage, monitoring equipment, and providing heating, ventilating, and air conditioning (HVAC).

Many times, the auxiliary systems receive less diligent thought and consideration than the major equipment. To have a well-functioning, easy to operate, and easy to maintain power plant, all of the auxiliary systems have to be carefully considered and conceived when designing the plant. A major design concern for auxiliary installations (such as compressors, strainers, and pumps) is their location with respect to performance considerations and maintenance access. Air intakes for compressors should minimize introducing dust and moisture into the compressed air. If pumps are to be self-priming, their orientation, suction line adit, and elevation should not prevent an effective start. Some pumps with suction line sources in the draft tube will take in air and become air bound; therefore, an air release valve should be provided at the appropriate location. Strainers should be arranged so that the baskets may be withdrawn readily without losing the entrapped material. If a large amount of debris is expected, pumps should have automatic self cleaning strainers. As much as possible, items requiring maintenance or monitoring should be installed at or slightly above a level available for normal walking access. Equipment should be located either on the floor or low on walls in roomy, easy-to-access areas so that monitoring and maintenance will be convenient and therefore not overlooked.

Cooling and Service Water

Cooling Water Uses

All plants, except small installations where self-cooling is part of the equipment design, require a cooling water supply. In determining cooling system flow and pressure, requirements for the following items should be considered: air compressors, auxiliary power generators, generator air coolers, transformers, turbine and generator bearing oil, turbine-runner seal lubrication, shaft-packing box, and heating, ventilating, and air conditioning requirements. Figures 9-1 and 9-2 show schematics of cooling water systems. Such systems should be designed with sufficient capacity to meet any overload operating condition that could reasonably be encountered. Water for firefighting purposes is considered separately in the "Fire Protection" section.

Sources for the Cooling Water Supply

A source upstream of the turbine usually can supply cooling water for plants with heads of 90 meters (300 feet) or less. For lower head plants, it is preferable to pump cooling water from the draft tube or tailrace. To some extent, economics may influence the choice between a penstock or draft tube supply (or a combination of both), but the supply reliability always should be the prime consideration.

Water from the upstream side of the unit can be taken from upstream or downstream of the turbine intake closure device, such as the turbine shut-off valve. Cooling water for turbine wearing rings, bearings, and other critical equipment should be taken from an area upstream of the closure device. This ensures a continuous cooling water flow if the unit should remain on the line with the device closed. Less critical flows can be taken from a place downstream of the flow closure device. When cooling water comes from upstream of the turbine in plants with heads exceeding 90 meters (300 feet), either the cooling water equipment must be designed for the higher pressure or some form of pressure reduction must be included in the pipeline before the water reaches the equipment.

It is also important to minimize the number and size of taps made upstream of the turbine to limit exposure to powerhouse flooding if a water line fails. Pressure water connections, such as cooling water supply, should be connected within 30 degrees of the horizontal centerline to avoid excessive debris and air. Guard and service valves should protect the connections.

System Design

The cooling water system design determines the number of pumps required for the water supplied from downstream of the turbine. A plant with multiple generating units and a common cooling water system should have a minimum of three pumps each sized for 50 percent of the total maximum plant requirement. Two pumps are for normal operation with the third pump on standby. Therefore, if two pumps are out of operation at the same time, the plant can still be operated

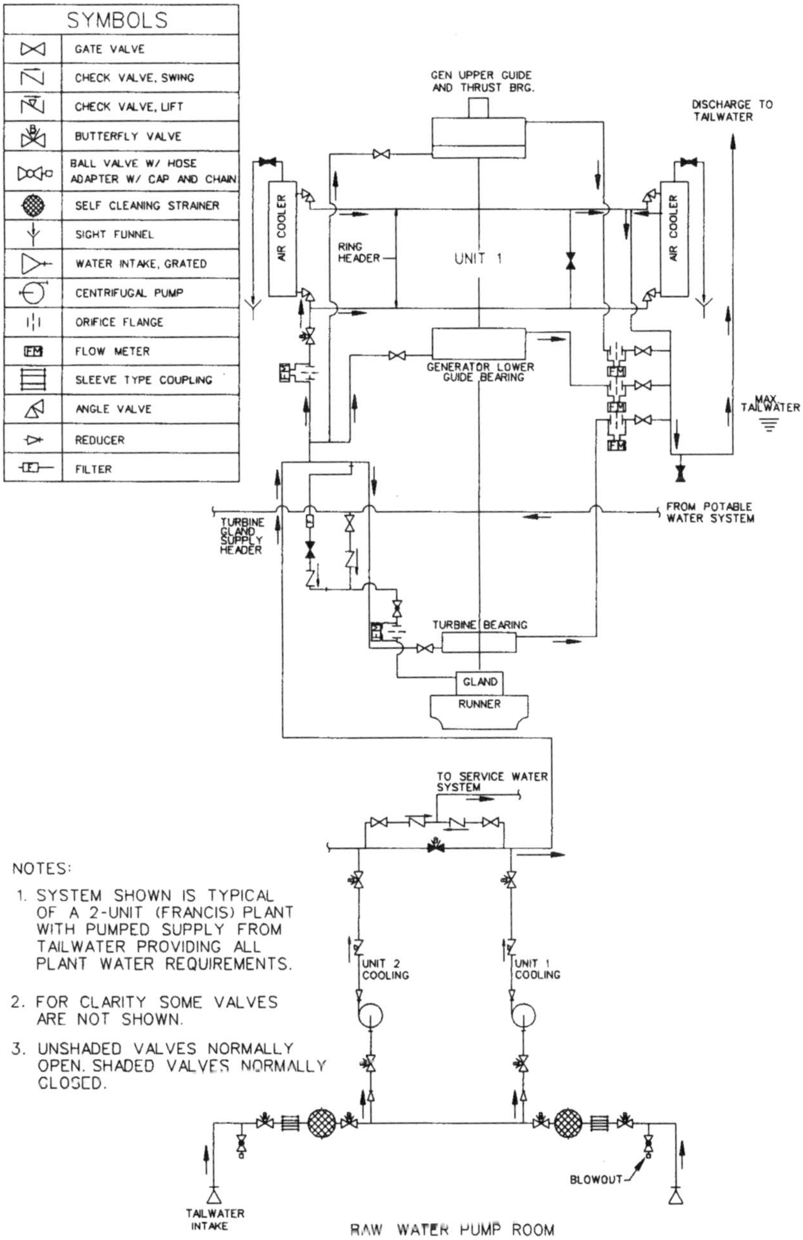

FIGURE 9-1: ***Unit Cooling Water Piping Schematic.*** *This schematic shows the cooling water piping for a unit. The cooling water system design determines the number of pumps required for the water supplied from downstream of the turbine. In a single-unit plant or a multiple-unit plant designed for independent cooling water systems at each unit, the plant should have a minimum of two pumps for each system with each pump sized for 100 percent of the maximum unit requirement.*

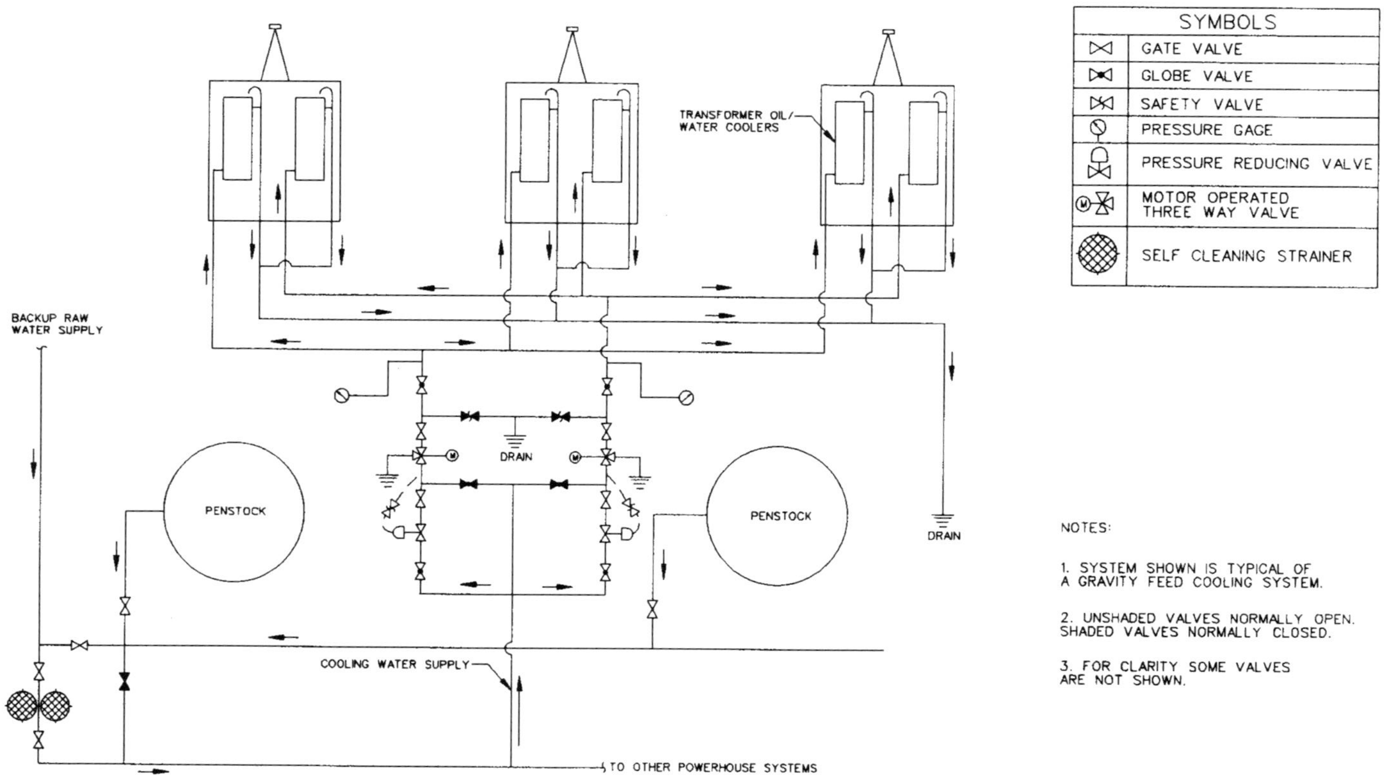

FIGURE 9-2: ***Transformer Cooling Water Piping Schematic.*** *This figure shows the schematic for water cooling a transformer. Transformer cooling water supply pressure has to be very carefully controlled to eliminate the possibility of contaminating the transformer oil.*

at reduced capacity with the third pump. In a single-unit plant or a multiple-unit plant designed for independent cooling water systems at each unit, the plant should have a minimum of two pumps for each system with each pump sized for 100 percent of the maximum unit requirement. Calculating pipe sizes should allow for deterioration with age because scaling, corrosion, or bacterial attack can reduce system flow capability.

Strainers should be located upstream of the pumps or pressure-reducing equipment. Often twin basket strainers are used since one basket can be backwashed while the other basket filters the water. Consideration also should be given to using automatic backwash strainers, particularly if the plant is in an unattended location or water conditions require frequent cleaning. Strainers should have a differential pressure switch to warn personnel if the strainer becomes plugged. Particular attention should be given to the equipment manufacturer's requirements for straining. For example, water to the turbine seal rings and shaft packing requires a higher degree of straining or, in some cases, filtering, than does water supplied to the generator air coolers.

The characteristics of the water supply should be considered in choosing piping materials. These include the water's chemical properties, scaling tendencies, total suspended solids (TSS), corrosive properties, and biological factors such as bacterial fouling. Butt-welded, black steel pipe often is used for exposed lines larger than 100 millimeters (4 inches). Large embedded pipe may be of cement-lined, ductile iron. Embedded pipe should have cleanouts to allow for removing foreign material. Corrosion-resistant materials should be used for smaller lines. Acceptable materials include stainless steel, brass, copper, fiberglass, PVC, and cement-lined ductile iron. Dielectric isolation should be provided between different metallic materials to prevent galvanic corrosion.

Lines up to 50 millimeters (2 inches) in diameter may use screwed- or flanged-gate or globe valves with union bonnets and rising stems. Globe valves should have renewable tapered stainless-steel seats and discs. Composition seats or disks do not hold up well in raw water service. Larger valves in critical locations should be plug or gate valves with bronze trim, outside screw and yoke, and rising stem—all with flanged ends. Valves in non-critical locations, such as isolating valves for system servicing, should be butterfly valves. All valves should be in accessible locations or be able to be operated remotely.

Locations subject to freezing may need heat tracing or provisions for draining. Heat-traced lines should have insulation. Pipes carrying water should be insulated to prevent sweating in locations where water condensation would be harmful or annoying.

Turbine bearings, runner seals, and shaft-gland or packing seals usually require cooling water. Other potential cooling-water demands include air compressors and aftercoolers, auxiliary power generators, and air conditioning equipment. Sight flow indicators and flow switches or flow meters with alarm contacts should be provided to monitor flow. Unattended plants, particularly, require automatic tripping and alarms. Provisions should be made to backwash coolers to control accumulation of biological growth.

Generator Cooling

The generator cooling is among the major uses of the cooling water system. An ample water supply should be provided, but overcooling should be avoided to prevent condensation from forming on the stator winding, especially when a generator is shut down. Automatic open/close valves downstream of the cooler should control the cooling water flow to make sure the coolers remain full of water and the generator maintains an even temperature. Small air-cooled units frequently require cooling water for lubrication oil.

Transformer Cooling

Transformer cooling water supply pressure has to be very carefully controlled to eliminate the possibility of contaminating the transformer oil. Transformer coolers should be double-wall tubes so that any leakage from the water side will not contaminate the transformer oil and will be readily detected. A vent to atmosphere on the water side of the cooler should be provided to eliminate any damaging pressure surges in the transformer cooler, as in Figure 9-2.

Service Water

Station washdown and other applications where potable water is not needed may require a service water system. A readily available source may be a connection on the cooling water supply header with a booster pump, if required. Figure 9-3 shows a typical service water system.

Intake and draft tube decks and pipe galleries may require service water outlets. The service water outlets should be located in accordance with the requirements and practices of the operating agency or company. Within those parameters, the outlets should be spaced to allow delivery of water to all needed locations using available hose lengths. As an example, outlets spaced at 25-meter (75-foot) intervals would be appropriate for 15-meter (50-foot) hoses. Service water outlets should be avoided in areas where possible damage may occur to moisture sensitive electrical equipment or controls.

Service water lines are not used frequently, and the choice of materials may reflect the limited service life. Consideration should be given to water conditions such as scaling, corrosion, or bacterial attack when evaluating material selection. Butt-welded steel pipe normally is used for lines over 75 millimeters (3 inches). Carbon steel (black) pipe can be used for sizes from 35 to 75 millimeters (1.5 to 3 inches). Smaller lines should be from brass pipe or copper water tubing. Dissimilar metals should be isolated to prevent galvanic corrosion. As service water lines may not be used for extended periods, the lines may have valved flushing connections to allow periodic flushing of stagnant lines. Valve selection and provisions for freeze protection are similar to the requirements for cooling water systems.

	SYMBOLS
	GATE VALVE
	GLOBE VALVE
	3-WAY 3-PORT PLUG VALVE
	BUTTERFLY VALVE
	BALL VALVE W/ UNIVERSAL AIR HOSE CONNECTION
	MOTOR OPERATED BALL VALVE
	PLUG VALVE 3-WAY 2-PORT 180°
	CHECK VALVE
	INTAKE, GRATED
	SELF-CLEANING STRAINER
	Y-TYPE STRAINER
	SIGHT FUNNEL
	PRESSURE GAGE
	GATE VALVE W/ HOSE CONNECTION
	ELECTRIC THERMOSTAT
	ROTOR VALVE 3 WAY 2 PORT 90°
DPS	DIFFERENTIAL PRESSURE SWITCH
BFP	BACK FLOW PREVENTOR

NOTES:

1. SYSTEM SHOWN IS TYPICAL OF AN 8-UNIT KAPLAN PLANT.
2. ALTERNATE INTAKE FOR DECK WASH AND TRANSFORMER SPRAY PUMPS WOULD BE RAW WATER HEADER.
3. DECK WASH HOSES PROVIDE BACK-UP FIRE PROTECTION. TRANSFORMER SPRAY PUMP PROVIDES BACK-UP FOR DECK WASH PUMP.
4. UNSHADED VALVES NORMALLY OPEN, SHADED VALVES NORMALLY CLOSED.
5. FOR CLARITY SOME VALVES ARE NOT SHOWN.

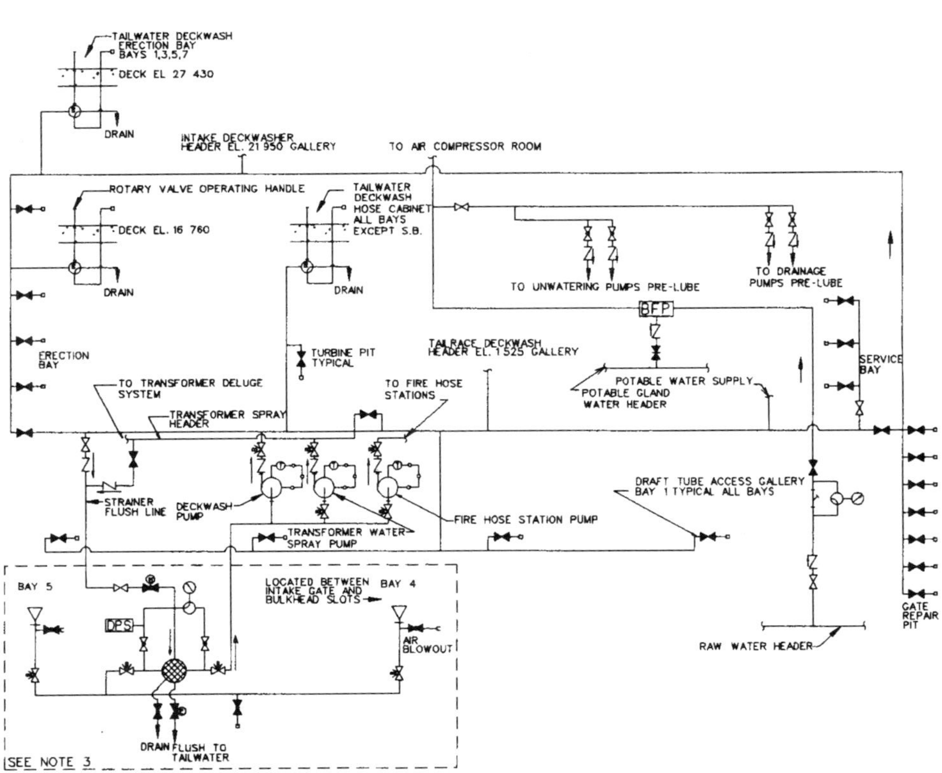

FIGURE 9-3: *Service Water Piping Schematic.* *This figure shows a typical service water system. Station washdown and other applications where potable water is not needed may require a service water system.*

Chapter 9

Potable Water

Need for Potable Water

Drinking, washing, and sanitation require potable water. Gland seals for pumps may need strained water or water with solids centrifugally separated to potable water standards for suspended solids. The volume rate of flow needed and the source of the potable water will depend upon the size, staffing, and location of the plant, climate, and other facilities that may need to be supplied from the system. Figure 9-4 shows a typical potable water system.

Supplies of Potable Water

Possible sources include wells, surface supply, and adjacent sanitary or municipal water districts. The supply chosen must meet standards for bacteriological quality and physical, chemical, and radiological characteristics. State or local regulations may limit the available choices.

Pumping, filtration, or other treatment may be required if a municipal source is not available. Water may need to be stored if peak demand exceeds the supply capacity, particularly if a fire protection reserve must be maintained. Supplying the fire protection system from the service supply may eliminate this problem if an acceptable supply can be obtained.

Fire Protection

Fires are more common at hydroelectric plants than is often realized. Fires can occur at places wherever there is a source of fuel for combustion. Fuel sources can be oil, combustible cable coatings, electrical equipment, chemicals, wood forms and skids used during construction and later stored for reuse, and gasoline—among others. Transformer fires are more common than fires in other areas. Figure 9-5 shows a typical fire protection water system.

The damage caused by a fire at a hydroelectric plant is more serious than just the cost for repairs and replacements. Human lives and safety may be involved, and the revenue lost because of an outage may exceed the costs for repairs and replacement. Much of the equipment at a hydroelectric plant is specially designed and cannot be replaced from a supplier's stock. Some equipment destroyed in a fire may require more than a year to replace.

The National Fire Protection Association (NFPA) Technical Committee on Fire Protection for Non-Nuclear Power Plants, Subcommittee for Hydroelectric Plants, has developed a Recommended Practice for Fire Protection of Hydroelectric Generating Plants, NFPA 851.[1] Any designer of a hydro plant should be familiar with and have a copy of NFPA 851 available for reference. NFPA 851 gives extensive additional references and outlines how to identify and protect individual fire hazards. Insurance requirements and state and local codes may have requirements more stringent than NFPA 851.

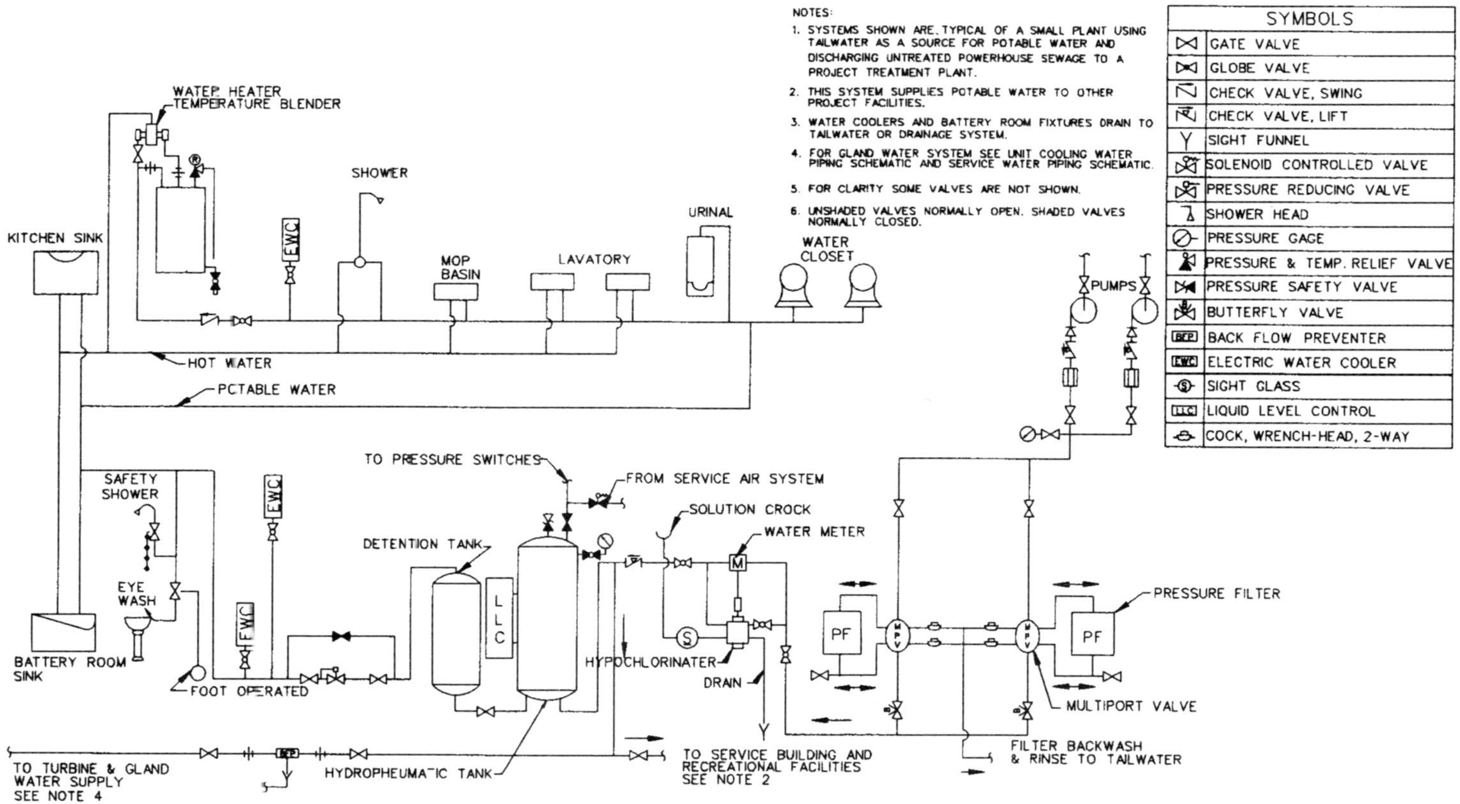

FIGURE 9-4: *Potable Water Piping Schematic.* *Drinking, washing, and sanitation require potable water. The volume rate of flow needed and source of the potable water will depend upon the size, staffing, and location of the plant, climate, and other facilities that may need to be supplied from the system. This figure shows a typical potable water system.*

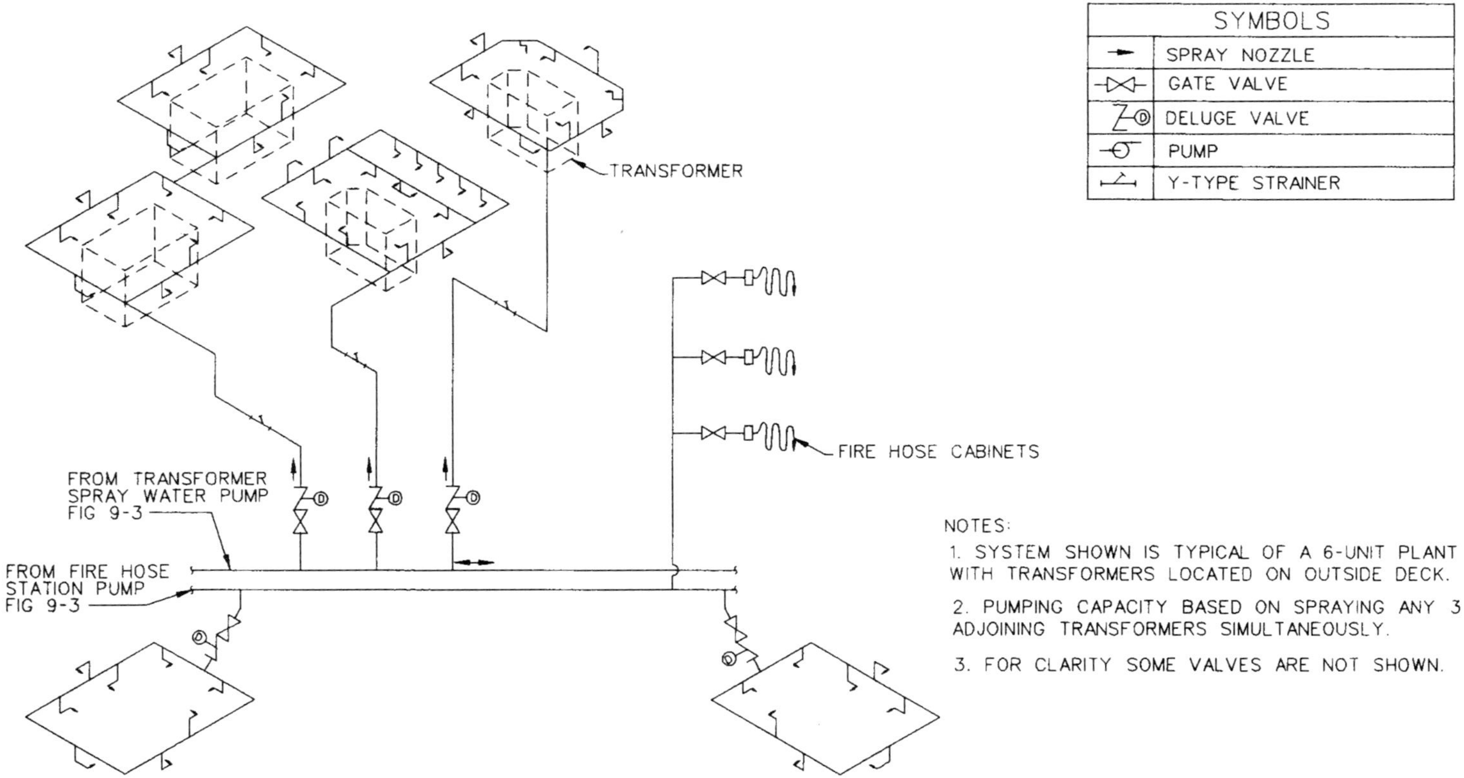

FIGURE 9-5: ***Fire Protection Water Piping Schematic.*** *Fires can occur at places wherever there is a source of fuel for combustion. Fires are more common at hydroelectric plants than is often realized because plants have oil, gasoline, combustible cable coatings, electrical equipment, chemicals, and wood forms and skids used during construction and later stored for reuse. Actually, transformer fires are more common than fires in other areas. This figure shows the layout for a typical fire protection water system.*

Controlling Fires in Rotating Equipment

Much research has been conducted in recent years regarding ways to control and quench fires in rotating electrical equipment. There has been considerable experience with quenching such fires at hydroelectric plants. The NFPA 851 Subcommittee has collected reports of fires at hydroelectric plants and how they were extinguished. Water deluge seems to be the most successful, safest, and, from most aspects, the preferred method for fighting such fires. Some owners prefer carbon dioxide (CO_2) for quenching fires in generators. CO_2 systems will require operator training and specific equipment for the selected system, and may require special venting. Figure 9-6 shows a typical fire protection CO_2 system.

In today's environmentally conscious world, disposing of water that contains oil, products of combustion, and firefighting chemicals and the gaseous products of combustion require special consideration. The environment inside underground power plants and inside confined spaces, including inside the bulb housing for a bulb turbine-generator unit, requires special consideration.

Plant Layout for Fire Safety

The plant's mechanical design should recognize the need for firefighting equipment, alarms, and sensors for detecting fires, heat, and smoke. Adequate water supply, including storage tanks and piping for delivering the water used for fighting fires, will be required. Assuring adequate water supply at the site and at elevations affording sufficient pressure is particularly important for plants at remote locations and in extreme climates.

Provisions for fire avoidance and firefighting affect the hydroelectric generating plant design and the arrangement of the spaces and equipment within the plant. Barriers should separate areas that have fire hazards. Sprinkler systems may be considered. Outdoor oil-insulated transformers should be separated from adjacent structures and from each other. Barriers should separate indoor oil-insulated transformers from adjacent areas and from each other. Similarly, oil-insulated circuit breakers and other oil storage and oil-filled equipment are fire hazards in hydroelectric plants and should be treated as such.

It is necessary to train and organize people to control the use of the fire fighting equipment and to organize fire brigades.

Many hydroelectric plants are far from municipal firefighting facilities. Some are in relatively isolated and inaccessible places. Normally, many hydroelectric plants do not have personnel at the plant who could fight a fire. A way to unlock facilities at an unattended hydroelectric plant must be planned in advance if it is intended that people be able to enter to fight a fire.

Smoke and Heat Venting

When there is a fire, those responsible for controlling and quenching the combustion often are unable to even locate the source of the fire. The smoke and products of combustion obscure the fire from view. The smoke and heat make access difficult and dangerous. Underground galleries in the powerhouse and

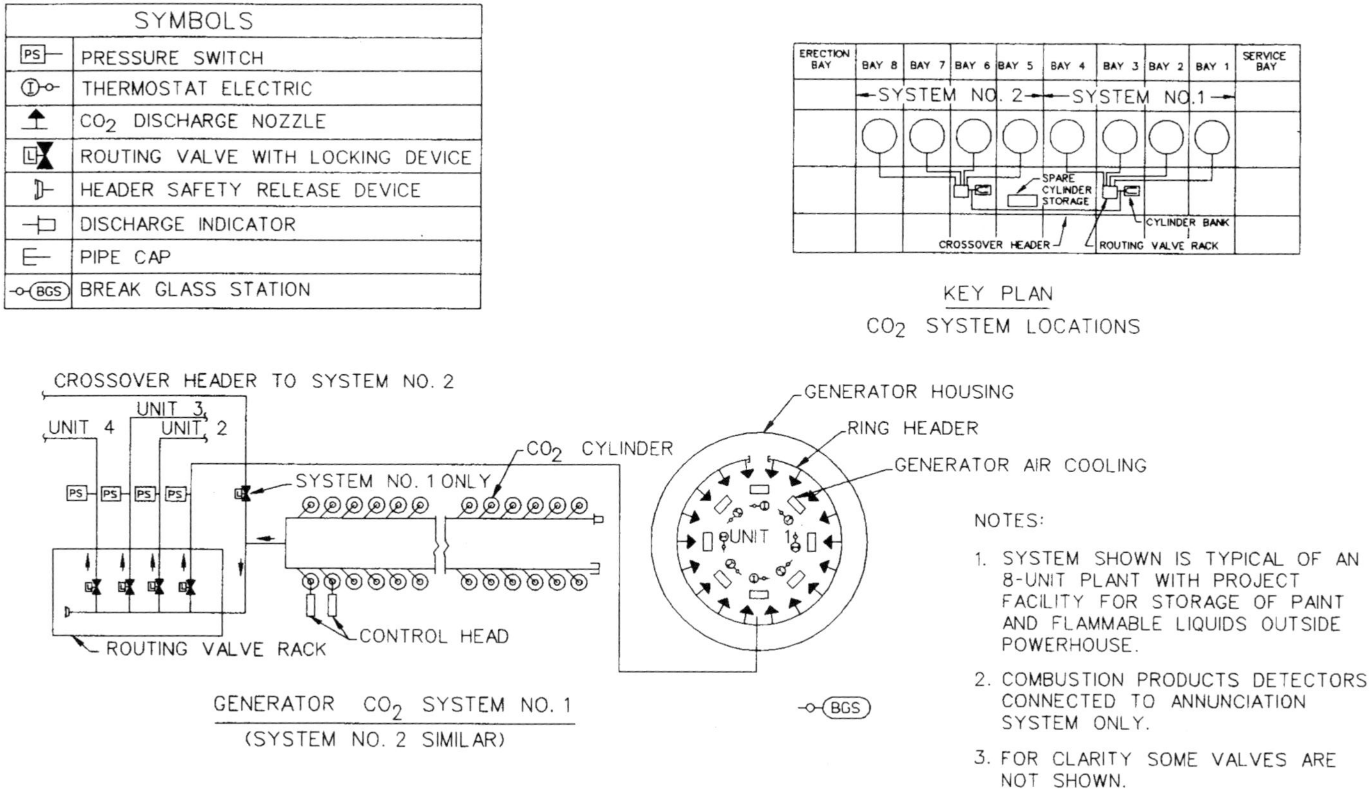

FIGURE 9-6: ***Generator Fire Protection Water Piping Schematic.*** *Some plant owners prefer carbon dioxide (CO_2) for quenching fires in generators. This figure shows a typical fire protection CO_2 system protecting two generator units. CO_2 systems require operator training and specific equipment for the selected system and may require special venting.*

underground cable tunnels are places where smoke and heat make it particularly difficult to control a fire. Evacuating smoke and heat from the power plant is essential before a fire can be controlled.

Fans and ducting must be provided for venting smoke and heat from a hydroelectric plant. This is a special problem in underground facilities. When fire heats air and puts products of combustion into the air in tunnels and in underground hydroelectric plants, the ventilation conditions that existed while the air was cold are altered. The designer must consider air flow stratification, the numerous nodes or junctions between tunnels and shafts, the likely frictional resistances with and without fire, and the placements and capacities of the fans and fire stops.

Plant ventilation should never be dependent on the functioning of a single fan and ventilation system. The ventilation fans should be of sufficient capacity to meet the ventilation requirements for venting heated air containing products of combustion with the largest fan out of service.

Water Supply and Drainage

Where multiple fire pumps are required, the pumps should not be subject to a common failure, such as a single electrical power supply or a single water supply valve. The pumps should have the capacity to meet the fire suppression flow requirements with the largest pump out of service.

All fire areas of the plant should have ways for drawing all spilled liquids directly to safe areas or for containing them in the fire area without flooding equipment and endangering other areas. For environmental reasons, liquid discharges from oil spills or from operating a chemical fire suppression system may have to be treated.

Consideration should be given to dedicating fire-protection systems, equipment, and installations solely for fire-protection purposes. Fire-protection provisions should cover the fire-protection equipment, also.

The largest, fixed fire-suppression system demand plus the maximum hose stream demand determines the water supply for the permanent fire protection installation. If water supply tanks have a dual purpose, a standpipe or similar arrangement should be provided to dedicate the amount required for fire protection use only. Where tanks are used, the water supply design should consider the time required for filling the tanks. In some rivers and tributaries, the existence of microorganisms limits the use of raw water without treatment for fire protection. For example, some rivers can contain small snails and mussels that can grow in stagnant water lines and plug them. Consideration of water quality may prevent long-term problems related to the supply of fire protection water.

Compressed Air System

Compressed Air Requirements

Hydroelectric power plants frequently require using compressed air for generator braking, governor oil pressurization, and maintenance. Also, draft tube water-level depressing may need compressed air for operating the turbine in condensing mode, for pump turbine starting, or for maintenance. Another use for compressed air is for depressing water below vertical Pelton wheels during periods of high tailwater. Typically, this application uses low pressure, high volume blowers. Figure 9-7 shows a typical compressed air system, and Figure 9-8 shows a compressor room for a powerhouse.

A small compressor can maintain compressed air storage tanks and receivers at normal working pressure. Compressed air is stored in several receivers from which it can be withdrawn when needed. Larger compressors may be required for draft tube depressing and for operating pneumatic tools or equipment for maintenance purposes. Aftercoolers remove moisture from the compressed air and prevent high temperatures in the receiver in order to avoid oil vapor explosions.

Generator Brakes

An independent distribution system, complete with receiver, supplies the generator brakes. The brake air receiver is supplied with air through a check valve from the main receivers. This ensures an adequate pressure and air quantity for braking regardless of pressure variations in the main system. The volume of air per brake application includes the initial volume, the displacement volumes of the brake cylinders with the brake lining worn to the rivet heads, plus all the piping between the air brake control valve and brakes.

Governor Oil System Requirements

The governor oil pressure tank will have to occasionally be recharged with compressed air to maintain the necessary air cushion. A governor air tank typically is provided for recharging the air in the governor oil pressure tank. The governor air tank should be maintained above normal governor oil tank pressure. Passing the high pressure air through a pressure reducer and throttling valve will allow refilling a governor tank with minimal moisture absorption. The system should have an air compressor with the required pressure rating and a capacity sufficient to charge the governor oil pressure tank in a reasonable time. This compressor should produce the maximum working pressure required by the governor system with a capacity balanced between the needs for quick recovery and compressor minimum running time. Some pressure oil systems now operate at 7,000 or 14,000 kilopascals, kPa (1,000 or 2,000 pounds per square inch, psi). The valves in the piping at the governor oil pressure tank should be installed in such a way that a partially closed or defective valve will be detected immediately.

Draft Tube Depression Systems

A draft tube depression system must be designed to meet plant-specific requirements. Figure 9-9 shows a typical draft tube depressing system. Depression air receivers and compressors will not necessarily be the same pressure as the service air system. The depression air pressure is dependent on the submergence of the

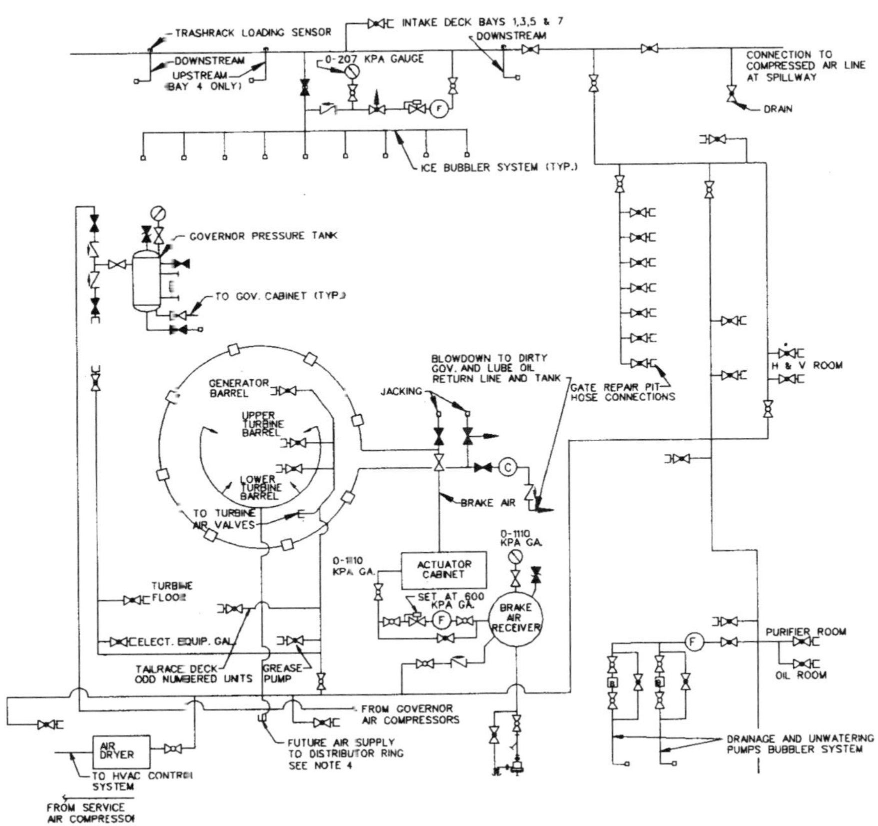

SYMBOLS
GLOBE VALVE
GATE VALVE
SWING CHECK VALVE
BALL VALVE
SAFETY VALVE
NEEDLE VALVE
PRESSURE REDUCING VALVE
PRESSURE GAGE
AIR FILTER
BUBBLER NOZZLE
FLOW DIRECTION
INVERTED BUCKET TRAP
CYLINDER (SIGHT) GLASS
SIGHT GLASS W/SHUT-OFF VALVE
PIPE CAP
OPEN HUB
BUBBLER CONTROL
UNIVERSAL AIR HOSE CONNECTOR
HOSE ADAPTER

NOTES:

1. SYSTEMS SHOWN ARE TYPICAL OF AN 8-UNIT POWERHOUSE WITH KAPLAN UNITS, INTAKE GATES, TRASHRACKS, GATE REPAIR PIT, AND WITH SPILLWAY AIR CAPACITY AVAILABLE FOR ICE BUBBLERS AND MAJOR MAINTENANCE.
2. ONE BRAKE AIR SYSTEM FOR EACH PAIR OF UNITS.
3. UNSHADED VALVES-NORMALLY OPEN, SHADED VALVES NORMALLY CLOSED.
4. MINIMUM PROVISIONS FOR FUTURE AIR SUPPLY TO DISTRIBUTOR RING IS NECESSARY FOR ALL UNITS.
5. FOR CLARITY SOME VALVES ARE NOT SHOWN.

FIGURE 9-7: ***Powerhouse Compressed Air System Schematic.*** *Hydroelectric power plants require compressed air for generator braking, governor oil pressurization, and maintenance. Also, draft tube water-level depressing may need compressed air for operating the turbine in condensing mode, for pump-turbine starting, or for maintenance.*

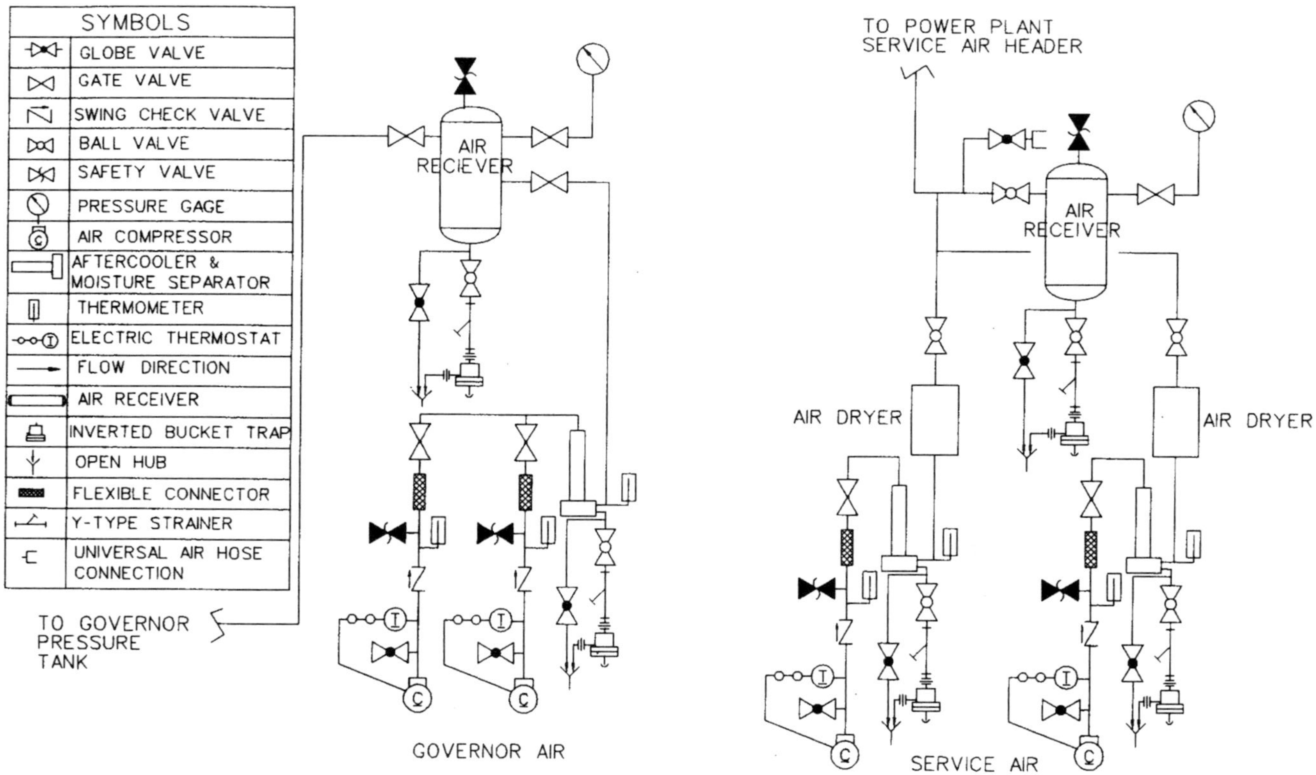

FIGURE 9-8: *Powerhouse Compressor Room Schematic.* *Compressor units should have adequate space, allowing for proper installation, servicing, and maintenance. The compressor area should have ample cooling and ventilation to remove heat produced by compressor operation. A small compressor can maintain receivers at normal pressure. Large compressors will be required for depressing draft tube water surface. Aftercoolers should cool the compressor discharge and remove moisture.*

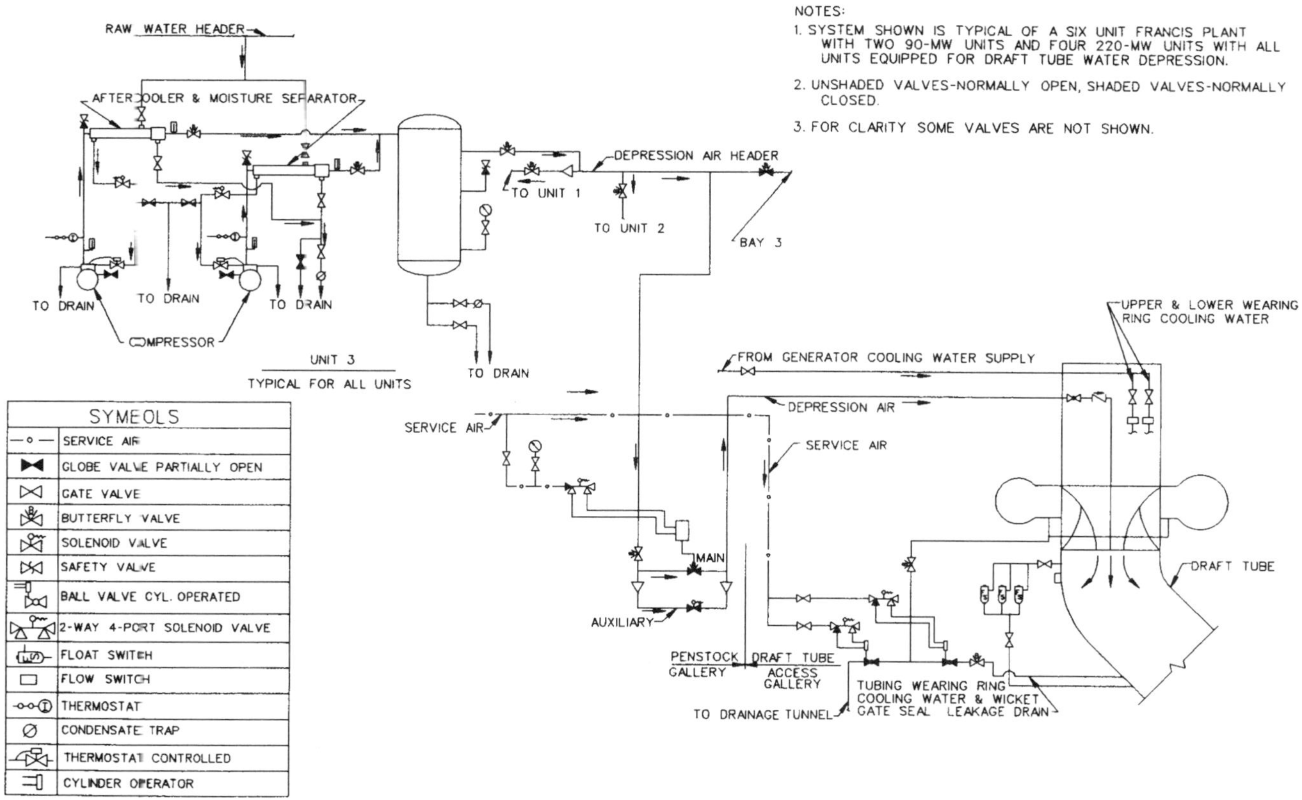

FIGURE 9-9: ***Draft Tube Depression System.*** *A draft tube depression system must be designed to meet plant-specific requirements. This figure shows a typical draft tube depressing system. Depression air receivers and compressors will not necessarily be the same pressure as the service air system. A draft tube depression system design should consider the solution of air into the water and leakage past the turbine wicket gates.*

turbines. A draft tube depression system design should consider the solution of air into the water and leakage past the turbine wicket gates. Normally, the compressors should be under start-stop control: starting when the pressure drops below the normal minimum and stopping when the pressure reaches maximum.

Installation Guidelines

All air receivers for the compressors should conform to the ASME Boiler and Pressure Vessel Code and have standard connections and outlets. Air receivers should have safety valves of a size and setting that comply with the code.

Compressor units should have adequate space allowing for proper installation, servicing, and maintenance. While the arrangement may be modified to suit the requirements of a particular plant, the clearances around the equipment should be adequate. This should include space for removing piston rods, intercooler tubes, and other large parts. A typical clearance of 1 meter (3 feet) around each machine and 2 meters (6 feet) between machines should ensure safe operation. Guards and enclosures should fully protect all moving parts, belts, and flywheels. On small self-contained units, the belt side should be located next to the wall. When within reach from the floor, discharge pipes to receivers should be heat insulated. The compressor area should have ample cooling and ventilation to remove heat produced by compressor operation. The floor should have openings or pipe trenches near the cylinders and air receivers for small cooling-water and pressure-control pipes. The exact arrangement of such small piping will have to be determined after the equipment has been selected.

Large reciprocating air compressors should be located on mass concrete to avoid transmitting loads and vibrations through the power plant structure. Care should be taken to provide a location where the equipment is not subject to flooding.

The air intake should connect to an outdoor space, free of spray, where the air is as cool as possible. The intake opening in the building must be screened properly, and the compressor intake may require an air silencer and filter. If the ventilating air system to the compressor room has sufficient capacity, the intake air may be taken directly from the compressor room.

Turbine Dewatering System

General Requirements

The turbine dewatering system may be a pressure-piping system drained by gravity to an open or a closed sump. The system should permit draining the spiral case and draft tube to tailwater level and then, if necessary, pumping the water remaining below the tailwater level to the sump. Usually, dewatering the penstock requires a separate system. Figure 9-10 shows a typical gravity drainage system. (This figure logically connects to Figure 9-11, Dewatering Piping Schematic. Traditionally, in hydropower plant design, the dewatering system and the gravity drainage system are shown on the same schematic drawing. The two systems

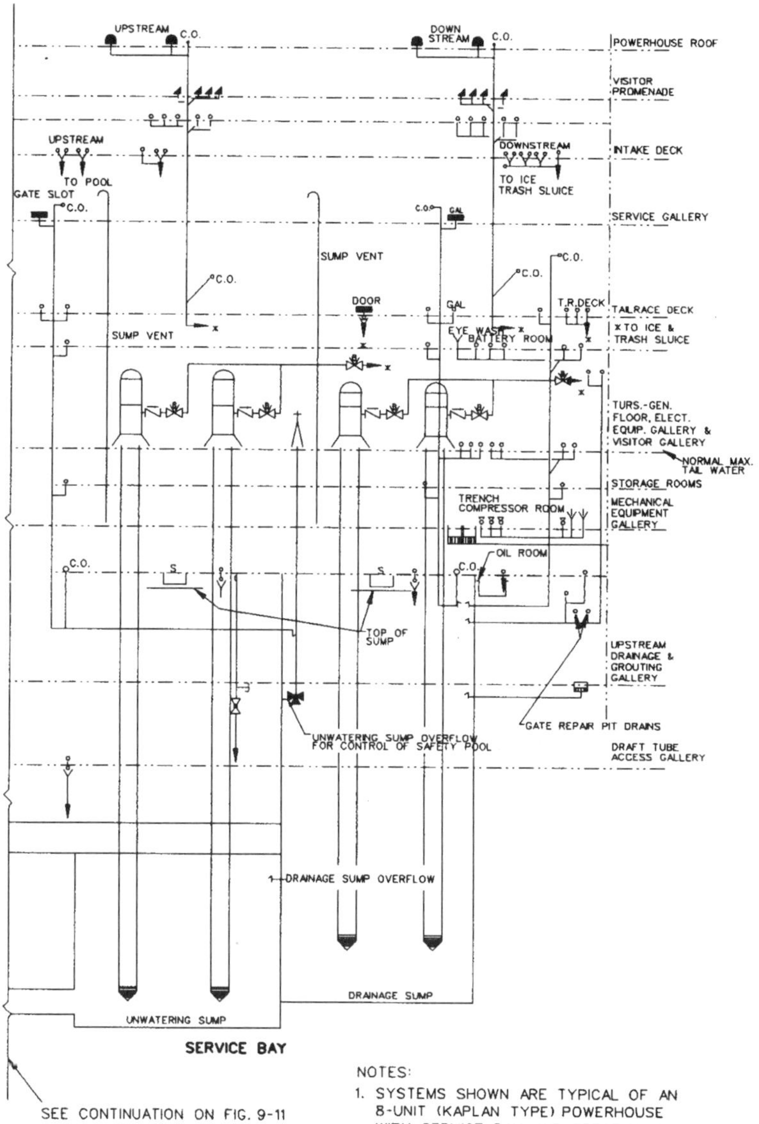

NOTES:

1. SYSTEMS SHOWN ARE TYPICAL OF AN 8-UNIT (KAPLAN TYPE) POWERHOUSE WITH SERVICE BAY AND ERECTION BAY (NOT SHOWN) AT OPPOSITE ENDS, FISH COLLECTION SYSTEM, OIL-WATER SEPARATOR AND ICE-TRASH SLUICEWAY.
2. UNSHADED VALVES-NORMALLY OPEN. SHADED VALVES NORMALLY CLOSED.
3. SEE FIG. 9-11 FOR SYMBOL DEFINITIONS.
4. FOR CLARITY SOME VALVES ARE NOT SHOWN.

FIGURE 9-10: ***Gravity Drainage Piping Schematic.*** *This figure shows a typical gravity drainage system. A gravity drainage system consists of floor drains and various equipment drains that drain to a sump. The drainage system could carry waste water, oil from a spill, or a combination of both. Usually, the water in the sump is pumped into the tailrace, but if it is mixed with oil, the oil must be separated from the water to avoid contaminating the tailrace.*

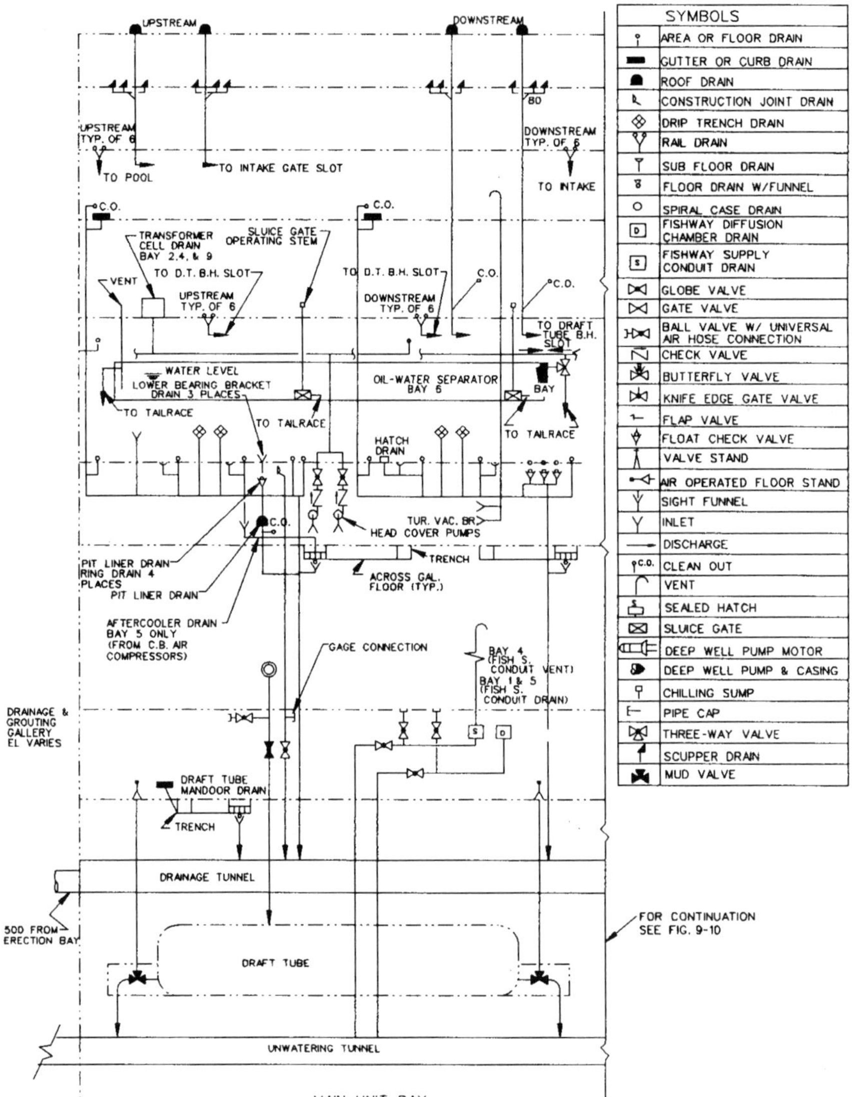

NOTES:

1. SYSTEMS SHOWN ARE TYPICAL OF AN 8-UNIT (KAPLAN TYPE) POWERHOUSE WITH SERVICE BAY AND ERECTION BAY (NOT SHOWN) AT OPPOSITE ENDS, FISH COLLECTION SYSTEM, OIL-WATER SEPARATOR AND ICE-TRASH SLUICEWAY.
2. UNSHADED VALVES-NORMALLY OPEN. SHADED VALVES NORMALLY CLOSED.
3. FOR CLARITY SOME VALVES ARE NOT SHOWN.

FIGURE 9-11: ***Dewatering Piping Schematic.*** *The turbine dewatering system may be a pressure-piping system drained by gravity to an open or a closed sump. Usually, dewatering the penstock requires a separate system. This schematic shows a typical dewatering system. The dewatering sump may act as a reserve drainage sump in case of an accident or flood. If two pumps are used, as in this figure, one of two pumps should handle the maximum drainage to the sump, pumping to the highest tailwater level.*

often use common structures and equipment, and sometimes may use the same sump or empty into the same drainage tunnel.)

The spiral case drain line design should dewater the case from the centerline of the distributor to the invert, including head gate leakage, in a reasonable time allowing for leakage through the head gate or turbine shut-off valve. For low-head machines, the penstock and the spiral case should be dewatered to the tailwater elevation through the turbine wicket gates before opening the spiral case drain line.

Each draft tube drain line under normal tailwater should discharge about 1 percent of the rated turbine discharge to help seal the draft tube bulkhead gates. With the water level at the floor of the tube, the line should be capable of removing about 1 percent of the rated turbine discharge for leakage through the draft tube bulkhead gates after sealing.

Dewatering Drainage Header

The dewatering drainage header is considered a low-pressure line except in some pumped-storage plants having deep submergence, 60 meters (200 feet) for example, of the pump-turbine. The exposed portions of the pressure drains and the exposed dewatering system downstream from the shut-off valve should be designed for the pressure conditions required. The pipe and fittings upstream from and including the shut-off valve should be suitable for the maximum expected pressure including waterhammer. Embedded seal strips or joints permitting slight movement should seal off the dewatering header from expansion or contraction joints. To minimize the danger of flooding the sump, the header should have valves with operating procedures to control the pressure within the spiral case within a safe value. The valve should be located in a vertical line, if possible, and the line from the case to the valve should be as short and as direct as possible to prevent clogging by silt. Also, it may be desirable to provide connections for an air hose to be used to break up the silt or to provide drain valves in the valve bonnet and the bottom of the slot.

Spiral Case Connections

All water take-offs from the spiral case should have suitable valves placed as close as possible to the casing to provide a safe, positive means of shutting off the water in case of emergency. Drains should take off from the low point of the invert and have suitable valves in the vertical run. To prevent large objects from entering and obstructing the piping and valves, inlets to the spiral case drain lines should have gratings.

Spiral cases that are preceded by a turbine shut-off valve, such as a butterfly valve, require an air and vacuum relief valve to permit dewatering and filling. This valve should be large enough to pass sufficient air to prevent an excessive vacuum inside the case during dewatering and should be able to open and release air when under maximum forebay pressure. Air relief vents should take off from the top of the casing, and guard valves should protect them.

Dewatering Sump

The dewatering sump may act as a reserve drainage sump in case of accident or flood. It should have automatic float controls and a high-level indicator similar to those of the drainage sump. In small plants, the dewatering sump may be combined with the drainage sump, or portable pumps may be used. Usually, large plants have a dewatering sump separate from the drainage sump.

Open or closed sumps may be used. A sealed, pressurized dewatering sump offers greater protection against powerhouse flooding should the dewatering pumps fail. If the sump is open, entrance hatches should not be bolted down, and they should open if the sump overflows. The sump should have suitable vents to the outside. The vents should be large enough that they do not choke with supersonic flow at the maximum dewatering rate.

Two or three dewatering pumps, similar to the drainage pumps, are desirable. One of two, or two of three, pumps should handle the maximum drainage to the sump, pumping to the highest tailwater level. The dewatering header should have an orifice plate in its discharge pipe to an open sump that will limit the inflow to an amount not exceeding the capacity of the dewatering pumps. The penstock and spiral case should be drained to the tailwater level by some means before draining water into the sump. Draining to tailwater level may be done in various ways and is discussed in the "Drainage Mechanisms" section. The orifice plate should be designed for a head that is the maximum tailwater elevation minus the sump ceiling elevation and a flow equal to the sump pump capacity at this head.

Tunnel and Penstock Dewatering

General Requirements

Usually, penstocks and tunnels are constructed of unlined rock, concrete-lined rock, concrete, or steel. Concrete-lined penstocks and tunnels often have weep holes through the liner to equalize the pressure inside and outside of the liner. Higher pressures on the outside of the liner can create buckling failures.

Penstocks and tunnels must be drained occasionally for inspection and maintenance purposes. During the draining operation, care must be taken to avoid developing differential pressures across the concrete liners or against the rock in unlined penstocks and tunnels. In unlined rock, water will penetrate cracks during normal operation and, if the penstock or tunnel is rapidly drained, this trapped water can cause rock to be broken from the walls.

Penstocks and tunnels must have adequate vent valves to admit air as the water is being drained. Failure to admit sufficient air can result in the collapse of the penstock. The Pacific Gas and Electric Company provides three to four times as many air vent valves as theoretically required along its penstocks to allow for any that may stick closed or malfunction during draining operations.

Establishing maximum allowable penstock or tunnel draining rates for each powerhouse usually provides protection against these consequences. The penstock or tunnel designer should establish these rates.

Drainage Mechanisms

Since the draining of penstocks and tunnels involves dissipating substantial potential energy stored in the water, the powerhouse design must have mechanisms to dissipate this energy. It is important to note that the pressure head in the water will range from maximum static to zero during the draining operation.

The goal of the drainage mechanism is to release water in a controlled manner at the flow rate corresponding to the allowable vertical distance per hour criteria over as much of the head range as possible. Often, various powerhouse drain valves will need to supplement the primary drainage mechanism during the later stages of draining when the pressure head is reduced.

The turbine can be used as the drainage mechanism for horizontal tunnels with large volumes whose draining causes little head change on the powerhouse. Once the water level reaches the penstock, the allowable drainage flow rate may be below the controlled ability of the turbine—especially if it remains connected to the electrical system at synchronous speed.

Schemes have been developed to use the turbine to drain the penstock by operation off-line at speeds less than synchronous. Such schemes must be planned carefully to provide speed control over a wide range, avoid "wire drawing" of the wicket gates, and avoid loss of oil film on the generator thrust bearing owing to low speed operation. (Wire drawing is a description of a surface that has been eroded by high-pressure, high-velocity water passing through a very narrow opening to a low-pressure area, such as between wicket gate sealing surfaces. The surface is characterized by rounded valleys between sharp ridges, oriented parallel to the flow of the water. The surface of the wire-drawn material looks like half-round wires had been drawn out of the surface at periodic intervals, thus the phrase "wire drawn.") For turbines so equipped, pressure regulator valves, such as Howell-Bunger valves, are more suitable since they do not have that problem.

Pelton turbines can drain the penstock or tunnel with the jet(s) operating against the deflectors inserted into the stream. The turbine should be blocked against rotation to avoid operating the generator thrust bearing at slow speed.

Drainage mechanisms also can be designed specifically for the purpose of draining penstocks or tunnels. Chapter 5, Major Valves, discusses valves capable of energy dissipation that can be used for drainage purposes.

Filling Rates

Penstock or tunnel filling rates can be much faster than draining rates. The initial filling rate during powerhouse commissioning should be slow. This allows for inspecting joints, couplings, and valves. Subsequent filling rates for penstocks can be about 30 vertical meters (100 vertical feet) per hour.

POWERHOUSE DRAINAGE

Drainage System Requirements

A gravity drainage system consists of floor drains and various equipment drains that drain to a sump. Because of the nature of these drains, the drainage system may carry waste water, oil from a spill, or a combination of both. Usually, the water in the sump is pumped into the tailrace. The oil must be separated from the water to avoid contaminating the tailrace. Figure 9-10 shows a typical gravity drainage system.

All walls in contact with water should have drain trenches along their interior faces. The trenches should slope to suitable floor drains. This includes outside walls retaining fill that may be saturated or ones that are below the annual high tailwater. Drain trenches should not cross expansion joints. Drains from expansion or contraction joints should discharge into the drainage system.

Oil storage and purifier rooms should have ways to contain spilled oil and carry the flow from the fire-protection sprinkler and deluge systems.

Drain Piping Design

All embedded pipe in the drainage system should be cast iron soil pipe with long sweeps at all horizontal turns. Quarter bends may be used in vertical turns. Horizontal lines should slope 20 millimeters to 1 meter (0.25 inch to 1 foot), if practical, with a minimum of 10 millimeters to 1 meter (0.125 inch to 1 foot). Drain lines less than 100 millimeters (4 inches) in diameter are not recommended.

Bell-trap drains should be provided in drain lines to a main drainage header or sump to prevent escaping odors. The discharge into the oil/water separator pit should be below the low-water level. This reduces the mixing of oil and other contaminants that are floating on the water surface. The vertical down spout in the sump should have a 20-millimeter (0.75-inch) vent hole above the maximum sump water level to prevent air lock. Vertical drains discharging into floor drains on lower floors need not be trapped.

Bell traps with hinged strainers should have the hinge pins removed so that the strainers will not become broken when left upright.

Drains from the battery room floor and sink should comply with local codes and requirements for dealing with an acid spill. Some locations may require a moat and other containment to confine a spill to the battery room for treatment.

Where embedded pipes cross expansion or contraction joints, provision should be made to permit minor movements without damage. Pipes crossing construction joints between first- and second-stage concrete should have a hub flush with the surface of the first-stage concrete to prevent damage during erection. Floor drains projecting above the subfloor should be installed when the finish floor is placed. Embedded pipe should be laid parallel to building lines, if practical, to avoid interfering with reinforcing and other embedded material.

Drainage Sumps

Each drainage sump should have enough volume to permit the minimum required running time of the sump pump during each cycle of operation. Sizing the pumps may require estimating seepage. Two sump pumps are preferred with each pump capable of pumping the estimated sump inflow. The pumps should operate alternately so they wear equally. Deep-well type pumps, with the motor above annual high tailwater and the discharge at normal tailwater, are preferred, but in some cases may be impractical. It may be desirable, for repair purposes, to use scavenging eductors or portable, submersible pumps for drying out the sump after bypassing the flow to the dewatering sump. Water-level switches start and stop the first pump as required, with the second pump starting if the first pump fails to handle the inflow. The sump should have a separate float control for a high-water level alarm. Penstock water-driven sump pumps or eductors are secure backups for electric motor-driven sump pumps if the electrical supply is vulnerable to interruption.

Oil entering the sump should be prevented from being pumped into the tailrace. The sump should have a sump dividing wall with horizontal holes in the bottom to establish a "clean" and "dirty" side of the sump. All drain lines should discharge on the dirty side and the pumps should be on the clean side. The dirty side should have ways to remove oil as needed. The sump should be large enough that the dirty side could hold the entire oil volume of the largest transformer or oil storage tank with the water level in the sump initially at the pump shut-off level.

SEWAGE AND WASTE TREATMENT

Sewage Treatment

Domestic sewage treatment is most conveniently handled by connecting to an adjacent sanitary district or municipality's system. In locations where this service is not available, provisions will have to be made at the plant. Figure 9-12 shows a typical sewage treatment system.

Larger plants may need to construct a privately owned sewage treatment facility. Such facilities may require that the treatment plant operators be licensed by the state.

Smaller plants may be able to use a septic tank for sewage disposal. A septic tank is constructed usually of coated metal or reinforced concrete. Both sedimentation and sludge digestion take place in a single compartment. The settled sludge is in contact with the sewage flowing through the tank. Because of these characteristics, the effluent is odorous, high in biological and chemical oxygen demand, and may contain pathogenic organisms. However, a septic tank may be applicable for a plant with very limited requirements. Effluent from a septic tank is disposed of in a tile field. Soil and ground water conditions required for an acceptable percolation test for the drain field may preclude use of a septic tank at the plant site. In this event, a holding tank may be the only practical alternate for a small plant.

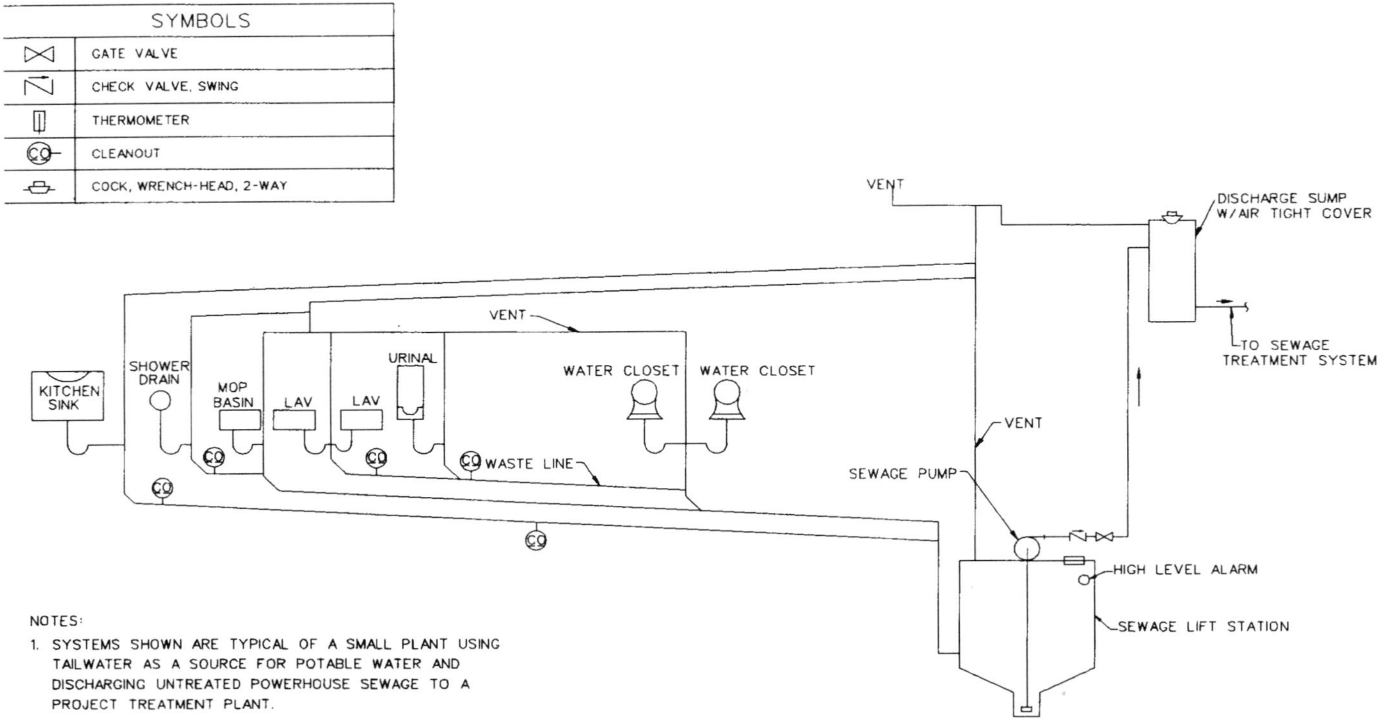

NOTES:

1. SYSTEMS SHOWN ARE TYPICAL OF A SMALL PLANT USING TAILWATER AS A SOURCE FOR POTABLE WATER AND DISCHARGING UNTREATED POWERHOUSE SEWAGE TO A PROJECT TREATMENT PLANT.
2. WATER COOLERS AND BATTERY ROOM FIXTURES DRAIN TO TAILWATER OR DRAINAGE SYSTEM.

FIGURE 9-12: ***Sewage Treatment Schematic.*** *Large plants may need to construct a privately owned sewage treatment facility. This figure shows a typical sewage treatment system. Water discharges may require a National Pollutant Discharge Elimination System (NPDES) permit or treatment prior to discharge. The major process pollutant from a hydroelectric facility is grease and oil.*

Holding tanks may be subject to strict state regulation, such as requirements for a maintenance contract with a licensed disposal operator, before a permit will be granted.

Water Discharges

Water discharges may require a National Pollution Discharge Elimination System (NPDES) permit or treatment prior to discharge. The major process pollutant from a hydroelectric facility is grease and oil. Typical outfalls and parameters to be monitored include:

- Sewage treatment plant for flow, total suspended solids, BOD, fecal coliform, pH, turbidity, odor, and color;
- Monitoring of oil/water separator or sump discharge for oil, grease, or suspended solids; and
- Cooling-water discharges for temperature.

Normally, turbine discharge has not been subject to such requirements. However, a few states have required an NPDES permit for turbine discharge. Quantities that may need monitoring include dissolved oxygen and pH.

Oil Handling Systems

System Requirements

Two basic categories of oils are encountered in hydro plants. One is lubricating-type oils used in the bearings of generators and turbines and in the hydraulic systems of governors. The other is insulating-type oils used in transformers and circuit breakers. Some lubricating applications in the plant also may use grease. Mainly, equipment size determines the quantities of each type of oil, and equipment design and the preferences and maintenance practices of the owner affect the method of handling. Figures 9-13 and 9-14 show typical lubricating and transformer insulating oil systems.

In general, small hydro plants require at most only a few hundred gallons of lubricating oil and often no insulating oil. Large, multiple-unit plants can require tens of thousands of gallons of oil.

Consideration must be given to handling the oils for initial equipment filling and subsequent oil changes. Usually small plants can handle oil using 55-gallon drums and portable pump and filter units for transferring the oil. When oil quantities reach thousands of gallons, however, handling with 55 gallon containers becomes impractical. One alternative method is to deliver oil in mobile rail or road bulk tanks and remove oil for outside treatment by the same means. Portable or permanently installed transfer equipment can transfer the oil within the plant. Another method is to install oil storage tanks at the plant complete with permanently installed oil transfer and treatment equipment. The latter method provides the greatest convenience for maintenance purposes, but incurs a greater initial cost, including the fire protection requirements described in the

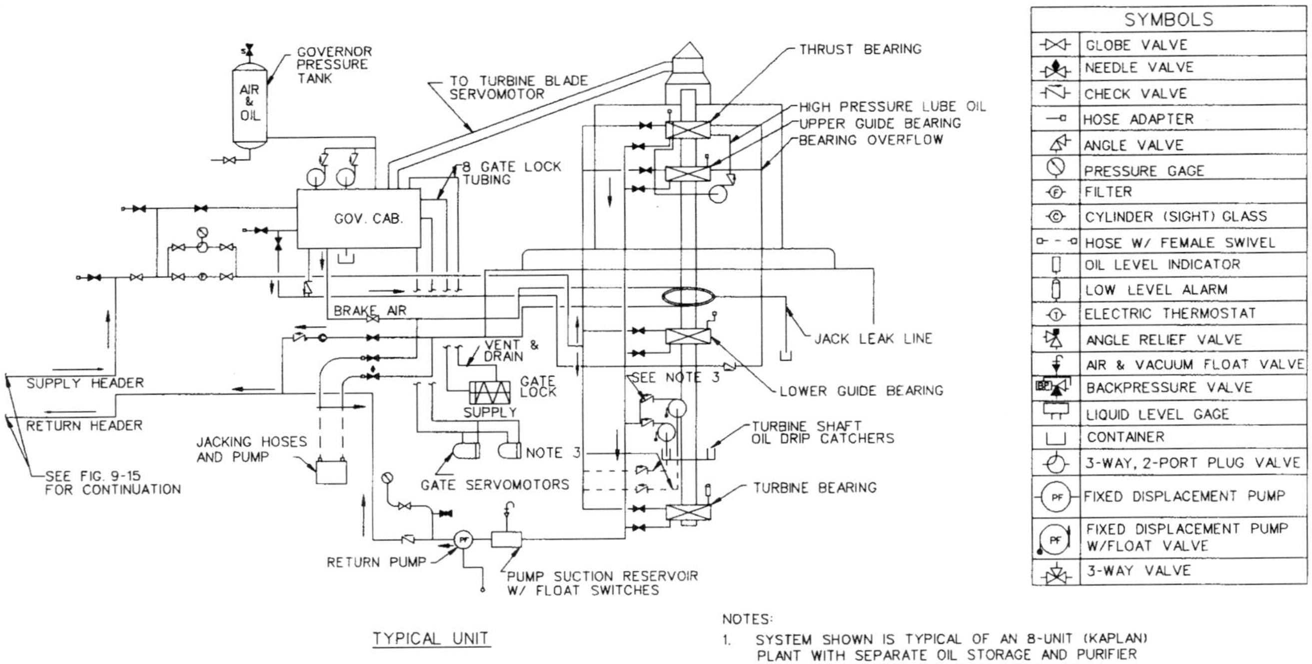

NOTES:

1. SYSTEM SHOWN IS TYPICAL OF AN 8-UNIT (KAPLAN) PLANT WITH SEPARATE OIL STORAGE AND PURIFIER ROOMS, OIL DELIVERY BY TANK TRUCK, AND A POWERHOUSE OIL-WATER SEPARATOR FACILITY FOR SPILLAGE.
2. WASTE OIL PUMP HANDLES GOVERNOR-LUBE OIL AND TRANSFORMER OIL.
3. USE GRAVITY RETURN FROM UNIT TO STORAGE ROOM WHEN TURBINE CONFIGURATION PERMITS.

FIGURE 9-13: ***Typical Lubricating and Governor Oil System Schematic.*** *The generator and turbine bearings and the governor's hydraulic system use lubricating-type oils. Shop-fabricated storage tanks for installation at the plant should meet applicable local codes. Tanks should have gauge glasses or gauge boards as needed.*

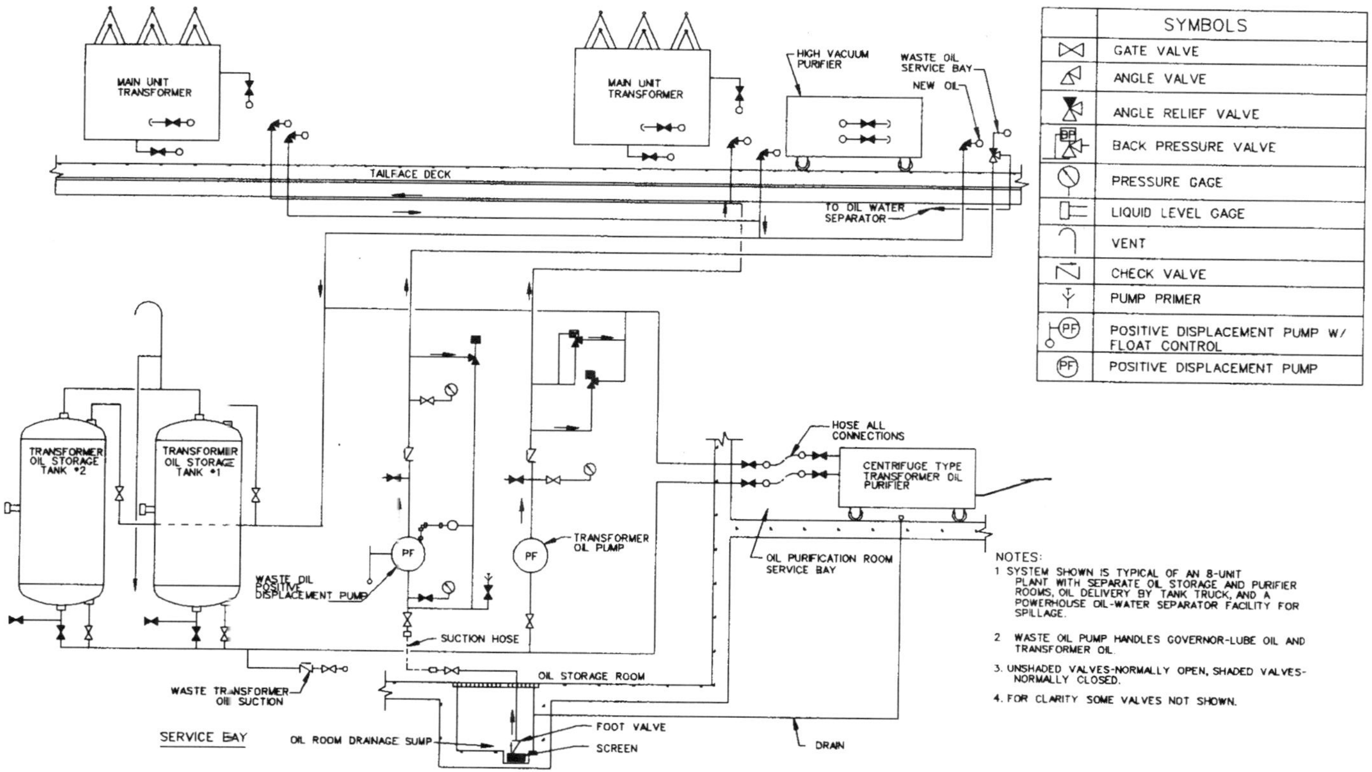

FIGURE 9-14: ***Transformer Oil Piping Schematic.*** *In a hydro plant, transformers and circuit breakers use insulating-type oils. This figure shows the piping in a typical transformer oil system. Tanks for insulating oil may require special breathers to prevent moisture contamination.*

"Fire Protection" section. Figure 9-15 is a schematic of typical oil storage and transfer facilities.

Design Guidelines

Shop-fabricated storage tanks for installation at the plant should be constructed to meet applicable local codes. Open vents/overflow are acceptable where oil is normally exposed to air. Tanks for insulating oil may require special breathers to prevent moisture contamination. Larger tanks will have to be field erected. Tanks should have gauge glasses or gauge boards as needed. Gauge glasses should have excess flow check valves and isolation valves to prevent leakage if the glass is broken.

Pumps for an oil transfer system usually are of self-priming, positive displacement design. Pump suctions should have fine mesh strainers, and discharges to the clean oil tanks and equipment should be through filter presses, cartridge filters, or centrifuges. Some owners require two of the three types of filters. Relief valves should be provided for sections of lines that may be isolated and become over-pressurized due to thermal expansion.

Usually, butt-welded steel transfer piping is used to minimize the risk of leakage, reduce friction losses, and reduce contamination. Expansion loops and offsets should be provided to allow for expansion. Lubricated plug valves offer the advantages of good sealing ability, low flow resistance, and quick operation. Gate valves can be used if tight shut-off is not needed, such as with fill or drain connections, which are usually plugged when not in use.

Air Conditioning Design for Power Plants

General Requirements

Heating, ventilating, and air conditioning (HVAC) for a hydroelectric power plant is similar to an industrial building, except for the special requirements that underground construction or large heat loads from electrical equipment may impose. Many hydro powerhouses are designed for unattended operation and do not require air conditioning for the personnel's comfort. In such cases, air conditioning costs can be avoided if electronic control equipment and communications and computer equipment can operate at ambient powerhouse temperatures. Figure 9-16 shows a typical heating, ventilating, and air conditioning system. In larger powerhouses, HVAC installations should normally move cool, moist air from the lower portions of the powerhouse to the generator floor. Brushes in electrical equipment perform better with a higher relative humidity. A modern plant design should have positive air pressure within the plant to exclude adverse environmental intrusion. This would keep out bugs, rain, and dust.

Heating and Cooling Requirements

Generally, the superstructures of power plants are ventilated at all times but are

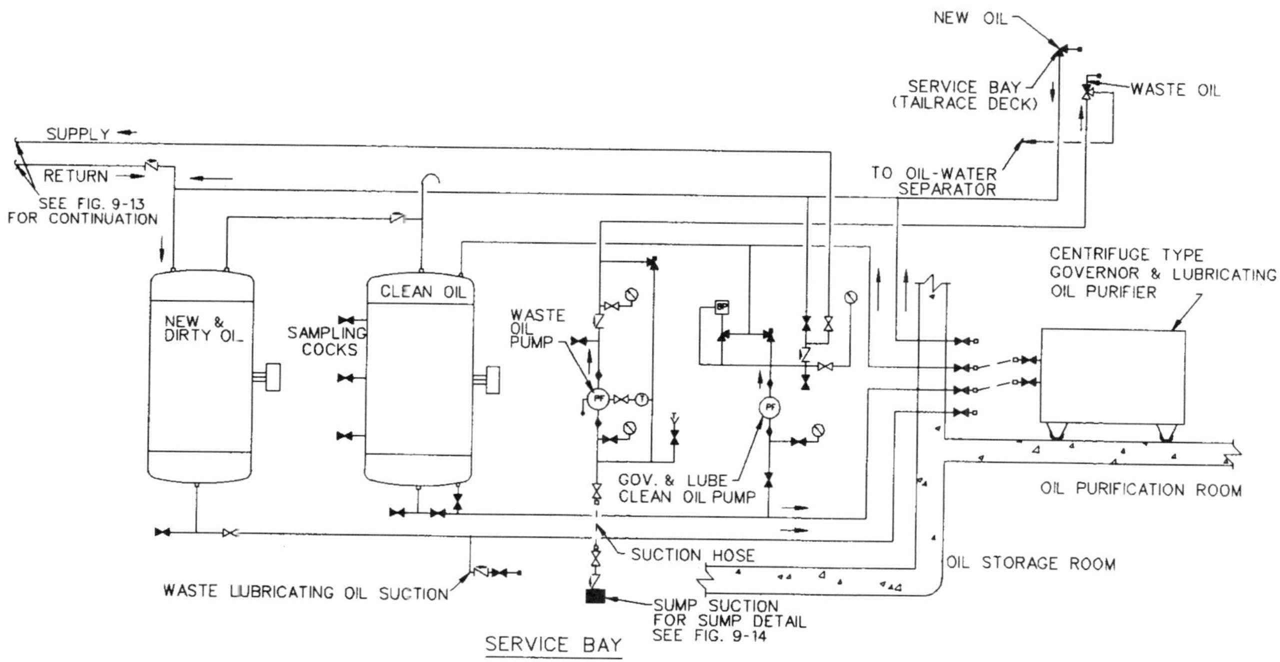

NOTES:

1. FOR SYMBOLS AND NOTES SEE FIGURE 9-13 LUBRICATING OIL PIPING SCHEMATIC.

FIGURE 9-15: ***Typical Oil Storage Room Schematic.*** *This schematic shows the typical oil storage and transfer facilities. Plants, except very small ones, should have adequately sized clean oil and dirty oil storage tanks. Positive displacement pumps should be provided to move oil through filters to storage tanks or bearing sumps.*

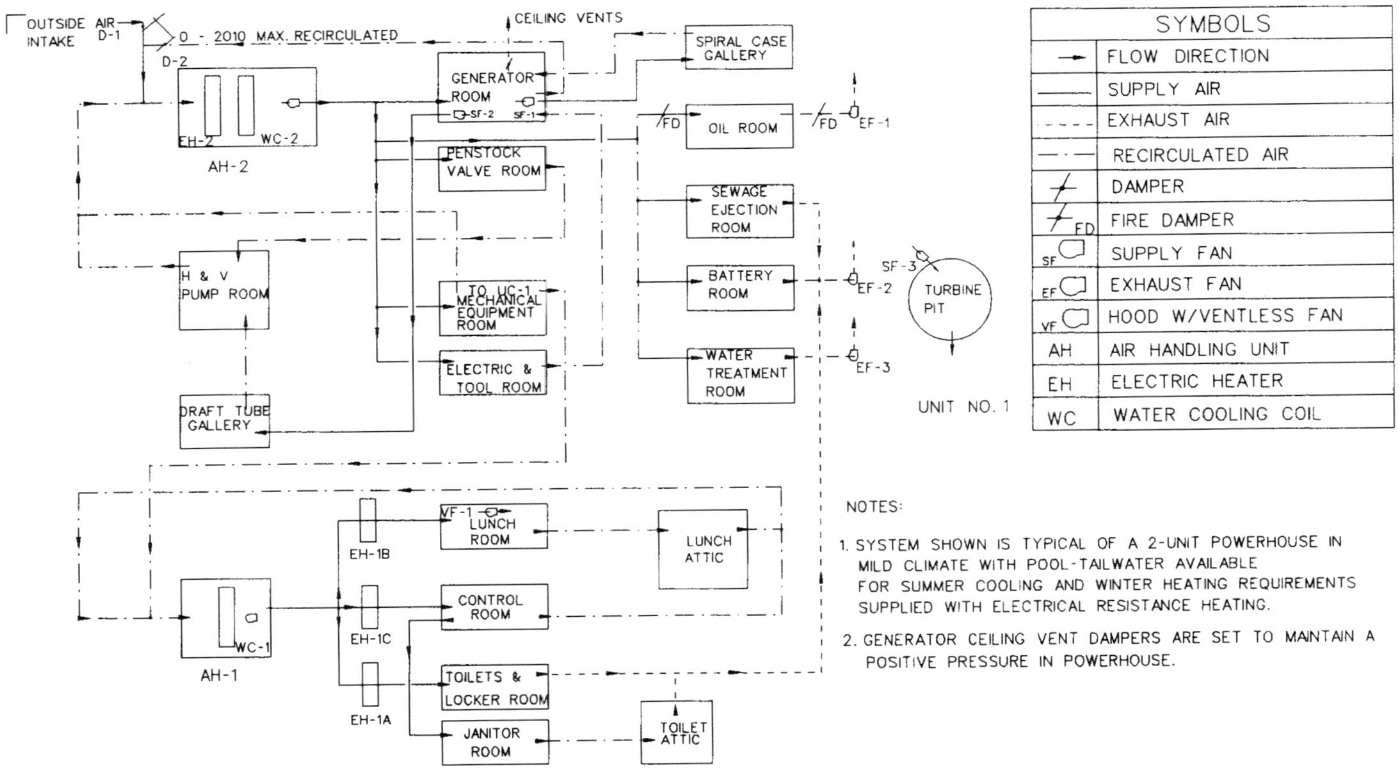

FIGURE 9-16: ***Air Conditioning Schematic.*** *Heating, ventilating, and air conditioning (HVAC) for a hydroelectric power plant is similar to an industrial building. This figure shows a typical heating, ventilating, and air conditioning system. In larger powerhouses, HVAC installations would normally move cool, moist air from the lower portions of the powerhouse to the generator floor. A modern plant design should have positive air pressure within the plant to exclude adverse environmental intrusion.*

not cooled or heated except in cases where:

- Temperatures must be maintained to protect equipment from freezing or excessively high temperatures; or
- Water condensation on metal work will cause damage to the metal work, or water droplets will cause damage to equipment below.

The substructures of power plants generally are heated and cooled to maintain minimum and maximum temperatures for control equipment, freeze protection, and to maintain minimum temperatures for the comfort of maintenance personnel. The minimum temperature is usually 7 to 10 degrees Celsius (45 to 50 degrees Fahrenheit) for equipment protection and 20 degrees Celsius (68 degrees Fahrenheit) for areas occupied by personnel and control rooms.

Generally, electric heaters provide heat. Where possible, the heat produced by electrical equipment should be available to adjust the air used for heating the surrounding areas.

All electric space heating is done by finned resistance coils instead of open element (blast) coils. The heaters generally are installed in the main air-conditioning units or in supply air ducts. When spot heating is necessary, a packaged unit containing both a heating coil and a fan is recommended. Toilet areas can use radiant ceiling heaters, and office spaces can use convection heaters.

Using water directly from the plant cooling water system or electrically driven refrigerating equipment provides cooling for the plant. Cold surfaces of non-embedded water conduits should be brushed with circulating air to remove excess moisture from the air and to cool the air so that the air may be used for cooling areas with high heat sources.

Oil storage tanks, oil transfer rooms, and DC-system battery rooms must have air exhausted continuously to the outdoors. Waste heat from adjacent rooms should heat the rooms since space heaters are seldom installed because of the explosive nature of the stored materials.

Power busses for large generating units may require separate cooling systems designed to handle their heat losses. The plant's air-conditioning units should handle all other loads within the plant.

Air Flow and Air Change Requirements

The following are recommendations for airflows, air changes, and temperatures for different areas in power plants.

Air intakes and exhausts:

Outdoor air intakes should have stationary louvers sized for a maximum air velocity of 2.5 meters/second (8 feet/second) across the gross area of the louver. Exhausts in walls should have stationary louvers sized for a maximum air velocity of 4 meters/second (12 feet/second) across the gross area of the louver.

Airflows for plant air-conditioning supplies:

- Main ducts up to 15 meters/second (50 feet/second), not to exceed a friction loss of 6-millimeter (0.25-inch) water gauge per 30 meters (100 feet);

• Branch ducts up to 10 meters/second (35 feet/second), not to exceed a friction loss of 5-millimeter (0.2-inch) water gauge per 30 meters (100 feet). Normally, the duct size is based on 3-millimeter (0.1-inch) water gauge loss per 30 meters (100 feet); and

• Discharge velocities into a room vary from 3 to 9 meters/second (10 to 30 feet/second) depending on the pattern and desired throw.

Plant exhaust velocities:

• From chlorine rooms, up to 18 meters/second (60 feet/second);
• From oil storage rooms, up to 15 meters/second (50 feet/second); and
• From toilet rooms, up to 6 meters/second (20 feet/second).

Air Changes

Table 9-1 shows the recommended air changes per hour for various areas of a power plant. Also included are some conditions that may have a bearing on the air change requirements. The information in this table evolved at the Bureau of Reclamation Denver Office. For all floors and galleries, the heat gain to each space, the air volume available, and the air temperature determine the air changes per hour.

Temperatures

Table 9-2 shows the recommended maximum and minimum temperatures that the designer should provide in various areas of a power plant. The information in this table evolved at the Bureau of Reclamation Denver Office.

TABLE 9-1: ***Suggested Air Changes for Various Power Plant Areas***

Space	Air Changes per Hour	Conditions When Air Changes Should Be Applied
Oil storage and transfer	6	Continuous
Toilets	12	When occupied and 15 minutes thereafter
Paint storage	6	Continuous
Paint booths	30	When painting
Welding rooms	6 to 12	When welding
Battery rooms		Minimum or sized to maintain hydrogen gas concentration below 1 percent by volume during maximum gas generation conditions
Chlorine storage indoors (Not allowed inside power plants in the U.S.)	6	Continuous
Generator floors	1	Dependent on heat gain from equipment
Sewage handling rooms	3	Continuous

TABLE 9-2: ***Maximum and Minimum Temperatures For Various Power Plant Areas***

Space	Minimum °C	Minimum (°F)	Maximum °C	Maximum (°F)
Oil storage and transfer	7	45	26	80
Toilets	18	65	29	85
Paint storage	7	45	26	80
Paint booths	18	65	26	80
Welding rooms	18	65	26	80
Battery rooms	10	50	26	80
Chlorine storage outdoors		*No requirements*		
Chlorine storage indoors	7	45	32	90
Generator floor floor level	7	45	26	80
Generator floor ceiling level	—	—	48	120
Pipe galleries	7	45	26	80
Electrical galleries	7	45	40	104
Visitor facilities	20	68	26	80
Control rooms	20	68	26	80
Computer rooms	22	72	25	78
Uninterruptible power supply rooms	18	65	32	90
Lunchrooms	20	68	26	80
Offices	20	68	26	80

TABLE 9-3: ***Suggested Maximum and Minimum Humidity for Various Power Plant Areas***

Space	Minimum percent relative	Maximum percent relative
Pipe galleries	—	80
Electrical galleries	40	80
Visitors facilities	30	80
Control room	40	80
Computer rooms	45	55
Offices	30	70
Lunchrooms	30	70
Intake air supply	—	80

Humidity

Table 9-3 shows the recommended maximum and minimum humidity that the designer should try to provide in various areas of a power plant. The information in this table evolved at the Bureau of Reclamation Denver Office.

The air-conditioning systems should pressurize the structure to restrict dust and drafts from entering through door cracks and other openings. To pressurize the building, an amount of outside air should be introduced into the building through the main air conditioning units that is equal to the amount of contaminated air being exhausted plus a minimum of one air change per hour for the generator floor. The amount of outdoor air should be balanced to give a gauge pressure difference of between 2 and 4 millimeters (0.08 and 0.16 inch) of water.

It may be desirable to pressurize the control room more than the rest of the plant to minimize dust accumulation.

Design Standards

The current Handbook of Fundamentals from the American Society of Heating, Refrigerating and Air Conditioning Engineers (ASHRAE) should be used as the reference for designing the HVAC system in the power plant.[2] This handbook contains information on many subjects that play a part in designing air-conditioning systems. These subjects should be reviewed, as necessary, to develop the background for good air-conditioning design. Also, the other ASHRAE volumes (such as Applications, Equipment, or System) can help in designing the power plant's HVAC system.

Controls

The heating, ventilating, and air-conditioning equipment should be controlled automatically to operate as an integrated system. If the design treats each unit bay as a system, the supply air equipment, all circulating air equipment, and all exhaust air equipment other than contaminated air exhaust equipment should be interlocked to operate with the others. This means that the exhaust fans and exhaust dampers should be controlled to exhaust an amount of air equal to the amount of outdoor air being introduced into the structure by the intake air dampers. Additionally, when the supply air unit is operating, the circulating fans should operate.

One sensor (thermostat) should control the heating, ventilating, and cooling stages of the main air-conditioning unit so that manual adjustments of separate sensors cannot cause overlapping of the stages. The controls should allow for using outdoor air to heat and cool the structure whenever possible to reduce the amount of energy required.

Air-conditioning controls should be integrated with the fire protection system to provide smoke and heat venting and safe conditions for exiting under fire conditions.

Monitoring Equipment

The usual mechanical variables that need to be measured in a hydroelectric plant include level, flow, pressure, speed, vibration, and temperature. Typical instruments used for these measurements are thermocouples, resistance temperature detectors (RTDs), dial thermometers, Bourdon tube pressure gauges, manometers, pressure transducers, and flow elements.

Level Measurement

The headwater and tailwater gauge wells should be located where they can provide readings for the full range of fluctuation. Typical ranges for headwater would be about 0.6 meter (2 feet) above maximum to 1.2 meters (4 feet) below minimum headwater. For tailwater, the corresponding range would be about 1.5 meters (5 feet) above maximum to 0.6 meter (2 feet) below minimum. The gauge wells and piping should be located where the water is free from debris and excessive turbulence. Instruments used for this measurement may include float gauges, pressure transmitters (possibly in a bubbler configuration), and various types of electronic level transmitters, such as capacitance or "sonar" devices.

Other places for using level instruments may include hydraulic oil tanks, bearing oil reservoirs, oil storage tanks, draft tubes, and sumps. Float switches often will be used for control of pumps or alarms in these applications.

Flow Measurement

Traditionally, flow has been measured with a meter calibrated to differential pressure taps or measured outside the powerhouse at a gauging station near the power plant. Recently, acoustic flow meters have been used to measure flow. The data from these meters are stored and used as evidence of water delivery.

Pressure Measurement

A Bourdon tube pressure gauge or pressure transducer connected to a digital display commonly measures pressure. The pressure gauges normally are located close to the penstock net-head piezometric taps or on the governor control panel. The digital output of the pressure transducer is easier to transmit long distances from the measuring point. The display usually is located in the control room.

Speed Measurement

Speed measurement in large powerhouses is usually a direct measurement of electrical frequency and is read out in percent of synchronous speed or rpm. Some smaller plants have mechanical tachometers reading directly in revolutions per minute. The speed readout is normally located on the governor control panel and also in the control panel in the control room.

Provisions for Index Testing

Hydraulic turbines often have provisions for index testing, usually with Winter-Kennedy taps in the spiral case. When specified, the turbine vendor will determine the locations of the taps. Care must be taken to ensure that the piezometric taps are surrounded by a smooth surface and are located away from changes in contour that may affect the flow. Suction elbow taps may be desirable for a pump turbine. Additional details concerning index testing will be found in Chapter 13, Inspection and Testing.

Temperature Measurement

Temperature elements often will have to be placed in wells. A well provides a pressure seal and mechanical protection for the measuring device. Design of thermometer wells involves some compromise because the factors required to produce adequate strength tend to reduce the accuracy and response of the temperature element. Temperature elements used in hydro plants may include dial thermometers, vapor- or gas-filled bulbs, thermocouples, and resistance temperature detectors.

Runout or Vibration Monitoring

Proximity sensors mounted near the shaft or seismic pickups or vibration switches mounted on bearing caps can monitor shaft runout or vibration. If the equipment is unattended, the designer should give extra consideration to equipment monitoring.

Some turbines experience a rough running condition because of draft tube pulsations at part load. This can damage the unit. Often, the load at which this occurs will vary with head. An accelerometer or pressure transducer mounted on the draft tube can detect this condition. The ability to telemeter this data to the remote operator of an unmanned plant may be an aid in loading the machine to avoid difficult operation conditions.

References

[1] *Recommended Practice for Fire Protection in Hydroelectric Generating Plants,* NFPA 851, National Fire Protection Association, Quincy, Massachusetts, current version.

[2] *ASHRAE Handbook of Fundamentals,* American Society of Heating, Refrigerating and Air-Conditioning Engineers, Atlanta, Georgia, current version.

Chapter 9

Bibliography

ANSI/ASME B31.1, *Code for Pressure Piping-Power Piping,* American National Standards Institute, 1977.

ASHRAE Handbook of Applications, American Society of Heating, Refrigerating and Air-Conditioning Engineers, Atlanta, Georgia, current version.

ASHRAE Handbook of Equipment, American Society of Heating, Refrigerating and Air-Conditioning Engineers, Atlanta, Georgia, current version.

ASHRAE Handbook of Systems, American Society of Heating, Refrigerating and Air-Conditioning Engineers, Atlanta, Georgia, current version.

ASME Boiler and Pressure Vessel Code, Section VIII, American Society of Mechanical Engineers, New York, current version.

ASME PTC 19.2-1964, *Supplement on Instruments and Apparatus,* Part 2, Pressure Measurement, American Society of Mechanical Engineers, New York, 1964.

ASME PTC 19.3-1974, *Performance Test Codes Supplement on Instruments and Apparatus,* Part 3, Temperature Measurement, American Society of Mechanical Engineers, New York, New York, 1974.

Bean, H.S., *Fluid Meters,* Chapter II-II, "General Requirements for Fluid Metering: Installation," American Society of Mechanical Engineers, New York, New York.

Flow of Fluids through Valves, Fittings, and Pipe, Technical Paper No. 410, Crane Co., New York, New York, 1969.

IEC Publication 41, *International Code for Field Acceptance Test of Hydraulic Turbines,* International Electrotechnical Commission, Geneva, Switzerland, current version.

IEC Publication 198, *International Code for Field Acceptance Test of Storage Pumps,* International Electrotechnical Commission, Geneva, Switzerland, current version.

Murdock, J.W., "Power Test Code Thermometer Wells," *Transactions of the ASME Journal of Engineering for Power,* American Society of Mechanical Engineers, New York, New York, October 1959.

Piping Design and Engineering, ITT Grinnell Industrial Piping, 1976.

Westway, C.R., and A.W. Loomis, editors, *Cameron Hydraulic Data,* 15th Edition, Ingersoll-Rand, Woodcliff Lake, New Jersey, 1977.

Winter, I.A., "Improved Type of Flow Meter for Hydraulic Turbines," *ASCE Proceedings,* Volume 59, No. 4, Part 1, April 1933.

Checklist for Chapter 9

Safety

1. Are adequate pressure relief devices provided for tanks, receivers, and isolated sections of piping?
2. Are high energy sources (for example, penstock connections) provided with means for isolation at the source?
3. Are fire protection, fire protection water supply, and ventilation systems adequate for emergency conditions?

Environmental

1. Are means provided to collect and treat potential sources of pollution (such as oil spills)?
2. Are adequate provisions available to contain major contamination (such as a tank or transformer failure or contaminated water from fire fighting)?

Location

1. Is equipment located with adequate room for ventilation, operation, and maintenance?
2. Is equipment located where it will not be subject to damage from flooding?
3. Is equipment located and mounted to avoid vibration problems?

Capacity

1. Do auxiliary systems have adequate capacity for both normal and emergency requirements?
2. Is the capacity of pumps and air compressors coordinated with the volume of sumps and air receivers to allow long enough running time for proper lubrication and avoiding short-cycling of equipment?
3. Are backup power supplies and equipment provided where continued operation of a system is vital?

Compatibility

1. Are materials used for piping suitable for use with the material handled?
2. Are motor and equipment enclosures appropriate for their environment?
3. Are motor electrical ratings (voltage, frequency, phase, and starting requirements) correct for the auxiliary power system at the plant?
4. Is electronic equipment suitable for the environment or located in air-conditioned areas?

Equipment Protection

1. Are strainers provided, as needed, to protect equipment from contaminants?
2. Are sight gauges or flow switches provided to monitor critical flows?
   Is freeze protection or heat tracing provided where required?

Equipment Monitoring

1. Are provisions made for contractual performance testing?
2. Can the equipment be operated remotely?

Chapter 10
Equipment for Environmental Control

This chapter discusses the available equipment for monitoring or mitigating the effects of hydroelectric projects on the environment. This chapter does not address the methods used to evaluate environmental effects and the appropriate mitigation measures of hydroelectric project construction and operation. State and federal statutes cover the methods used to evaluate environmental effects and include review by state and federal resource agencies. While limited to a discussion of available equipment, this chapter contains a brief discussion of hydroelectric-related environmental effects to clarify the purpose of the equipment.

Effects to Streamflows

Hydroelectric dams can significantly affect streamflows both upstream and downstream of the dam site. Dam construction typically requires constructing a temporary cofferdam, which can block or divert flows from the main river channel. Wetlands, aquatic organisms and their habitat, and water quality (through increased turbidity and siltation and, in some instances, the release of toxic sediments) can be adversely affected by such activity. The effect can depend on the length of the bypass reach, the extent and duration of the diversion, and the contribution of local tributaries and ground water to flows in the bypass reach.

Hydroelectric projects also affect streamflows by impounding upstream waters and, depending on the design and operation of the project, alter the downstream flow regime. Run-of-river projects are operated such that incoming flows are not stored in large quantities. Peaking projects typically store flows for subsequent generation release during periods of high electricity demand. The operation of peaking projects can result in sudden fluctuations in water levels downstream, scouring of the stream bottom, bank erosion, and increases in downstream turbidity. Furthermore, drastic fluctuations in peaking flows can adversely affect downstream aquatic and riparian habitats and, subsequently, fish and invertebrate populations.

Regardless of the type of project, project operation mostly affects stream reaches where flows are bypassed for power generation. As indicated previously, operation can affect biological and physicochemical parameters. However, other factors that could be affected include agricultural water use (for example, irrigation), recreation, aesthetic quality (due to the loss of waterfalls, exposure of the stream bottom, and eroded stream banks), and cultural resources (erosion and destruction of archaeological sites).

Establishing a minimum flow in the affected stream reach can mitigate the effects of project-related changes in the streamflow regime. The needs of the downstream resources determine the required timing, duration, and flow rate. In any case, the needs downstream of the diversion structure are evaluated against the project's power generation benefits to determine the best means of using the available streamflow.

EFFECTS TO AQUATIC COMMUNITIES

Hydroelectric projects can affect aquatic communities in a number of ways. The primary effects are:

- Fish entrainment and impingement;
- Effects from turbine;
- Reservoir fluctuations;
- Blockage of fish migration routes; and
- Flow-related changes to aquatic habitats.

Fish Entrainment and Impingement

The entrainment of fish that are drawn into hydraulic turbines by the water flowing into the turbine intake affects fish and fish populations in two ways. First, large numbers of juvenile fish may be discharged from an impoundment into the downstream waters through hydraulic turbines. This is usually restricted to a short time in the spring and early summer. As summer progresses, the number of fish that are susceptible to entrainment declines rapidly due to natural mortality and the increased capability (associated with growth to a larger size) of the fish to avoid environmental hazards.

The second potential effect on fish is individual injury. Fish entrained in the water passing through a turbine may encounter several potential hazards. The potential for a hydroelectric project to entrain fish depends on several variables. The main factors include the inlet velocity, volume of flow, hydraulic head, depth of the turbine intake, and the size and nature of the fish population. Other features of the unit, such as the spacing of the trashrack bars, may also influence the size of the fish that may be entrained. Fish small enough to pass between trashrack bars may be able to sense the turbulence caused by the water passing through the rack and avoid being entrained.

Impingement or collisions of fish on screens or trashracks is not a problem at most hydroelectric plants simply because most of them do not have screens fine enough to cause impingement. Usually, fish impingement is associated with small mesh screens. In the Pacific Northwest, a number of hydroelectric units have been equipped with flat tilting, wedgewire panels to screen out downstream-migrating salmon smolts. These facilities have successfully diverted the smolts into a bypass conduit that discharges to the tailwater area downstream without impingement.

Turbine Effects

Several factors cause injuries to fish passing through hydraulic turbines. Strikes by turbine blades, collisions with support members, abrasion, cavitation, changes in direction of flow, and pressure changes are the major sources of injury to entrained fish. The type of turbine, number of blades, speed of rotation, cavitation, hydraulic head, depth of intakes, and the species and size of the fish determine the frequency and severity of these injuries. Cavitation can be important because it may subject passing fish to sudden, intense forces.

Propeller-turbine runners have three or more blades, which are either fixed at one blade angle or adjustable over a range of blade angles (Kaplan turbines). Francis turbine runners have curved buckets instead of blades, and the number of buckets on Francis runners is greater than the number of blades on propeller runners. It is generally believed that propeller and Kaplan turbines pose less threat to fish than Francis turbines because they have fewer blades.

The survival rates for fish passing through hydraulic turbines also generally correlate with the efficiency of the turbines. Therefore, Kaplan turbines, which have a higher efficiency over the full turbine operation load range than Francis or fixed blade propeller turbines, pose the lowest threat to fish.

The hydraulic flow through axial bulb and tube-type turbines is more linear than for units having a spiral case, semi-spiral case, or integral intake. Consequently, there is a lower potential for injury of fish passing through axial turbines. While the fish passage studies conducted to date are not conclusive, the following general factors have been identified that reduce the threat that turbines pose to fish mortality:

- Axial turbines;
- Larger turbines;
- Slower, rotating speeds;
- Fewer blades or buckets; and
- Increased efficiency.

Reservoir Fluctuations

The water level in reservoirs formed by dams may fluctuate widely, especially for storage projects. Typically, run-of-the-river projects are operated with stable pools. However, dams with flashboards, which can collapse or be lowered when subjected to flood flows, require the pools to be drawn down below the crest of

Chapter 10

the dam for flashboard installation. These drawdowns are normally of minor magnitude and short duration.

The most probable effect of these maintenance-related drawdowns to fish populations is the exposure of fish eggs during the spawning season. Short-duration drawdowns of 1 to 1.2 meters (3 or 4 feet) appear to have no effect on the adult fish populations or the fishery.

Hydroelectric projects that store water in the spring for power generation during the ensuing year may cause dramatic drawdowns during drought years with some associated effects on aquatic ecosystems. However, these projects are designed and licensed to operate in this manner and any fish habitat, fish production, and fishery that results from this operation may be a net gain over conditions without the project.

Pumped-storage hydroelectric projects may cause water level fluctuations in two water bodies. Usually, an upper reservoir, or forebay, is constructed to receive water pumped from a lower body of water, or afterbay, during off-peak hours. The process is reversed to generate power during peak generation needs. Man-made upper reservoirs often are sized for one or two days of generation storage. Consequently, the man-made reservoirs do not have significant aquatic ecosystems established. The lower reservoir, or afterbay, also may fluctuate to some extent during pump-back/generation cycles. In general, the afterbays typically have large storage capacities, and pump-back does not significantly affect water level fluctuations. Pumped-storage project designs and license conditions typically include the limits of permissible reservoir fluctuations.

Blockage of Fish Migration Routes

One of the most recognized effects associated with hydroelectric projects is that of blocking fish migration routes. This typically involves the upstream migration of anadromous fish during the spawning season, but may involve other species that undergo extensive spawning movements.

Also, the presence of a hydroelectric facility can impede downstream movements. In some cases, the only means of downstream passage is through the facility's water intake system, resulting in fish mortality as discussed previously.

Several methods can facilitate the upstream and downstream passage of fish. These include various types of fish ladders, fish elevators, pumps, screens, and other devices. These measures are discussed later.

Flow-Related Changes to Aquatic Habitat

As previously indicated, the presence of hydroelectric dams can significantly affect the flow regime of a stream and aquatic habitat both upstream and downstream of the project dam. These effects are discussed below.

Upstream effects. The construction of a hydroelectric dam typically results in the inundation of terrestrial habitat upstream of the dam. The amount of habitat that is inundated depends primarily on the size of the reservoir and the topography and vegetative patterns of the flood plain. For example, a reservoir constructed in a relatively flat flood plain may result in the loss of more land than a similar sized reservoir built in a sparsely vegetated, steep sided canyon. However, flooded vegetation can serve as valuable aquatic habitat for food, cover, and reproduction for a variety of aquatic organisms. Although much of the organic material that is inundated decomposes within a few years after inundation, some materials such as tree stumps and logs can provide cover for a much longer time.

Creating a reservoir also can result in the inundation of existing wetland areas. In some cases, however, the reservoir creates new wetlands in areas adjacent to the reservoir that were formerly well drained.

Freshwater streams generally contain pool and riffle sequences that provide a wide variety of water depths, flow velocities, and substrate materials. Such variety in habitats gives rise to highly diverse communities of aquatic organisms. The inundation of a stream reach can drastically affect aquatic habitat by altering flow patterns, flow velocities, water depths, water temperature, dissolved oxygen (DO) concentrations, and other habitat characteristics. The comparatively slow flow velocities in reservoirs can allow sediments suspended in the water column to settle out and cover the stream bottom. Consequently, the composition of the aquatic community can change such that species adapted to lake environments are favored, while species that require stream conditions are reduced in number or eliminated altogether.

Although the impoundment of a stream can result in adverse environmental effects, some beneficial environmental effects also can be realized. If public opinion and the management goals of the resource agencies favor the development of "flatwater" recreational opportunities, stream impoundment may be viewed favorably. In reservoirs that thermally stratify, "two-story" fisheries can be established with warm-water species occupying the upper, warmer part of the water column and cold-water species (i.e., trout) occupying the lower part.

Downstream effects. In addition to affecting upstream conditions, stream impoundment can significantly alter downstream habitat conditions. Reduced streamflows in bypassed stream reaches can reduce the quantity and degrade the quality of available aquatic habitat. Rapid fluctuations in streamflows (such as those associated with peaking projects) can temporarily increase the amount of available habitat during power generation, allowing aquatic organisms to move into the newly inundated areas. Sudden reductions in streamflow associated with reductions in power generation can strand these organisms or their eggs on the exposed substrate and result in death due to desiccation, suffocation, or predation. Also, aquatic organisms can be trapped in isolated pools and exposed to overcrowding, predation, excessive temperatures, and low oxygen concentrations.

It is possible to operate some hydroelectric projects in such a way to stabilize downstream flows and reduce seasonal fluctuations in aquatic habitat. This can enhance downstream aquatic habitat during "dry" periods. In many streams, however, the naturally high flows occurring in the spring flush out sediment and other materials deposited during the previous summer and fall seasons. This exposes stream substrates and makes the substrates available for use as spawning habitat and cover by fish and invertebrates. Reduced streamflows can result in the accumulation of excessive amounts of sediment and organic material in the bypass reach and affect the amount of available spawning habitat and cover.

High spring streamflows also can carry sand and gravel from upstream which, when deposited further downstream, provide additional spawning substrate and cover or replace materials that were flushed further downstream. Sand and gravel flushed from upstream can be trapped in the reservoir, making these materials unavailable to downstream areas as spawning substrate or cover.

Stream impoundment also can benefit downstream resources. Reservoirs can act as sediment "traps" and reduce the effects of siltation on downstream aquatic habitat. Reservoir tailwater areas can serve to attract and to concentrate fish and provide additional recreational fishing opportunities. Large impoundments can thermally stratify in the summer, forming a warm-water layer on the top and a cold-water layer on the bottom. Water released from the lower, cold layer can provide an ideal downstream environment for cold-water species such as trout. In many cases, reservoirs have allowed the development of "world class" trout streams due to the ability to control the temperature of water releases. At the same time, however, downstream conditions may become less suitable for warm-water species and affect the viability of the local warm-water fish community.

Effects to Water Quality

For the most part, the most significant effects of hydroelectric projects on water quality result from temperature changes and dissolved oxygen (DO). The disturbance of contaminated sediments also may be of concern in some areas.

Temperature

Reservoirs typically increase the retention time of incoming flows, resulting in increased water temperatures. Depending on the depth of the reservoir, local climatic conditions, and other factors, the water column can thermally stratify, usually in the summer and early fall. Releasing these waters from the reservoir can, in turn, increase or decrease the temperature of waters in downstream reaches, depending on the depth at which the water is withdrawn from the reservoir. Thermal stratification persists until heat loss from the warm layer and wind-induced turbulence combine to cause a mixing of the water layers or "turnover." Then, the warm layer temperature is approximately the same as the cold layer temperature.

Reduced flows in the downstream reaches resulting from project-related flow diversions also can increase temperatures in downstream areas, particularly in shallow pools.

Dissolved Oxygen

As indicated previously, inundation of terrestrial vegetation usually accompanies dam construction. Localized depletions in dissolved oxygen (DO) concentrations can result as organic material decomposes. However, this effect is relatively short term in nature.

In many cases, depth-specific differences in DO concentrations accompany thermal stratification of a reservoir. The photosynthetic activity of phytoplankton and aquatic vegetation and the turbulence resulting from wind and wave action and incoming streamflows are important in maintaining DO concentrations in reservoirs. During periods of thermal stratification, these factors maintain relatively high DO concentrations in a reservoir's warm layer. The cold layer, however, is isolated from turbulence-producing atmospheric activities by the warm upper layer. Oxygen consumption continues in the cold layer because of the respiration of aquatic organisms and the decomposition of organic matter. Typically, the cold layer is deep enough to limit light penetration and to minimize photosynthetic activity and oxygen production. Consequently, DO concentrations in the cold layer are lower than those in the warm layer and are sometimes unsuitable for many aquatic organisms to survive. Following "turnover," DO in the reservoir generally returns to uniformly high concentrations.

Waters discharged from the lower cold layer can, in turn, affect downstream DO concentrations. Low DO concentrations can accompany cold layer discharges that provide suitable temperatures for fish species not normally occurring downstream as discussed above. Under such circumstances, resident as well as stocked species can be affected.

Gas Supersaturation

Gas supersaturation, often associated with high-head hydroelectric projects, occurs when air and water are mixed under pressure. This results in high concentrations of dissolved gases. Supersaturation at hydroelectric projects typically occurs when air is entrained in the spillway discharge. Spillway aeration (to reduce cavitation) and dispersions and stilling basin turbulence cause air entrainment. Air entrainment in unit intakes also can occur with lower reservoir levels and result in air supersaturation in the penstock.

Gas supersaturation can result in fish mortality and is of particular concern in areas that support salmonid populations. Potential mitigation measures may include altering spillway design and controlling the water level fluctuations at penstock intakes.

Contaminants

Project construction and operation activities can sometimes result in the spillage of oil, fuel, and other petroleum materials. Turbine wicket gate bushings often are lubricated with grease, and turbine bearings and servomotors use oil. As a result, the turbine may release small amounts of oil and grease during operation.

Other Effects

Older hydroelectric projects often contain products having materials that have been found to pose health risks to personnel. The two most common materials are asbestos and the polychlorinated biphenyls (PCBs). Mercury or other contaminates may have drained into project reservoirs.

In the past, asbestos was commonly used for cable insulation, generator stator coil insulation, ceiling insulation, gaskets, pipe insulation, brake shoes, and other similar applications. When asbestos products are identified in a power plant, utilities commonly hire experienced specialty contractors authorized by the responsible governmental agencies to remove it. It is preferable to remove asbestos to minimize any future risk to plant personnel and the surrounding environment.

Oil-filled electrical equipment such as transformers, breakers, and switches often contained a PCB additive as a flame retardant in case of combustion. As with asbestos, utilities generally have PCB-containing oil or equipment removed from the plant to prevent potential contamination.

Reservoir construction or sedimentation removal may contain contaminants such as mercury. Prior to reservoir construction or reservoir sediments removal, a sampling and testing program should be conducted to determine the presence of any hazardous materials. Governmental agencies specify the sampling procedures and disposition of hazardous materials in the United States and other countries.

Remedial Measures and Related Equipment

The previous section discussed the environmental effects of hydroelectric projects. The following sections present the measures and equipment that can mitigate these effects.

Stream Gauges and Streamflow Monitoring

Streamflow is the rate of flow of water in a stream or river. Measurements of stream stage or elevation usually determine the streamflow because of the difficulties in performing continuing measurement of streamflow. A current meter or other hydrology measuring methods correlate the stream stage to the streamflow.

Many hydroelectric projects are required to maintain specific minimum streamflows. While operators typically calculate the required flow from hydraulic turbine or gate discharge design curves, project owners often are required to monitor the project flows at a downstream location.

The simplest form of a water elevation or streamflow gauge can be a staff gauge and scale. Readings are taken manually at prescribed intervals. Staff gauges also are used to monitor the elevation of the headwater pool.

It is often desirable or required that streamflow or water surface elevations be monitored and recorded continuously. Float-type water elevation measuring equipment typically measures and records tailwater and headwater elevations.

These devices often are installed by governmental agencies to provide hydrologic records of the flow in rivers and streams.

Fish Protection Systems

Currently, several types of fish protection systems are being used or tested at hydroelectric projects. The major categories of fish protection systems are:

- Fish ladders;
- Fish elevators and locks;
- Fish bypass systems;
- Fish pumps;
- Traveling screens; and
- Fish screens.

Fish Ladders

Fish ladders provide a "passive" means of passing fish upstream around a dam. A fish ladder consists of a series of pools, each of which is slightly higher in elevation than the preceding pool. The fish ladder is open at both the top (upstream) and the bottom (downstream) ends. Water is passed through the ladder allowing fish to enter at the bottom, swim through the series of pools, and exit at the top.

Two of the most critical factors of a fish ladder design are the location of the entrance of the ladder and establishing an appropriate attraction flow. Without proper planning, the most sophisticated fish ladder system is useless if the fish cannot "find" the ladder entrance. In cases where upstream fish passage is a concern, a hydroelectric developer should consult with a qualified consultant prior to proceeding with project design and construction.

Figure 10-1 shows the three most basic fish ladder designs: the pool and weir, vertical slot, and Denil fishways.

The pool and weir design, probably the oldest fish ladder design, cannot operate under fluctuating heads, but it is often fitted with a regulating pool at the upper end of the ladder to remedy this problem.

By contrast, the vertical slot design, frequently used at facilities along the U.S. West Coast, can operate under a wide range of streamflows and does not require a regulating section. The vertical slot design is more complicated than the pool and weir design.

The Denil design has been implemented successfully, but generally is effective only with limited reservoir fluctuations. The Denil fish ladder must be kept clean and free of debris accumulation and requires more frequent maintenance than the other designs.

Figure 10-2 shows the typical location of fish passage facilities.

Obviously, when selecting a fish ladder design, one must consider the species of fish to be passed upstream. Also, the state and federal fish and wildlife agencies may have specific preferences for fish ladder design and should be consulted prior to expending a large amount of resources on developing the facility.

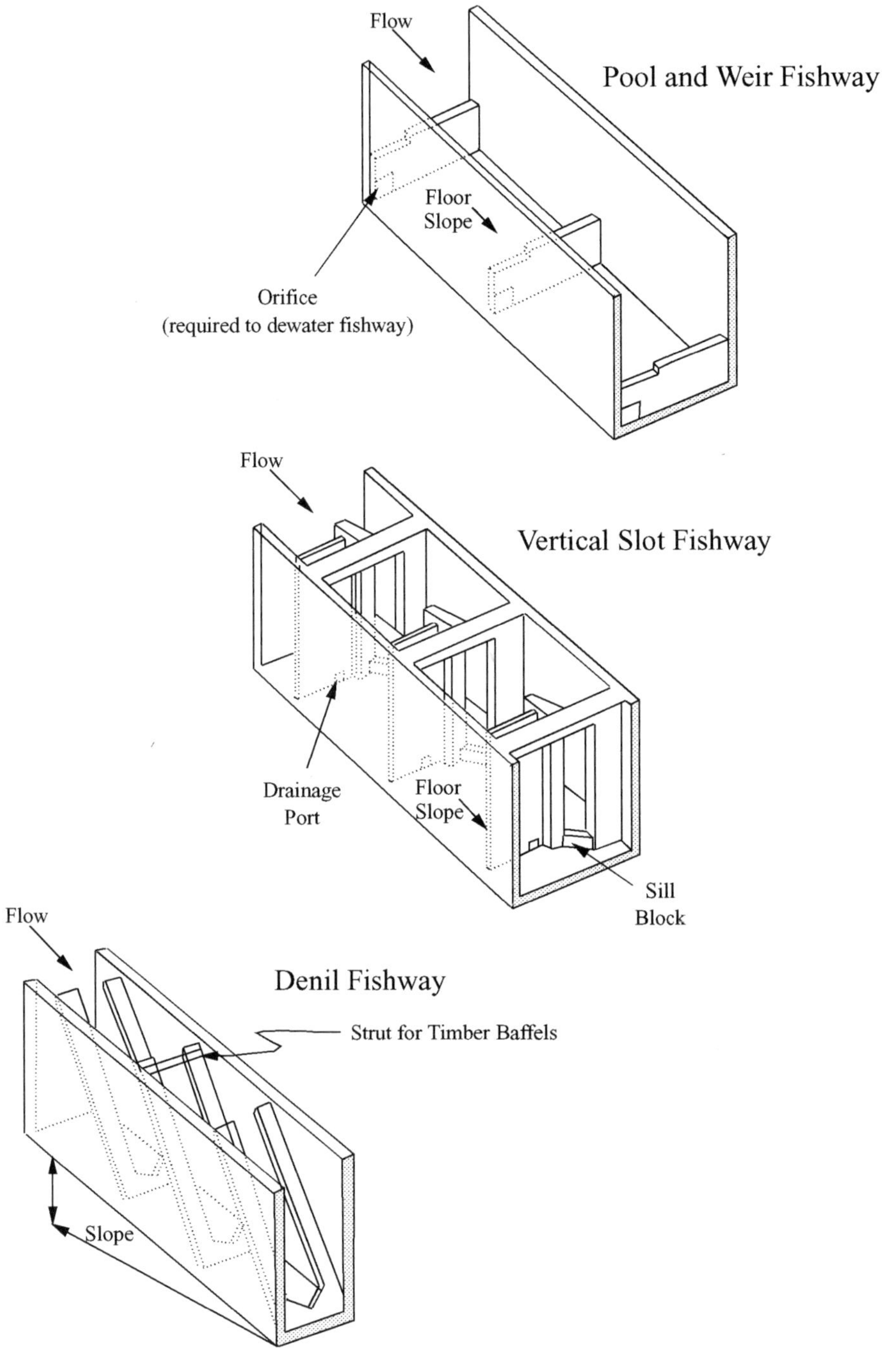

FIGURE 10-1: *Fish Ladder Designs.* *These isometric drawings show the three basic fish ladder designs: pool and weir fishway (top), vertical slot fishway (middle), and Denil fishway (bottom). (Source: S.G. Hildebrand, et al. Redrawn with permission from the Oak Ridge National Laboratory.)*

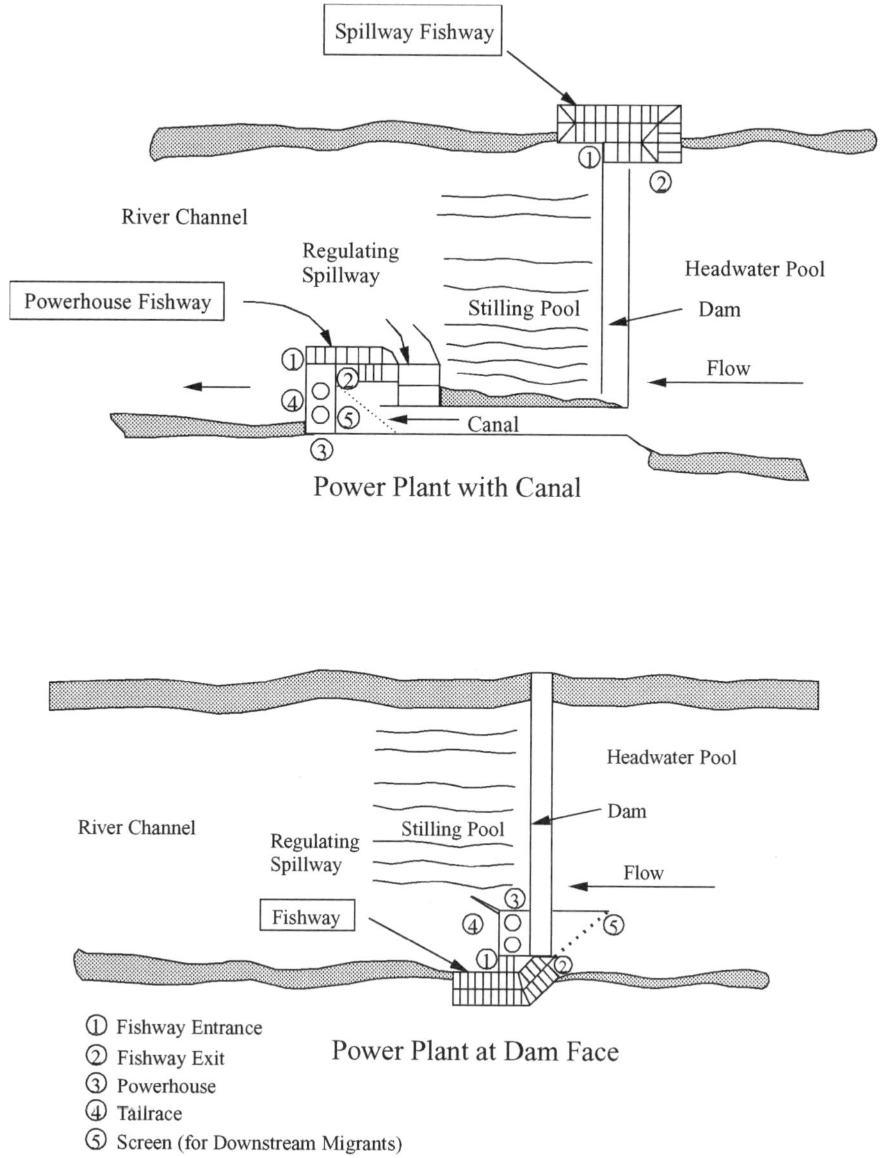

FIGURE 10-2: ***Fish Passage Locations.*** *The top figure, showing a power plant with a canal, has fishways adjacent to the powerhouse as well as on the other bank. The bottom figure, in which the power plant is at the face of the dam, has only one fishway.*

(Source: S.G. Hildebrand, et al. Redrawn with permission from the Oak Ridge National Laboratory.)

Fish Elevators and Locks

Fish elevators and fish locks are manual methods used to pass fish upstream past a dam. Attraction flows properly located and of the appropriate flow rate draw the fish into the collection chamber on the downstream side of the dam. Once an

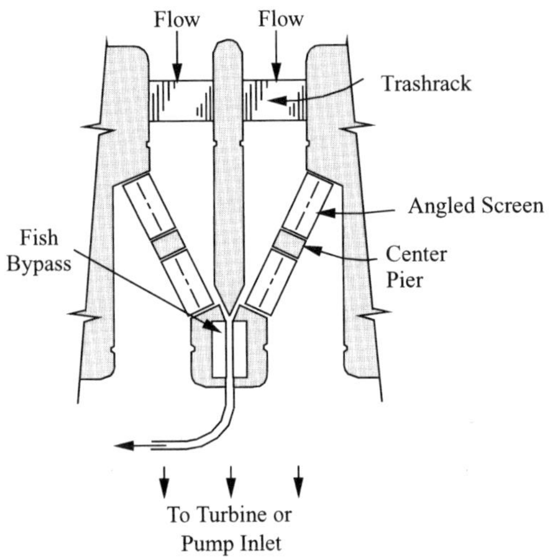

FIGURE 10-3: ***Conceptual Angled Screen Design.*** *As seen in this figure, angled screens are set at an angle to the streamflow and act to guide fish to the opening of the fish collection or bypass entrance.*
(Redrawn with permission from the Electric Power Research Institute.)

adequate number of fish have been collected, fish elevators use a water-filled hopper to raise the fish above the dam. The fish are then transferred to a holding tank for transporting to the release area or to a trough for releasing into the reservoir. Fish locks are operated by closing the downstream side of the collection chamber, flooding the chamber with water, and allowing the fish to pass through a gate at the top (upstream side) of the chamber.

Fish Bypass Systems

Fish bypass systems provide safe passage for fish migrating downstream from the vicinity of the project intake. Fish bypass systems generally consist of two components: a fish diversion structure to guide fish away from the project intake and a fish return structure to safely pass diverted fish downstream.

Fish diversion structures. The various types of fish diversion systems include angled screens, angled rotary drum screens, inclined plane screens, louvers, and removable inclined screens.

Angled screens are set at an angle to the streamflow and act to guide fish to the opening of the fish collection or fish bypass entrance, as shown in Figure 10-3. This system is effective in diverting numerous species and life stages of warm-water and cold-water fish downstream.

Rotary drum screens use a screen cylinder that rotates on a horizontal axis to remove materials from the incoming flow. Figure 10-4 shows a rotary drum screen that uses the outer periphery of the cylinder as its screening surface. It can be set at an angle to the flow to divert fish to a fish return system. These systems have been used in the western United States on some water diversion projects.

Inclined plane screens or skimmers, as seen in Figure 10-5, divert fish upward in the water column and toward the bypass outlet.

Louvers are a series of evenly spaced, vertical slats aligned across a channel at an angle to divert fish to one side or the other. Fish typically try to avoid the turbulence created by the louvers and, in this manner, are guided toward the fish return structure. Generally, louvers are effective in diverting fish, but are more

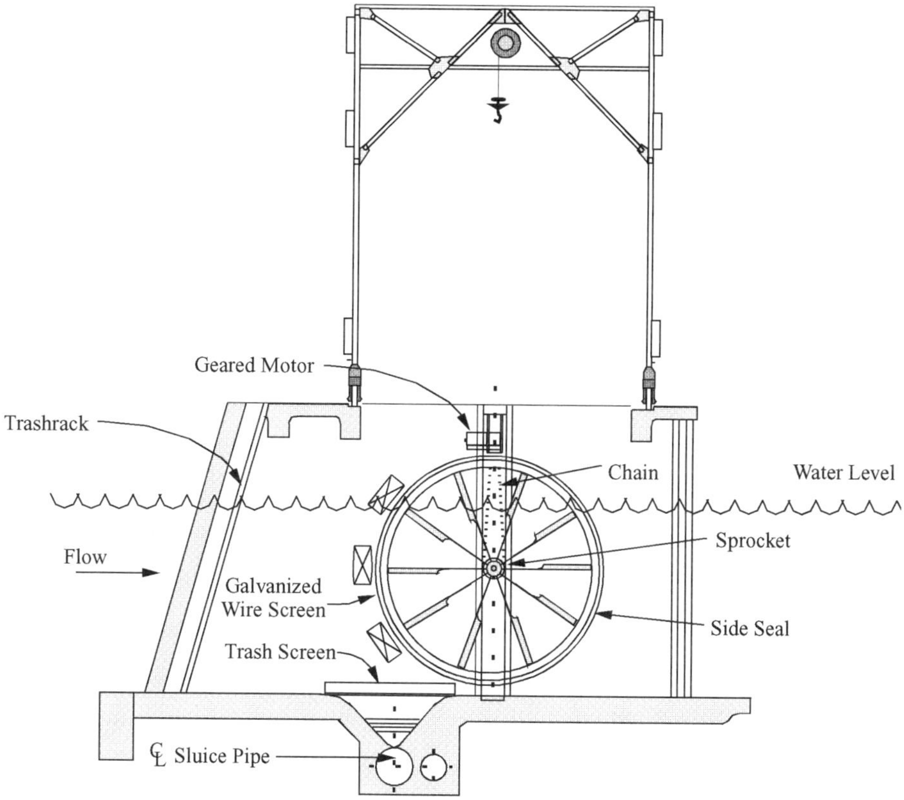

FIGURE 10-4: *Rotary Drum Screen.* *Rotary drum screens use a screen cylinder that rotates on a horizontal axis to remove materials from the incoming flow. This figure shows a rotary drum screen that uses the outer periphery of the cylinder as its screening surface. It can be set at an angle to the flow to divert fish to a fish bypass or collection system.*
(Redrawn with permission from the Electric Power Research Institute.)

species-specific than screens. These systems have been shown to require less maintenance than screen systems.

Some hydroelectric projects in the western United States have tested removable inclined screens. The design in Figure 10-6 screens only the upper portion of the intake area. Consequently, it is effective only for fish that concentrate in the upper levels of the water column.

Fish capture and transport. After fish are diverted away from the intake entrance, they must be collected and safely returned to their natural environment. A fish return system typically consists of a transition section and a pipeline or trough. In some cases, a pumping unit transports fish through the pipeline and into the stream.

Tapered transition sections between the diversion structure and the pipeline

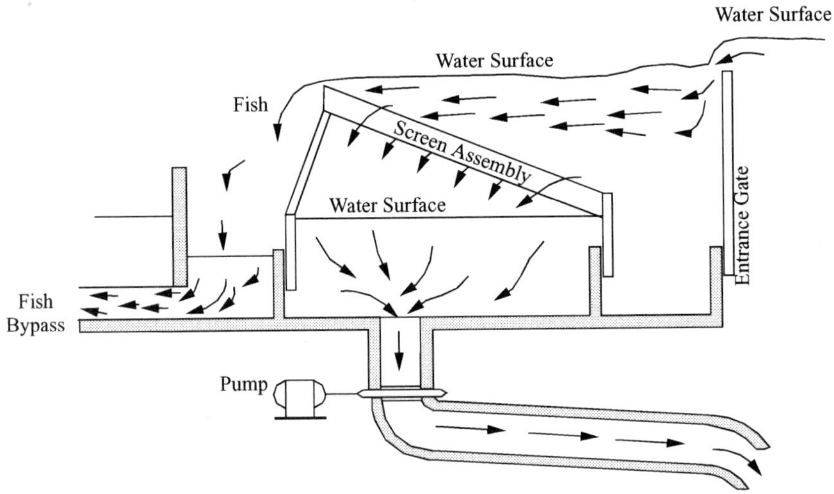

FIGURE 10-5: ***Inclined Plane Screen (Skimmer).*** *An inclined plane screen, as seen in this cross-sectional view, diverts fish upward in the water column and toward the bypass outlet.*
(Redrawn with permission from the Electric Power Research Institute.)

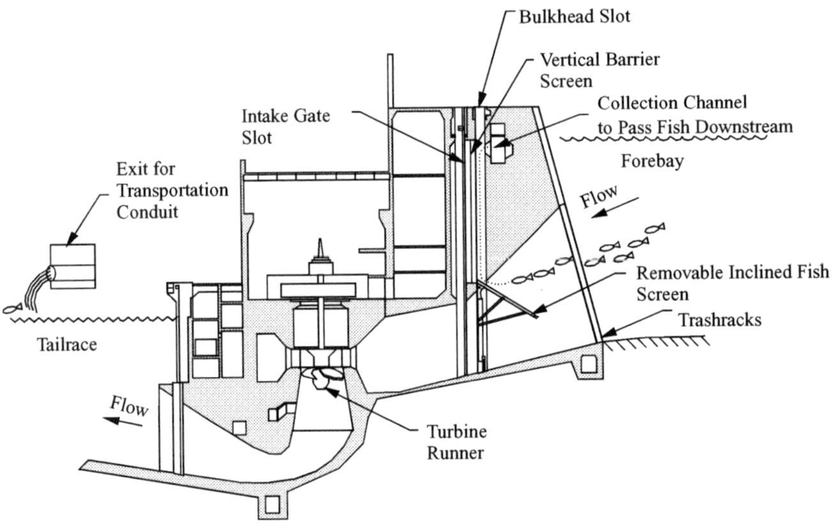

FIGURE 10-6: ***Typical Removable Inclined Screen.*** *The design shown in this figure screens only the upper portion of the intake area. Consequently, it is only effective for fish that concentrate in the upper levels of the water column.*
(Redrawn with permission from the Electric Power Research Institute.)

guide the fish to the pipeline opening. The pipeline, itself, is usually sized according to the size of the fish but must be large enough to avoid clogging from debris. Specific criteria for pipeline design are available.[1]

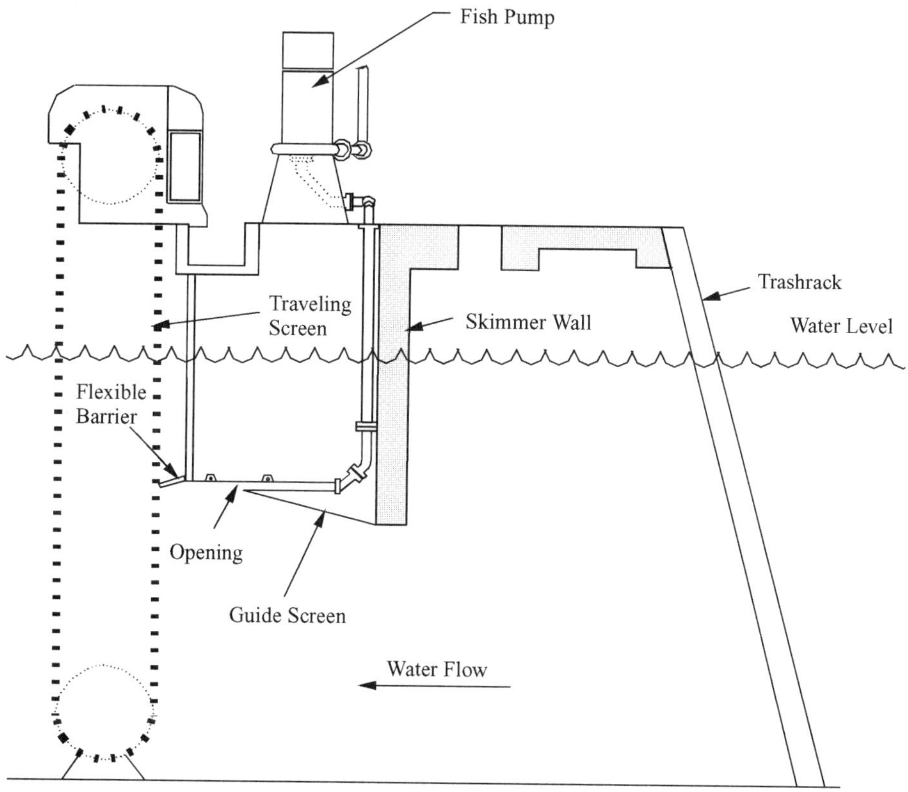

FIGURE 10-7: ***Fish Pump System.*** *This cross-sectional drawing shows a method to collect and pump fish from the intake to avoid turbine entrainment.*
(Redrawn with permission from the Electric Power Research Institute.)

Chapter 10

Fish Pumps

Various types of pumps have been considered for use at power plants to collect and transport fish entrapped in the intake screenwells. The fish pump system, shown in Figure 10-7, has been successfully tested. Critical factors appear to be fish length and pump speed.

Figure 10-8 shows two types of fish pumps: the peripheral-type jet pump (top) and the screw-propeller pump (bottom). Experiments with the peripheral-type jet pump have been successful in transporting fragile fish species. However, field application of this technology has been limited. Screw-propeller pumps are effective in transporting fish of various species and life stages. Critical factors are the pump speed (for example, low speeds result in higher fish survival) and size (for example, must be adequate to pass the largest fish to be transported).

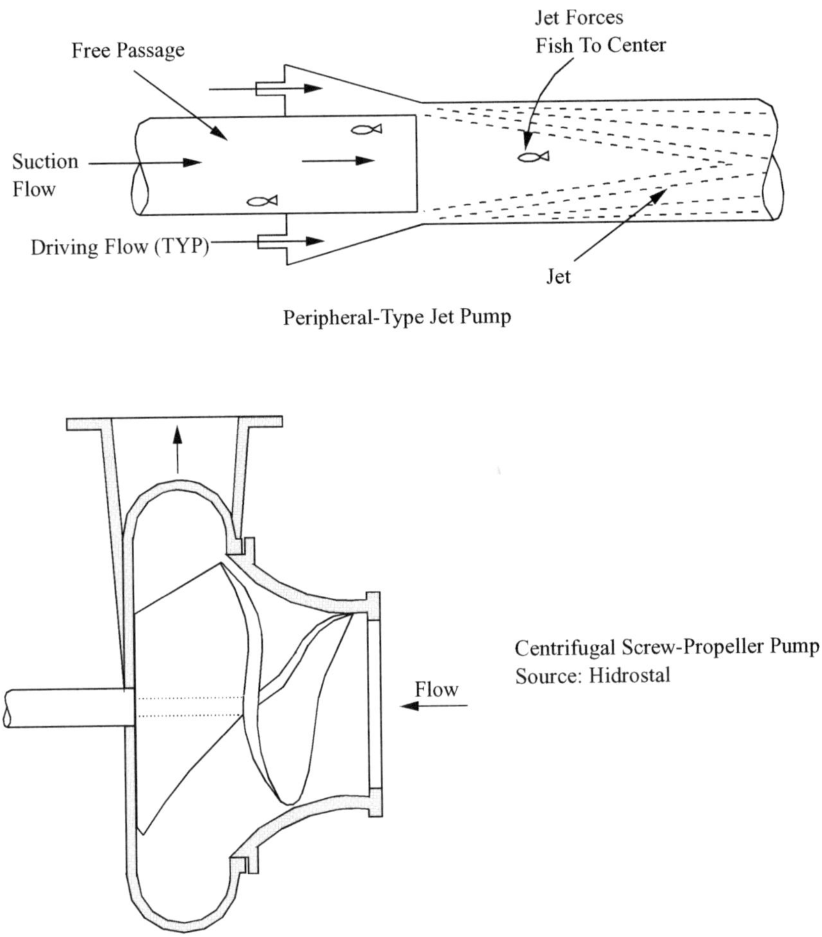

FIGURE 10-8: ***Fish Pumps.*** *This figure shows two types of fish pumps: the peripheral-type jet pump (top) and the centrifugal-type screw-propeller pump (bottom).*
(Redrawn with permission from the Electric Power Research Institute.)

Traveling Screens

Traveling screens consist of a series of wire mesh panels, which are fastened to steel frames and revolve vertically over head and foot sprockets. As water passes through the wire mesh, the upstream face of the traveling screen collects and retains debris and fish. When the rotating panels are rotated above the water surface, the screen is cleaned with a high-pressure water spray. The debris is collected for subsequent disposal. Traveling screen systems designed for fish protection usually include features for returning the collected fish to the stream below the dam.

One problem associated with the use of screens is the impingement of fish on the face of the screen. Fish mortality can occur owing to suffocation or physical damage. Frequent cleaning of the traveling screens can minimize impact-related fish mortality. Spray wash pressures must be low enough to minimize physical damage during cleaning, and the return line must be designed to safely collect and return fish to the stream. Traveling screen systems without fish collection buckets often are provided on intakes. These screens are self-cleaning since debris collected on the upstream face is removed on the downstream side by the flow in the intake. However, the maintenance problems with the operating mechanism remain.

Fish Screens

Several types of fish screening systems have been designed to minimize the passage of fish into water intakes at hydroelectric facilities. Primarily, the species and life stage of fish that are of concern determine the type of screen and the specific design features (for example, mesh size). Other factors, such as economic feasibility and structural compatibility with other project features, also must be considered. The principal types of fish screening systems are discussed below.

Bar racks. Bar racks prevent the entrance of large fish into the facility's intake system. However, bar-rack spacing small enough to prevent larger fish from entering may not prevent smaller fish. Intake velocities at many low-head hydroelectric projects are low enough that even small fish can usually escape if they pass through the bar racks.

Stationary screens. Both hydroelectric and other electric generation facilities have used stationary screens to successfully block fish entrainment without significant head loss if the screens are kept clean, the flow velocities are low, and the fish have an adequate escape route. Stationary screens, while effective in screening fish from the intake, can be difficult to keep clean. Therefore, stationary screens often are designed to be removable for cleaning or during periods when they are not needed; however, removable screens have higher initial and operating costs.

Cylindrical wedgewire. Thermal-electric stations have successfully used cylindrical wedgewire screens for water system intakes. The large screening surface and associated low intake velocity are effective in excluding various fish species and life stages. Because of the relatively high cost and the large screening area required for effective use, cylinder wedgewire screens are not feasible for turbine intakes, but may be used to screen cooling water intakes.

Barrier nets. Barrier nets, shown in Figure 10-9, prevent fish from entering into the water intake system. The size of the fish to be excluded dictates the mesh size of the net. The proper net design can minimize the numbers of fish that pass through the net, but the net may require frequent maintenance and repair if the stream has large amounts of debris.

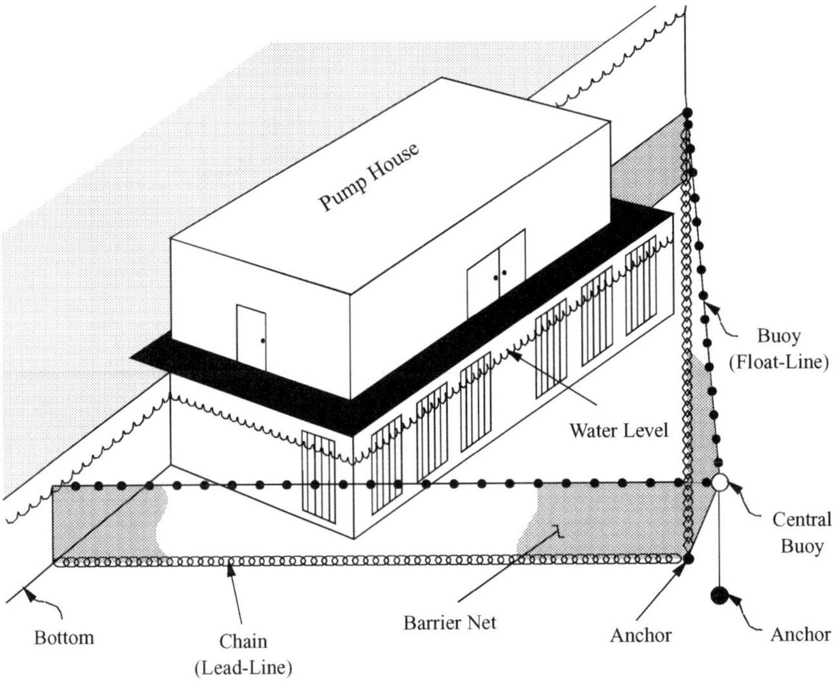

FIGURE 10-9: *Barrier Net.* *Barrier nets, shown in this figure, prevent fish from entering into the water intake system. The size of the fish to be excluded dictates the mesh size of the net. The proper net design can minimize the numbers of fish that pass through the net, but the net may require frequent maintenance and repair if the stream has large amounts of debris.* *(Source: 1977 Hudson River Aquatic Ecology Studies at the Bowline Point Generating Station. Redrawn with permission from the Electric Power Research Institute.)*

WATER QUALITY ENHANCEMENT SYSTEMS

Many methods have been used to improve low DO concentrations at hydro projects. The hydro industry practices and standards for improving DO are changing in response to the environmental concerns of our society. No single method of improving DO is successful and cost-effective for all applications; therefore, alternative methods are usually evaluated for each project's specific requirements and features.

Reservoir Destratification

The temperature of the water in a reservoir often varies over the depth of the water from the surface. In warm months, the water temperature near the surface is higher than water at lower depths, while during cold months the water near the surface may be colder than water at lower depths. Significant differences in the water temperature at different depths in reservoirs are referred to as thermal stratification.

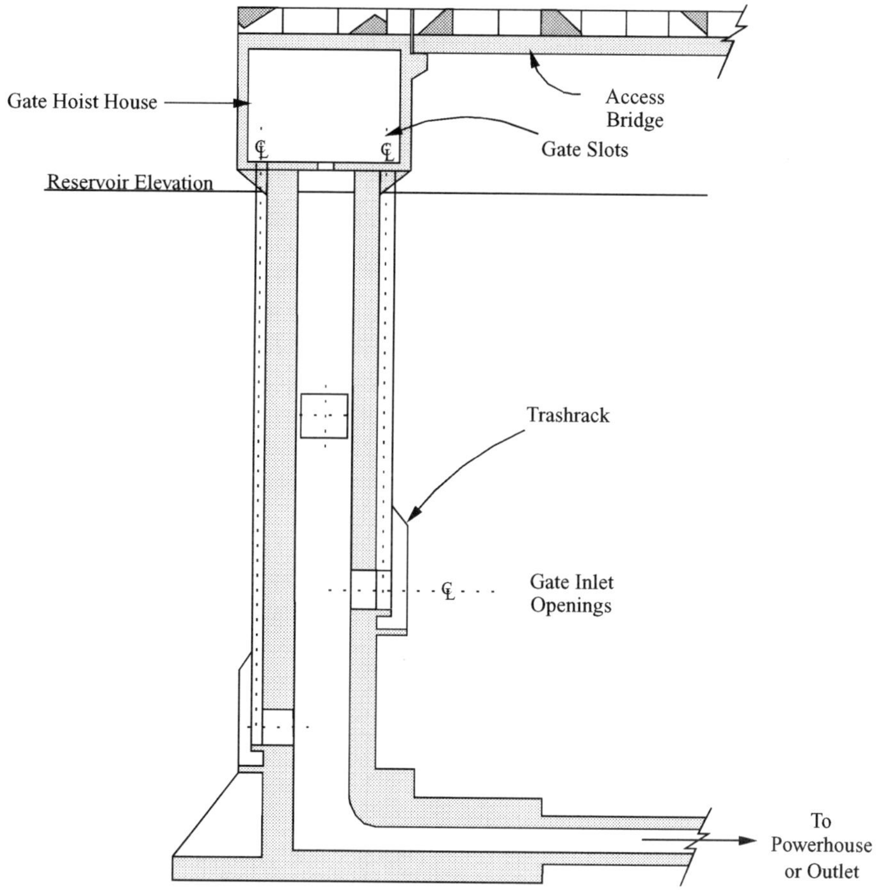

FIGURE 10-10: ***Multilevel Inlet.*** *This figure shows an example of a multilevel inlet that can select the elevation where water is withdrawn to modify project dissolved oxygen (DO) concentrations and temperature requirements.*

The DO content in reservoirs varies with water depth and temperature. For water having the same temperatures, the DO content increases with increasing depth because of the increasing pressure at lower depths. For water at the same depth, the potential DO content decreases with increasing temperature. Significant differences in DO content in reservoirs are referred to as DO stratification.

Since the DO concentration can vary depending on water depth in the reservoir, one method to improve DO concentrations is to mix or destratify the reservoir DO and thermal layers. Vigorous mixing moves the high DO water in the reservoir, thus equalizing both DO concentrations and differences in water temperatures. Except for smaller impoundments, complete destratification cannot be achieved since it would require excessively large mixing facilities and energy inputs. Most project purposes can be adequately met with partial destratification systems.

The two most common and efficient methods of destratification are pneumatic systems using compressed air bubblers and hydraulic systems using water circulating pumps. Currently, pneumatic systems are the most common type. They have the advantage of introducing additional oxygen into the impoundment through the bubblers. However, they also increase nitrogen saturation levels. By contrast, hydraulic systems avoid the nitrogen saturation conditions. Based on model comparisons of the two systems, hydraulic systems are considered more efficient destratifiers than pneumatic systems.

Multilevel Inlet Structures

Using intake structures that can select the elevation where water is withdrawn can modify project DO concentrations and temperatures. As a result, higher temperature water with relatively high DO concentration near the top of the reservoir can be mixed with low temperature water with lower DO levels. Figure 10-10 shows an example of a multilevel inlet.

Inlet Aspirator

The inlet aspirator system uses the low pressure at the penstock intake to draw water through a pipe extending from the intake depth having a low DO level to depths having a higher DO concentration. The upper section of the pipe can slide inside the lower section (telescope) to be adjusted to various reservoir levels. Figure 10-11 shows a conceptual drawing of an inlet aspirator system.

The system would increase the DO levels of powerhouse releases by mixing the reservoir thermal layers. The aspirator pipe is sized to provide a desired mix ratio at maximum flow conditions. Although the fixed-pipe size would limit operational flexibility, the system would not require the high number of gates of the multilevel inlet system. Minimum penstock approach velocities would be required to reduce the pressure enough to overcome the density difference of different temperature waters and draw water from a higher elevation into the penstocks. For this reason, the system may not be effective for low discharge conditions.

Turbine Aspiration

The low pressure that is present below Francis runners commonly is used to draw air into the turbine at part load to stabilize the runner vortex. While the air vent valve is normally closed above part loads, the valve could be opened manually to allow aspirating at all loads.

Turbine runners draw significantly less air when operating at rated capacity, but draw increasingly more air at lower loads. Baffles installed near the vent holes on the turbine hub could increase the vacuum and, thereby, maximize air aspiration. The vent baffles allow a larger volume of air to be aspirated over a wider range of operating conditions. This increases the air-to-water flow ratios for corresponding conditions. Removing restrictions that limit flow in the air vent piping and installing a bell-mouth intake on the air vent pipe can further increase natural aspiration.

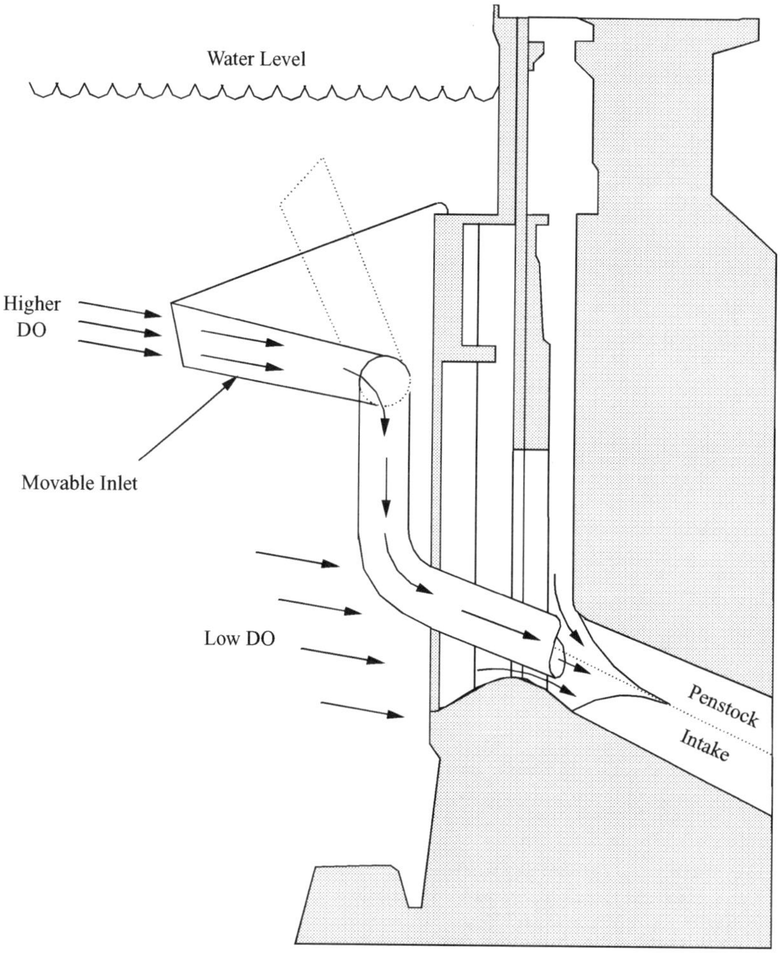

FIGURE 10-11: ***Inlet Aspirator.*** *This figure shows a conceptual cross-sectional drawing of an inlet aspirator system. This system uses the low pressure created at the penstock intake to draw water from a higher elevation having a higher dissolved oxygen (DO) concentration. The water is drawn through a pipe to the intake elevation having a low DO level. The vertical section of the pipe can "telescope" or be adjusted to various reservoir levels.*
(Courtesy U.S. Army Corps of Engineers.)

The success of turbine aspiration and improving DO levels varies widely depending upon the turbine design, operating head, and tailwater submergence. Another drawback of opening the air vent valve at higher loads is the reduction in turbine efficiency.

Draft Tube Venting Ring

A venting ring with a triangular cross-section was fabricated from sheet metal and installed in the draft tube at the Tennessee Valley Authority's Norris Dam. Air was aspirated through a pipe into the venting ring. DO concentrations increased significantly; however, adding the baffle ring caused a loss in turbine efficiency. An additional turbine efficiency loss occurred at peak efficiency from air induction through the baffle ring.

As with other alternatives that use air aspiration, the extent of aspiration varies with tailwater level and with turbine load. The aeration efficiency is lowest at the higher output levels where it is needed most.

Mechanical Aerators

Mechanical surface aerators are commonly used to increase the DO levels in bio-oxidation-type waste water treatment processes. Generally, floats or brackets support these devices over the water so that the impeller spins in a horizontal plane just below the surface. The ability to increase oxygen concentration depends on the initial DO concentration level. Standard lines of aerators, ranging in size from 0.75 to 75 kW (1 to 100 horsepower), are available from several manufacturers.

Mechanical surface aerators are not efficient for maintaining high DO levels. These systems have high power requirements, high initial cost, high maintenance costs, unsightly appearance, noisy operation, and create obstructions. Therefore, these systems are usually one of the less desirable alternatives for water quality enhancement.

Compressed Air Injection

Air compressors can be used to inject air directly into the flow passage of the turbine. It is possible to inject air in quantities up to 5 percent of water volume. However, such quantities have a variety of adverse effects. Potential injection points that can be used include the penstock fill line, piezometric taps, penstock drain line, turbine air vent, and draft tube drain line. Compressed air injection also has been used in reservoirs. When injecting air into the upper section of the penstock, the air often accumulates at the top of the penstock section and bubbles upstream to the intake. Similar problems can occur for other injection points when the unit load is changed. The U.S. Army Corps of Engineers has found air injected into the draft tube drain line to be relatively ineffective because of the low pressure in this region and the short contact time before the water is released into the tailrace. Air injection at higher pressure locations is more effective, but requires higher pressure compressors. Air injection also reduces the turbine efficiency much like turbine aeration.

Liquid Oxygen

Liquid oxygen (LOX) can be injected directly into the pressure taps located in the penstock or spiral case. Pipe diffusers can be added to the injection taps to improve the oxygen absorption efficiency. Injecting liquid oxygen is more effective in increasing the DO in the turbine discharge, and more efficient than injecting air into the system. The major disadvantage of oxygen injection involves safety and storage requirements. Oxygen injection has a negligible effect on power generation.

Spilling Weirs

Spilling water over a dam crest or a cascading natural stream bed can significantly increase the DO concentration in water. Stepped or cascading weirs can be designed and constructed to maximize air entrainment as water splashes down the steps. Water spilled over a bypass weir is not available for generation; therefore, other methods that are more energy efficient in improving DO often are selected.

Weirs could be installed in tailrace channels to improve DO concentrations. A weir's effectiveness in improving DO increases in proportion to the available head and subsequent spilling weir length.

Fixed-Cone Dispersion Valves

Fixed-cone dispersion valves described in Chapter 5, Major Valves, can control water discharges. The valve disperses the water and hydraulic energy into the air by spraying the conical discharge over a large area. Inherent in the design is the ability to significantly increase the DO concentration in the water that is released. However, power generation is lost from the dispersed water.

Methods to Reduce Petroleum Contamination

Oil Separators

Oil separators are designed with the water collection sumps of hydropower stations to remove oil and grease from the powerhouse drainage before it is discharged. Chapter 9, Auxiliary Systems, discusses the oil-handling system in a power plant. Oil and grease from the turbine pit and powerhouse floors are potential contaminants that could enter the powerhouse drainage system and eventually the station sump.

The station sump should have two or more compartments to allow the oil products to settle out and separate from the water. The sump discharge should be located at the bottom of the sump to minimize petroleum discharges that tend to float to the surface.

Oil separators skim the oil and petroleum products from the water surface. One common type uses a stainless steel tape to transfer oil from the top of the sump to a storage container for removal.

Self-Lubricated Wicket Gate Bushings

The headcover and bottom ring can have self-lubricated wicket gate bushings to eliminate the potential minor source of water pollution from the wicket gate bushings and grease buildup on the headcover. The self-lubricated bushing should be considered to replace the classical bronze wicket-gate bushings for new turbines and when upgrading existing turbines. Previous wicket-gate bushings required lubricating grease to minimize friction, wear, and corrosion of the wicket gate stem. By eliminating high viscosity grease, reduced bushing clearances are possible. This can enhance governor control. Additionally, eliminating grease and grease buildup can help in identifying water leakage above the headcover as well as in reducing maintenance costs.

Self-lubricated bushings fall into two main categories: polytetrafluoroethylene (PTFE) fluorocarbon and carbon bronze bushings.

PTFE bushings are water-lubricated, synthetic, frictionless bearing materials that are inserted into and supported by stainless steel or bronze bushings. The PTFE bushings do not require a lubricating medium other than water. It is recommended that PTFE bushings be used with wicket gate trunnions either constructed, sleeved, or overlaid with stainless steel to prevent corroding the wicket gate trunnion and increasing bearing wear. Installations that use self-lubricated bushing need to consider possible expansion from thermal changes and water absorption. PTFE bushings are prone to wear when the water has a high particle content, in which case bearing end seals are recommended. PTFE bushings also are brittle, and care during installation is required to prevent cracking during installation. Similarly, the trunnions should be provided with chamfers or relief areas to facilitate installation.

A second category of self-lubricated bushing is the carbon bronze or carbon-impregnated bronze bushing. Carbon bronze bushings contain circular inserts of carbon material incorporated in the bronze bushing. The carbon material lubricates the wicket gate stems during operation of the wicket gate. Care in handling, storage, and installation is required to prevent damage and having the carbon material crack and flake.

Carbon-impregnated bushings are bushings in which carbon particles are put in the molten bronze during the manufacturing process. The carbon modules lubricate the mating wicket gate trunnion as the bronze bushing wears.

Water-Lubricated Turbine Bearings

Turbine bearings usually are oil lubricated; however, smaller turbines can have "water-lubricated" guide bearings to eliminate the potential for oil spills that could potentially be discharged from the plant.

Double-Walled Heat Exchangers

Turbine and generator bearing oil reservoirs also commonly use single-pass, water-cooling coils. Most transformers are oil-filled and use oil as a heat transfer medium. Some transformers have an additional oil or water heat transfer system. Using double-walled heat exchangers in turbine, generator, and transformer cooling systems can reduce the risk that cooling coil failure will discharge oil into the water. Double-walled heat exchangers provide a means for leaks to be discovered and contained without contaminating the system. Dry-type transformers also are available for smaller capacities, but are physically larger and more costly.

Non-Petroleum-Based Lubricants

Food-quality grease and hydraulic oil products are available and can be used for greasing the classic bronze grease bushings and hydraulic systems. Food-quality grease and hydraulic oil often are used in potable water systems and also have been used successfully on hydraulic turbine grease and hydraulic fluid systems. Some current products may not be less toxic or more biodegradable than petroleum lubricants. Furthermore, higher volumes of food quality lubrication may be required to achieve the same level of lubrication as traditional lubricants.

Petroleum Storage

Petroleum products, typically including oils and painting supplies, often are stored in hydro powerhouses. The location of any storage facilities should be selected to minimize the potential for contamination and fire.

Storage locations should be provided with curb barriers to prevent any spills from entering the other powerhouse areas or the powerhouse drainage system. Also, the storage areas for petroleum products should have fire doors that close automatically and a fire detection and suppression system. The fire suppression system also should restrain any leaking petroleum products and prevent contamination. (See Chapter 9, Auxiliary Systems, for details on fire protection systems.)

Hydraulic Systems

Oil hydraulic systems are commonly used for wicket gate servomotors, runner blade servomotors, inlet valve operators, spillway gate operators, and other applications. While oil hydraulic systems are very reliable, they present a potential source of contamination that using water hydraulic systems often can eliminate. Water hydraulic systems have been successfully used for inlet valve operators, inlet valve seals, and spillway gate operators. Also, water hydraulic systems have been used—with more limited success—for turbine blade servomotors.

Reference

[1]Committee on Hydropower Intakes of the Energy Division of the American Society of Civil Engineers, *Guidelines for Design of Intakes for Hydroelectric Plants,* Chapter 7—Fish Passage and Protection, American Society of Civil Engineers, New York, New York, 1995.

Bibliography

Allen, H.H., and L.R. Aggus, *Effects of Fluctuating Reservoir Water Levels on Fisheries, Wildlife, and Vegetation; Summary of a Workshop,* Miscellaneous Paper E-83-2, U.S. Army Engineer Waterways Experiment Station, Vicksburg, Mississippi, February 24-26, 1981.

Assessment of Downstream Migrant Fish Protection Technologies for Hydroelectric Application, Prepared by Stone & Webster Engineering Corporation, Electric Power Research Institute, Palo Alto, California, September 1986.

Barnhart, G.A., *Proceedings of the 1982 Northeastern Division, American Fisheries Society, Cold-Water Workshop,* New York Department of Environmental Conservation, Ithica, New York, 1982.

Bell, M.C., *Fisheries Handbook of Engineering Requirements and Biological Criteria,* Volumes I and II, U.S. Army Corps of Engineers, North Pacific Division, Portland, Oregon, 1973.

Bell, M.C., *Fisheries Handbook of Engineering Requirements and Biological Criteria,* Fish Passage Development and Evaluation Program, U.S. Army Corps of Engineers, North Pacific Division, Portland, Oregon, 1991.

Bell, M.C., *Revised Compendium on the Success of Passage of Small Fish through Turbines,* Report for U.S. Army Corps of Engineers, North Pacific Division, Contract No. DACW-57-88-C-0070, Portland, Oregon, 1991.

Bioengineering Symposium, Proceedings of the Symposium held September 1-2, 1993, in Portland Oregon, American Fisheries Society, 1993.

Dissolved Oxygen Study, Table Rock Dam and Lake, U.S. Army Corps of Engineers, Little Rock District, Little Rock, Arkansas, February 1985.

Entrainment Impact Estimates Using the Equivalent Adult Approach, Prepared by the Office of Biological Services, U.S. Department of the Interior, Fish and Wildlife Service, Washington, D.C., September 1977.

Evaluating Hydro Relicensing Alternatives: Impacts on Power and Nonpower Values of Water Resources, EPRI GS-6922, Electric Power Research Institute, Palo Alto, California, August 1990.

Evaluation of Dredged Material Proposed for Ocean Disposal, EPA-503/8-91/001, Environmental Protection Agency, Washington, D.C., 1991.

Ferguson, J.W., "Analyzing Turbine Bypass Systems at Hydro Facilities," *Hydro Review,* Volume XI, Number 3, June 1992, pages 46-56.

Fish Entrainment and Turbine Mortality Review and Guidelines, EPRI TR-101231, Project 2694-1, Electric Power Research Institute, Palo Alto, California, September 1992.

Fish Protection/Passage Technologies Evaluated by EPRI and Guidelines for Their Application, EPRI TR-104120, Electric Power Research Institute, Palo Alto, California, May 1994.

Fourth Progress Report on Fisheries Engineering Research Program, 1966-1972, U.S. Army Corps of Engineers, North Pacific Division, Portland, Oregon, May, 1976.

Francfort, J.E., G.F. Cada, D.D. Dauble, R.T. Hunt, D.W. Jones, B.N. Rinehart, G.L. Sommers, and R.J. Costello, *Environmental Mitigation at Hydroelectric Projects: Volume II: Benefits and Costs of Fish Passage and Protection*, Contract DE-AC07-761D01570, Idaho National Engineering Laboratory, EG&G Idaho, Inc., Idaho Falls, Idaho, January 1994.

Hannan, H.H. and T. Cole, "Dissolved Oxygen Dynamics in Reservoirs," In Thornton, K., (ed.), *Perspectives in Reservoir Limnology,* John Wiley & Sons, Inc., New York, New York, 1985.

Hildebrand, S.G., et al., *Analysis of Environmental Issues Related to Small Scale Hydroelectric Development II: Design Considerations for Passing Fish Upstream Around Dams,* Publication No. 1567, Oak Ridge National Laboratory, Oak Ridge, Tennessee, 1980.

1977 Hudson River Aquatic Ecology Studies at the Bowline Point Generating Station, prepared for Orange and Rockland Utilities, Inc., Lawler, Matusky, and Skelly Engineers, September 1978.

Hydraulic Model Evaluation of the Eicher Passive Pressure Screen Fish Bypass System, EPRI AP-5492, Final Report, Prepared by Eicher Associates, Inc., Electric Power Research Institute, Palo Alto, California, October 1987.

Kennedy, R.H., R.C. Gunkel, and J.M. Carlile, *Riverine Influence on the Water Quality Characteristics of West Point Lake,* Technical Report, U.S. Army Engineer Waterways Experiment Station, Vicksburg, Mississippi, 1983.

Kennedy, R.H., K.W. Thornton, and R.C. Gunkel, "The Establishment of Water Quality Gradients in Reservoirs," *Canadian Water Resources Journal,* Volume 7, 1982, pages 71-87.

Loar, J.M., et al., *Analysis of Environmental Issues Related to Small Scale Hydroelectric Development I: Dredging,* Publication No. 1565, Oak Ridge National Laboratory, Oak Ridge, Tennessee, 1980

Loar, J.M., and M.J. Sale, *Analysis of Environmental Issues Related to Small Scale Hydroelectric Development V: Instream Flow Needs for Fishery Resources,* Publication No. 1829, Oak Ridge National Laboratory, Oak Ridge, Tennessee, 1981.

Martin, D.B., and R.D. Arneson, "Comparative Limnology of a Deep-Discharge Reservoir and a Surface-Discharge Lake on the Madison River, Montana," *Freshwater Biology,* Volume 8, 1978, pages 33-42.

Mathematical Methods to Evaluate Entrainment of Aquatic Organisms by Power Plants, Prepared by the National Power Plant Team, U.S. Department of the Interior, Fish and Wildlife Service, Washington, D.C., September 1977.

Neitzel, D.A., C.S. Abernethy, and E.W. Lusty, "Evaluation of Rotating Drum Screen Facilities in the Yakima River Basin, South-Central Washington State," *Fisheries Bioengineering Symposium,* American Fisheries Society Symposium 10, Bethesda, Maryland, 1991.

Pankratz, T.M., *Screening Equipment Handbook for Industrial and Municipal Water and Wastewater Treatment,* Technomic Publishing Company, Lancaster, Pennsylvania, 1988.

Pearce, R.O. and R.T. Lee, "Some Design Considerations for Approach Velocities at Juvenile Salmonid Screening Facilities," *Fisheries Bioengineering Symposium,* American Fisheries Society Symposium 10, Bethesda, Maryland, 1991.

Physical Impacts of Small-Scale Hydroelectric Facilities and Their Effects on Fish and Wildlife, Prepared by the U.S. Fish and Wildlife Service, U.S. Department of the Interior, Washington, D.C., September 1984.

Proceedings: Fish Protection at Steam and Hydroelectric Power Plants, EPRI CS/EA/AP-5663-SR, Electric Power Research Institute, Palo Alto, California, March 1988.

Proceedings of the Symposium on Small Hydropower and Fisheries, American Fisheries Society, Western Division and Bio-Engineering Section, Aurora, Colorado, May 1985.

Research Update on Fish Protection Technologies for Water Intakes, EPRI TR-104122, Electric Power Research Institute, Palo Alto, California, May 1994.

Review of Hydroelectric Projects Licensed by the Federal Power Commission, Prepared by the Mitre Corporation, U.S. Department of the Interior, Fish and Wildlife Service, Washington, D.C., September 1977.

Turbak, S.C., D.R. Reichle, and C.R. Shriner, *Analysis of Environmental Issues Related to Small-Scale Hydroelectric Development IV: Fish Mortality Resulting From Turbine Passage,* Publication No. 1597, Oak Ridge National Laboratory, Oak Ridge, Tennessee, 1981.

Turbine-Related Fish Mortality: Review and Evaluation of Studies, EPRI AP-5480, Final Report, Prepared by Eicher Associates, Inc., Electric Power Research Institute, Palo Alto, California, October 1987.

Turner, R.R., et al., *Analysis of Environmental Issues Related to Small-Scale Hydroelectric Development III: Water Level Fluctuation,* Publication No. 1591, Oak Ridge National Laboratory, Oak Ridge, Tennessee, 1981.

Updated Compendium on the Success of Passage of Small Fish Through Turbines, U.S. Army Corps of Engineers, North Pacific Division, Portland, Oregon, 1981.

Chapter 11
Equipment Layout Considerations

Chapter 11

When laying out equipment, the main concern is for the long-term reliability of producing either electrical or mechanical power. The equipment layout in a hydroelectric plant depends on many factors:

- Type of powerhouse;
- Kinds and sizes of the principal turbine-generating equipment;
- Length of the water conductors;
- Plant location (above ground or underground);
- Access provided for construction and future maintenance;
- Routing for connecting power busses or cables to the main step-up transformers;
- Maximum water levels to which the powerhouse structure will be subjected; and
- Safety of people and physical property.

In general, there are three main types of hydroelectric power plants: conventional plants on the surface, power plants in underground caverns, and plants in pits excavated into the ground and rock.

Conventional, surface plants may be totally enclosed (indoor) with no superstructure protecting the outdoor generators, or the plants may be semi-outdoor. In some cases, the upstream portion of the powerhouse may contain the intake trashracks, gates and hoists, and water passages leading to the turbines. In that case, the powerhouse also functions as part of the structure that retains the reservoir water. In other cases, the intake is separate from the powerhouse. Water flows to the turbines through a flume or canal, a wood or metal pipe (the penstock), a tunnel, an open forebay, or any combination of those.

A power plant in an underground cavern may be located anywhere between the intake and the tailrace—depending on geologic and on other conditions.

If there is a long tailrace tunnel, both upstream and downstream combination surge shafts and gate chambers may be needed to minimize waterhammer pressures and/or pressure surges. Some underground plants with moderate head have pressure shafts through rock that lead directly from a vertical elbow in the concrete intake to the turbines. The turbines are located in a vault in the rock of the downstream face of the power cavern.

Power plants in pits may have one or more units, but the excavation depth ordinarily does not exceed the distance from the lowest point of the substructure to the roof in a similar conventional plant. Plants in pits having one, two, or three units may have all units served by a hoist arm rotating on a pivot. In this case, access from the hoist arm to the main equipment is through large, watertight hatches in the roof.

Obviously, the kind of powerhouse, the head, the geographic elevation, the ground cover soils, characteristics of the rock, and the climate greatly influence the equipment requirements and layouts. In general, the equipment involved includes:

- Equipment for drainage;
- Equipment for ventilation and air conditioning;
- Equipment for heating and cooling;
- Water supply (other than the water released through the turbines);
- Sumps and sump pumps or eductors;
- Access into the power plant and to the equipment;
- Choices between central systems or unit-by-unit systems;
- The kinds and locations of hoisting equipment;
- The type of emergency power supply or service;
- The transformers and auxiliary electrical equipment; and
- The emergency and service turbine water supply shut-off gates and valves.

Closeness to or remoteness from sizable communities also influences the type and size of equipment needed for fire protection, sewage, potable water supply, machine shop, warehousing, and other mechanical systems of the hydroelectric plant.

The powerhouse exists as a place into which the principal equipment and all of the supporting auxiliaries are installed. The designer should keep in mind that selecting the main turbine-generating equipment and determining its location control much of the powerhouse design. The plant is arranged and designed so the main equipment can be installed, operated, and maintained safely and conveniently. The auxiliary mechanical equipment and systems, together with electrical equipment and systems and civil and structural facilities, exist for that purpose. This chapter does not discuss ways to determine the type and location of the power plant, but several of the references listed in the bibliography at the end of this chapter discuss this.

Laying out and designing the mechanical equipment and systems must not be done without coordinating with the civil and the electrical design engineers.

The mechanical designer of the plant must obtain information fundamental to the electrical and civil design and evaluate this information to carry out the plant design. This information should include: heat loads and personnel requirements; need for cooling electrical busses and cables; protection and utility services for the electrical equipment; and other matters. The following paragraphs provide a brief discussion of equipment design considerations. A checklist at the end of this chapter lists the primary information needed for designing the mechanical systems that affect the plant layout.

Runner Elevation

The requirements for establishing the runner elevation (setting) will vary depending on the type of turbine and how the turbine will be operated. Reaction machines (Francis, Kaplan, propeller, and Deriaz) usually are installed beneath the lowest tailwater level. However, geologic strata may cause turbine runners to be installed anywhere from below the tailwater to above the tailwater. Where cost-effective, it is desirable to install the turbine high enough that water drains below the wicket gates, or even to below the turbine runner, to accommodate maintenance and inspection. As discussed in Chapter 3, Hydraulic Turbines, the selection of runner elevation is dependent on the turbine-generator speed and on other turbine characteristics. The selection should be made in consultation with the turbine supplier. Deeper installations help avoid cavitation, but may increase powerhouse excavation cost and cost of the water conductors.

Pelton and other impulse-type turbine runners must be above the tailwater level in the turbine pit, so there will be no interference with evacuating water from the runner that will interfere with its rotation. Rotation of the air mass directly beneath the turbine runner and above the water will churn waves whose crests are above the static water level, and must be accounted for in determining the setting of the Pelton turbine runner. Manufacturers should be consulted regarding settings of Pelton turbine runners.

For Pelton turbines, if high tailwater occurs infrequently, such as during high flows during a spring runoff, it may be economical to provide an air depression system to lower the water level in the pit. Setting the turbine runner and the amount of air depression, if any, becomes an economic decision in which the following should be evaluated:

- Frequency of the need for depressing the water level;
- The cost of depression systems operating at different pressures; and
- The cost of lost energy owing to variations in the available gross head as a result of different runner placements above tailwater level.

In this situation, it will be cost-effective to provide a chamber in the tailrace to recover entrained depression air from the water. This minimizes net air consumption if the Pelton turbine is operated below prevailing tailwater level for a considerable amount of time.[1]

Closure Valves (Turbine Inlet Valves)

The function and location of closure valves depends on the type and arrangement of the power plant, the provision of upstream water control gates (see Chapter 7, Gates), the number of turbines supplied from a single principal water conductor, the means and access for servicing and maintaining the valves, and the length of the water conductors. Chapter 5, Major Valves, discusses closure valves (turbine inlet valves).

If a single water conductor supplies more than one turbine, each turbine should have a turbine closure valve. Otherwise, more than one unit would be out of service each time one of the units is inspected, maintained, or repaired.

The plant may need two valves upstream from the turbine—particularly if there is a remote intake where it could take time to lower stoplogs or intake gates. One valve normally would be open and would be closed only in an emergency, on failure of the downstream valve, or while the downstream valve is being maintained. The upstream valve would need minimum maintenance because it would be operated infrequently. At some plants, the upstream valve is a less expensive butterfly valve, if the head is not too great, while the service valve is a spherical valve.

Multi-stage reversible pump-turbines usually have no wicket gates to stop the flow of water. In that case, the second valve, which can shut off or open the flow of water, is located downstream from the maintenance valve. The downstream, or service, valve would be used for normal start up of the units and for normal closure. Some inlet spherical valves, as mentioned in Chapter 5, Major Valves, have both upstream and downstream movable seals. These seals allow maintenance on the downstream seals while the upstream seals are closed and retaining full head pressure.

Both conventional and underground hydroelectric plants may house the valves in the principal equipment enclosure. This avoids needing a separate crane. However, it is often more economical to house the valves upstream from the main enclosure—either outdoors or, for underground power plants, in a separate valve cavern. The desirability of a separate valve cavern (gallery) may be dictated by the difficulty and cost of creating the roof arch when the width of span in the machine hall is increased to provide for the valves and the crane access to them. Using hatches above valves for access limits the amount of usable floor space at upper levels.

Pressure Regulators and Turbine Bypass Valves

Chapter 12, Hydraulic Transients, discusses having valves serve as pressure regulators and protective devices for the machinery and the water conductors. Besides these uses, valves are used as either synchronous or non-synchronous bypasses. This allows water to be released downstream from the power plant even when one or more of the turbines is not operating.

The pressure regulator usually is linked mechanically to the turbine gate operating servomotor or the wicket gate operating ring. Sometimes, the valve also is used as a synchronous turbine bypass to release water for municipal purposes, for irrigation downstream, or for instream flow requirements. In this case, the valve may be at a different elevation than the centerline of the turbine distributor, and, then, the linkage to the turbine flow control may be hydraulic. For mechanically linked regulators, the linkage should be as short and as stiff as practical with a minimum of bearings and bell cranks, where backlash and lost motion can occur. Lost motion or flexibility in the linkage can cause excessive pressure increase on rejecting load.

A pressure regulator linked hydraulically to the turbine flow control should be as close as practical to the turbine. The hydraulic lines should be protected against damage or accidental leakage or loss of oil. For added security, the pressure regulator's hydraulic oil system should be separate from the turbine governor oil system.

The layout, setting, and downstream hood (if used) arrangement of fixed-cone dispersion-type free discharge valves, frequently used as pressure regulators and synchronous bypasses, merit careful consideration. Accidents have occurred when fixed-cone valves have been installed in inappropriate layouts or when they have been improperly hooded. Problems can result from vibrations of a fixed-cone dispersion valve set so deep that adequate quantities of air were not reaching the valve.[2] Vane failures have occurred on valves that have been hooded, possibly because the hoods may have interfered with admitting adequate air to the discharging jet of water.[3] Faulty air admission rather than the vane characteristics may have caused vibrations.

A fixed-cone dispersion valve should be situated at least ten pipe diameters downstream from a bend or from a vortex-shedding obstruction. At that distance, the stream lines will be nearly parallel to the vanes of the valve, and the pressure fluctuations of the vortices will be nearly damped by the turbulent flow. If the water approaches the vanes at an angle, unbalanced forces may cause the thin vanes of the valve to vibrate and crack at the welds. Water will approach at an angle if there is a bend in the conduit upstream from the valve. A vortex-shedding obstruction upstream from the valve, for example a butterfly valve disc, will cause unbalanced pressures on opposite faces of the vanes that also induces vibration in the vanes.

The vertical location of a fixed-cone dispersion valve relative to the downstream water surface also requires careful consideration. In theory, a column of air can be drawn to a depth as great as the barometric pressure. In practice, the maximum column will be less than the atmospheric head. The jet from a fixed-cone dispersion valve draws in air at its margins. The kinetic energy in the jet supplies the energy needed to draw in the air and mix it with the water. This process changes the destructive mass of high velocity water into a nondestructive spray of low velocity foam within a few valve diameters downstream from the valve discharge.

Tremendous quantities of air are drawn from the surroundings into the jet.

At some valve sleeve openings, the quantity of air equals the quantity of water. If the valve is set above the maximum water level and is not hooded, there is no problem with drawing air into the jet. If the valve is not submerged so deeply as to cause the air column to "make and break," the discharge jet will draw a column of air to itself. If the valve is submerged deeper, the column of air feeding the jet may "make and break," causing severe vibration and instability. Finally, if the valve is submerged more than 10 meters (34 feet) (at sea level), no column of air can be drawn to the valve, so there is no danger of a column collapsing and, consequently, no accompanying vibration or damage. There are many situations in which such valves are submerged deeply and used to disperse a jet so it will not impinge downstream onto a concentrated area.

Service and Erection Facilities

Service and erection facilities can be at either end or in the center of a multi-unit hydro plant. In underground plants, those facilities are usually near where the access tunnel or shaft enters the power cavern machine hall.

After the turbine and generator types and sizes have been selected, one of the earliest steps in laying out a hydroelectric plant is working out access for delivering the machine parts and assemblies into the plant and having space for erecting the units. In a surface power plant, permanently installed or temporary equipment can handle major loads. Permanent lifting equipment is needed in an underground plant. Chapter 8, Lifting Equipment, discusses permanently installed lifting equipment. Crane clearances must be worked out in three dimensions so the remainder of the plant can be designed with adequate dimensions for erecting and installing the principal machinery and the auxiliary equipment. Underground plants often have vaults opening from the main power cavern. These vaults have monorails that are adequate for assembling and maintaining the embedded and removable turbine parts.

Access

Provisions should be made for handling large or heavy components or assemblies during installation and dismantling. Cranes, hoistways, hatches, pulling eyes, trolleys, and jacking devices are commonly used. An excessive number of large hatches will use floor space that could otherwise be available for equipment or laying down parts. Alternate means for moving equipment laterally (for example, air cushions) to a place where access for lifting is most convenient have been successful.

The service bay of a plant often has space for stacking the generator core and rotor rim laminations and installing the pole pieces of a large generator rotor. Additional space may be needed for assembling a large Kaplan turbine runner and shaft. At a few plants with very large units and with limits on dimensions and weight of parts that can be transported to the site, the stators typically are stacked in the final stator location.

Provisions also should be made for bringing materials and maintenance supplies, such as barrels of oil and palleted materials, into storage rooms in the plant. A plant may have a separate jib crane and small hatchways for this purpose. The erection and service areas of a large powerhouse may have areas and base plates for rotor/runner erection, the transformer transfer track, railroad and truck access, and shop and storage areas.

On large turbines, it is convenient to have a hoist, traveling on a circular rail above the headcover in the turbine pit, for handling wicket gates, levers, and links. If possible, there should be two entrances to the turbine pit, so one can be used if the other is blocked.

Valves, especially critical shut-off valves, should be located where an operator can shut them in an emergency without having to bring a ladder or other access device. If a valve must be at a high elevation, a chain or motor operator should be provided. In some instances, removable handles or locks are needed for valves whose inadvertent operation would result in an unsafe situation.

For both safety and access considerations, piping should not be routed above switchgear, motor control centers, or other critical electrical components.

Access to Surface Power Plants That May Be Submerged

The elevation at which access to the powerhouse is provided depends on the maximum flood levels of the headwater and the tailwater. At some plants where the head is relatively low, 6 to 15 meters (20 to 50 feet) under normal flow conditions, the powerhouses have integral intakes. During floods on some rivers, the dams become little more than submerged overflow weirs with the difference between headwater and tailwater level being only a third of a meter (a foot) or so. Water may flow over the roof during great floods. To be above high water level, the gantry crane rails and the transformer for the units should be above the road access to the roof.

If no other access during floods were provided, a boat would be the only means of reaching the roof of the powerhouse. All, or almost all, of the upper portions of the powerhouse walls and the roof would be underwater. Obviously, no major access doors should be in the walls of the structure. Watertight hatches in the roof can provide access to the equipment. At those plants, the roof of the powerhouse has stairwell penthouses with watertight doors large enough for personnel access only.

Since such plants cannot have a safe road access to a large opening in the powerhouse wall, a gantry crane is used for bringing heavy equipment to the plant and erecting the equipment through roof hatches. It is practical then, since the powerhouse has an integral intake and a deck over the draft tube gate openings, to provide upstream and downstream extensions of the crane rail beams on the gantry. The single crane can then handle the intake gates, the trashracks, and the draft tube gates or bulkheads using either the main or auxiliary trolleys. The crane legs may, if necessary, be designed for the main hook to travel between the legs to either the upstream or the downstream extension. The gantry crane may travel from the nearest road access to the powerhouse roof by rails on beams that are supported by concrete piers. In other cases, a roadway may lead to the powerhouse roof deck. The roadway will be submerged during a major flood.

If the elevator extends to the roof, elevator hoist rooms also must be protected against flooding. Similarly, all supply and exhaust air ventilation openings must be above the maximum flood water level. In that case, no openings would exist in the powerhouse walls.

Main electric power busses and cables can be brought to the mainland from the powerhouse in watertight tunnels—usually of reinforced concrete. If the land end is high enough above the flood water level, a watertight tunnel can provide personnel access to the plant.

Elevators

Depending on the location, nature, and size of the plant and the installed equipment, the plant should have permanent passenger or freight elevators. Other aspects of the layout of plant equipment and auxiliaries determine the elevator locations for access to and egress from the occupied and working areas of the power plant. However, for safety, the travel distances to elevators, ladders, hoistways, stairways, and ramps should meet requirements of all applicable codes. For interpreting the code, general areas inside a hydroelectric plant are "special purpose industrial occupancies" for maximum travel distances to means of egress. In the code, "occupancies in unusual structures" are for temporary occupancies, means of egress from inside the structures, piers of large bulb units, open structures, and underground volumes (such as tunnels, surge shafts, power caverns, and galleries).

The most recent editions of ANSI A17.1 and ANSI A 17.2, along with applicable state and local codes, give guidelines for designing, constructing, testing, inspecting, and maintaining permanent elevators.[4,5]

Using Temporary Cranes

In some cases, providing permanent lifting equipment capable of handling the heaviest loads, often the assembled generator or the generator rotor, might not be justified. For future availability of lifting capacity, permanently installed, indoor cranes usually are more economical than relying on temporary cranes. Some engineers recommend using mobile cranes for erecting the principal machinery—usually at small plants. If difficulties are encountered during erection and commissioning or if the heaviest components of the machines need to be removed for repair or replacement later, the cost of bringing a mobile crane to the site or the costs of prolonging an outage may surpass what was thought to be saved by not having permanent cranes. Mobile cranes discussed in this section do not include powerhouse gantry cranes, which are movable. They are discussed in Chapter 8, Lifting Equipment.

If the power plant does not have permanent lifting equipment capable of handling the heaviest loads during erection, temporary cranes must be brought to the site. Lifting equipment of lesser capacity should, however, be permanently installed for maintaining the plant. Before selecting a power plant design using temporary equipment for installing and removing the major loads, investigate thoroughly the costs, including evaluation of the costs of time delay and potential increased time during major outages.

Mobile cranes include:

- Wheel-mounted cranes (usually for relatively light lifts);
- Crawler-mounted cranes;
- Barge-mounted cranes; and
- Cranes supported on structures whose frames can move on rails, such as gantry cranes (lateral motion of the wheels supporting the fixed structure) and ringer, or whirly cranes (lateral motion of the wheels and rotational motion of the crane bed).[6]

Determining the kind of temporary crane to use depends on the magnitude of the load and the maximum radius at which the load must be handled. If the combination of load and radius is not too large, a wheel-transported crane may be used. This crane is common at small hydroelectric plants. In the United States where suitable cranes can be rented at any sizable city, the time needed for bringing a wheel-transported crane to the site and preparing it for use is relatively short. However, in some countries and in remote locations, delivery time may be long. If the crane is needed for maintenance or repairs after an emergency shutdown, the time needed for delivering the crane to the site and preparing it for use prolongs the time a unit is out of service.

Rental costs for wheel-transported cranes are low compared with rental costs for larger-capacity crawler cranes or for ringer cranes. The cost for using a wheel-transported crane (for example, one capable of a 45,000 kilograms (50-ton) load at a radius of 9 meters (30 feet)) for several months will include costs of: the monthly rental; transporting the crane, the operator, and the oiler to the site; preparing the crane for use; preparing for transport from the site after use; transport from the site; crane operator's and oiler's salaries and living expenses; and the crane owner's overhead and profit for the entire time the crane is at the site.

The cost for renting a crawler crane, capable of handling the same 45,000 kilograms (50-ton) load, but at a radius of 18 meters (60 feet), would likely be considerably greater than for renting a wheel-mounted crane. The crawler-crane transport and preparation consume much more time, and the investment of the crane owner in the larger capacity crane is greater. Transport and preparation costs are the same whether the crane is at the site throughout the initial period for installing the machinery or whether the crane is at the site only a short time for performing maintenance or repair on the principal machinery.

Because rental costs are less for a smaller crane, the cost of lifting can be reduced if the crane can approach closely to the locations where the major load is delivered or assembled and where it is to be installed. At many sites, and particularly when there is more than one unit, the access road or siding might not closely approach all of the units. It may be possible to bring a roadway to the roof of the powerhouse from where heavy loads can be brought to and lowered through a hatchway in the roof. Barge-mounted cranes can be used where the tailwater is at a river or lake.

When a mobile crane is considered for use at an existing hydroelectric plant, the crane selection may depend on the safe maximum loads that can be supported by the roof deck and by the walls and frame supporting the roof. The crane itself, the space needed for rotating the crane cab, and ballast weights and base of the boom, the hatchway, and the lowboy truck would all have to fit on the roof if the radius at which the load is handled is minimized. In some cases, sufficient space can be provided in advance by designing and building the roof with an extension downstream from the generator hall. Space also must be provided for the crane outriggers, which spread wider than the truck wheels. In use, the entire crane is jacked up on its outriggers so the wheels do not touch the ground or deck. All of the load is transferred from the outriggers to the roof. The maximum load on any one outrigger must be determined since the loads are not the same on each outrigger. Plates and timbers can be used to spread the maximum load transferred to the supporting deck or ground to an area greater than the area beneath the outrigger. Doing that, however, requires additional space.

An important disadvantage of planning to support the crane and its load on the powerhouse roof is that the walls and roof must be built and cured long enough to support the design loads. That may delay the time when the crane can first be used for installing the embedded portions of the machines. If it were not for depending on the roof to support the crane, the embedded parts could be safely installed earlier without the protection provided by the powerhouse walls and roof. The revenue lost by delaying the initial erection and operation of the principal machinery until the roof can safely support the crane and the load may exceed the possible savings of using a wheel-transported crane rather than a crawler-crane capable of handling the load at a greater radius.

Installing the embedded parts with a temporary crane and providing a permanent bridge crane or gantry crane for installing the heavier components later may cost less than using a temporary crane for the whole installation. Having permanent cranes thereafter will be convenient. Future maintenance costs and loss of revenue during time out of service will be reduced, even though the owner must pay the annual cost for maintaining the permanent crane. Revenue will be lost while waiting for transporting and preparing a temporary crane for use—particularly if a crawler crane is needed. Providing a permanent powerhouse bridge crane or gantry crane, depending on the layout and available accesses, may be less costly.

Each hydroelectric plant has its unique combination of access, possible powerhouse structures, maximum load and radius for handling the load, number of units, and circumstances related to proximity and availability of suitable rental equipment. In some locations, rental equipment of adequate capacity is not available in reasonable time—at any cost. Planning installation and future major dismantling of the equipment affects the layout of the power plant and its environs and the design of the structures. Installing the equipment is part of the overall planning of the project design and its construction. How the equipment will be installed and maintained should be worked out before final details of design are undertaken.

Special Use of Cranes

A large, multi-unit plant that is outdoor or semi-outdoor may not have an enclosure large enough for disassembling and working on the turbine or the generator. Most large plants have a separate erection and maintenance bay. Where winter weather may be very severe, a gantry crane can be fitted with panels that slide down between the gantry legs to enclose a large working space. The interior space can be heated, and personnel can carry out major maintenance on the unit protected from the weather.

Maintenance Provisions

Consideration should be given to where the spare parts supplied with the turbines and generators will be stored. The turbine floor level should have space for tool boards for the units. At multi-unit plants, it is often convenient if the wrench boards supplied with the turbines are mounted on wheels so they can be easily moved from one unit to another.

Spacing between the equipment in the plant should consider the clearances needed for removing individual components and the working space needed for the people and the tools required for maintenance. There should be temporary storage space for laying down dismantled parts. If the plant has stoplogs instead of draft tube or intake gates, there should be a storage rack for the stoplogs. Provisions should be made to allow convenient access to the turbine runner for inspection and to make weld repairs in place, if possible. Vertical turbine units should have a platform of removable planks or plates installed that permits measuring clearances, centering, and repairing the turbine runner from below. Providing slots, covered by removable plates, for the platform supports in the walls of the upper part of the draft tube will allow this.

Many owners use removable lower sections of the draft tube cone on vertical reaction turbines to allow removing the runner from below without removing the generator or the bearings, the distributor mechanism, or the headcover. Special runner handling cars with hydraulic jacks permit removing the runner without relying on a powerhouse crane. The powerhouse crane can be used together with a runner handling cart, with wire rope cables passed through the hollow center of the shaft. The cables support a cross frame that is installed beneath the runner.

However, using cables through the shaft is no longer recommended because of the danger to human life if the cables fail. Large Francis and Pelton turbines often are dismantled from below.

Miscellaneous Considerations

Because powerhouse walls may be damp, especially below water levels in the substructure, mechanical and electrical components should not be mounted directly to the walls. Equipment mounted on standoffs from the walls will be less likely to suffer damage from moisture and corrosion. It is helpful to use moisture-tight enclosures for critical instruments and controls.

Control rooms should be designed to minimize noise transmission from the equipment in the power plant. This is important to protect the hearing of plant operators and to meet Occupational Safety and Health Administration (OSHA) standards.

Oil pipes between the governor or hydraulic controller and the wicket gate (or needle or Kaplan blade) servomotors should be as short as possible to minimize time lags and improve response time. The designer should arrange the plant galleries so the governor oil system components can be placed near the turbine oil-hydraulic servomotors. Piping for auxiliary systems in the hydroelectric plant is discussed in Chapter 9, Auxiliary Systems.

Large plants often have storage rooms for combustible materials such as paints, oils, greases, and cleaning compounds. Large plants often have machine shop facilities in the plant or in a separate machine shop and maintenance building. This is particularly true at sites that are isolated from industrial machine shops and in countries where local machining facilities are expected to be inadequate for the large equipment.

The drawings in the equipment procurement specifications should give the maximum outline dimensions and recommended equipment locations, so the powerhouse design may continue without waiting for the final dimensions from the manufacturers.

References

[1]Grein, H. and R.K. Holler, *Operation of Pelton Turbines Under Back Pressure Conditions,* Escher-Wyss.

[2]Broehl, D.J., and J. Fisch, *Solution of Vibration Problems Experienced with Howell-Bunger Valves at Round Butte Dam,* International Congress on Large Dams, Ninth Congress, Istamboul, Question 33, Response 20, 1967.

[3]Mercer, A.G., *Vane Failures of Hollow-Cone Valves,* International Association of Hydraulic Research Symposium, Stockholm, 1970.

[4]*Safety Code for Elevators, Dumbwaiters, and Moving Walks,* ANSI A 17.1, American National Standards Institute, current version.

[5] *Practice for Inspection of Elevators,* ANSI A 17.2, American National Standards Institute, current version.
[6] Shapiro, Howard I., *Cranes and Derricks,* McGraw-Hill, Inc., New York, New York, 1980.

BIBLIOGRAPHY

Fire Protection for Hydroelectric Generating Plants, ANSI/NFPA 851, American National Standards Institute, 1987.

Goncharov, A.N., *Hydropower Stations Generating Equipment and Its Installation,* Energiya, Moskva, 1972, translated from Russian by Th. Pelz, Israel Program for Scientific Translations, Keter Publishing House Jerusalem, Ltd., Jerusalem, 1975, and available from the U.S. Department of Commerce, National Technical Information Service, Springfield, Virginia.

Hydraulic Institute, *Standards for Centrifugal, Rotary, and Reciprocating Pumps.*

Life Safety Code, ANSI/NFPA 101, American National Standards Institute, 1985.

Mechanical Design of Hydro Plants, Design of TVA Projects, Technical Report No. 24, Volume 3, Tennessee Valley Authority, U.S. Government Printing Office, Washington, D.C., 1960.

Mosonyi, Emil, *Water Power Development,* Publishing House of the Hungarian Academy of Sciences, Budapest, Hungary, 1957.

CHECKLIST FOR CHAPTER 11

1. Have access and lifting requirements for equipment used during construction been evaluated?
2. Have permanent entrances and exits during normal and maximum flood conditions been adequately evaluated considering life safety code requirements?
3. Has the turbine placement been evaluated with respect to minimum and maximum tailwater levels considering local geology and temporary and permanent access?
4. What structural protection is needed for the main machinery and the auxiliary systems?
5. Where will the intake and the draft tube closure devices be located? How have they been selected?
6. Has the choice between permanent and temporary lifting equipment for maintenance been adequately evaluated?
7. What type of turbine guard and service valves will be used and where will they be located?
8. What type of pressure regulators and turbine bypass valves will be used and where will they be located?
9. Is there a need for permanent service and erection facilities inside or near the power plant?

10. Is there a need for elevators for personnel or equipment? If so, do they comply with life safety codes?
11. Have unit-by-unit auxiliary systems or central systems been adequately evaluated?
12. Have fire protection systems, equipment locations, and the water supply and conduits for fire water been selected?
13. How will power be supplied in emergencies? How will fire protection and other systems be activated or shut down if there is loss of power from both the generating units and the transmission line?
14. Is there a need for machine shop facilities at or near the plant?
15. Have heat loads been evaluated to determine permanent and temporary heating, ventilating, and cooling systems?
16. Are the air supply and exhaust vent openings above the maximum flood water level?
17. Has the need for and details of draft tube or tailwater depression systems been evaluated?
18. Should the turbine supplier furnish a hoist inside the turbine pit?
19. Have valves been specified, which have accessible operators as well as locks or removable handles?
20. Does the layout avoid routing piping above switchgear, motor control centers, or other electrical equipment?
21. Are adequate, small lifting devices provided at convenient locations for normal plant operation and maintenance?
22. Has adequate space been provided within the power plant for laying down parts during assembly and disassembly?
23. Has adequate space been provided around each piece of equipment to permit personnel to operate, lubricate, and repair the device?
24. Are there provisions for access beneath the turbine runner for inspection, taking measurements, maintenance, and making repairs?
25. Would removing the turbine runner from below be desirable? If so, has space been provided for moving the upper portion of the draft tube cone (if there is one) and the runner to places where they can be out of the way or repaired?

Chapter 12
Hydraulic Transients

Hydraulic transients are disturbances that occur in a fluid during a change from one steady state to another. The disturbances considered in this chapter are pressure changes caused by pressure waves propagating throughout the hydraulic system. Figure 12-1 shows a comparison of the predicted and the actual field pressure changes at a well-designed hydroelectric plant. This chapter emphasizes designing a plant in such a way that no damage occurs from the inevitable transient conditions.

Various references provide descriptions of the basic equations of continuity and motion needed for calculating the pressure changes associated with hydraulic transients. These references include Wylie and Streeter (method of characteristics), Bergeron (graphical method), and Rich (arithmetic method).[1,2,3] Any of these methods will achieve comparable results. Originally, designers used these methods manually. Proven, reliable computer programs are now commercially available that calculate these values electronically. However, the buyer and user should be aware of how the programs operate and decide whether the programs are applicable to the case at hand. It is unwise to use such tools unless the user is thoroughly familiar with the subject.

Among those methods that use elastic (nonrigid) water column theory are those of Allievi, as explained in documents such as Parmakian's.[4] The simplified methods usually yield safe estimates provided that the designers uses conservative parameters. Additional guidelines to use in the various methods for computing the pressures are given by Pejovic, Boldy, Obradovic, Chaudry, and Zipparro and Hasen.[5,6,7] Those references are from among many available works detailing the methods of calculation. This chapter describes what cases should be considered and does not repeat details of the rigorous calculations given in the various references.

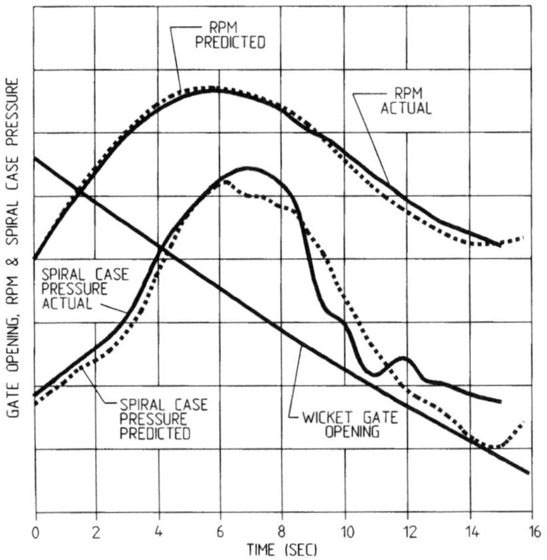

FIGURE 12-1: ***Transient Calculations versus Field Results.*** *This graph shows the transient calculations (predictions) versus field (actual) results obtained following a load rejection. The data is from Unit 1 at the Cerron Grande Dam.*

DESIGNING FOR TRANSIENT CONDITIONS

Transients will occur in every hydro plant because changes must be made starting up from standstill, varying load, and shutting down. Also, transients result from human errors, malfunctioning equipment, transmission line disruptions, accidents to the water conductors, and earthquakes. The engineer must focus attention on those transients whose magnitudes, if not otherwise mitigated or avoided, might endanger the plant's safety and its personnel or might damage the equipment or cause operational difficulties. The engineer analyzes the transients to determine the most severe operating conditions. The designer of the machines and the water conductors will then be able to provide a safe, undamaged hydroelectric system.

A hydroelectric unit contributing to speed or frequency regulation of a connected load must be controlled to adjust flow through the turbine as rapidly as practical, to minimize changes in speed or frequency when the load changes. However, as the time interval allowable for changing flow decreases, the magnitude of the transient changes in pressure in the water conductors increases. The practical limit for the rate at which water flow can be changed then becomes the ability of the water conduits to withstand the magnitudes of the transient pressures. If the generator's overspeed is not a concern, the governor could shut off the flow very slowly, wasting a small amount of water but causing only minor pressure transients.

Hydraulic transient pressures can be addressed by providing sufficient strength in the water conduits to accommodate them, by limiting their magnitude through measures described later, or by a combination of both. With increased strength, the water conduits can withstand higher maximum and lower minimum pressures. However, water column separation as described below must be avoided under all circumstances. Increasing the wall thickness of the water conduit or using a higher strength material (for example, higher strength steel, hoop wound concrete, or hoop wound steel) or combinations of higher strength material will allow larger transients without damage.

One way to minimize hydraulic transient pressures is to maintain a constant flow of water through the turbine. Smaller turbine-generator units can have a "water rheostat" control instead of a traditional governor. This avoids the need to change flow through the turbine as the electrical load changes. The water rheostat is a variable resistance placed in the tailwater to dissipate a portion of the generator output. As the electrical load required from the plant changes, the turbine control device simply assigns any excess power being generated to the water rheostat for dissipation as heat in the water. Thus, the generator output and corresponding turbine flow remains constant.

Pressure transients exceeding design amounts may destroy penstocks, tunnels, valves, or other system components. Strong increases in pressure may crack internal linings of the water conductors; may damage connections between steel and concrete; or may damage the hydraulic machines, the thrust bearings, air valves, penstock valves, or gates. The damage from transient pressure may not be evident immediately but, after repetitive transients, the machinery or the water conductors may fail.

Minimum pressure during transients, below design amounts, may cause mechanical damage to water conductor supports or collapse penstocks or concrete water conductors that are not designed to withstand such strains. Water column separation and the subsequent rejoining may create forces that damage turbine runners and non-rotating parts. This condition occurs when the local pressure drops below the vapor pressure of water or the partial pressure of dissolved gases. The gases may promote corrosion and weakening of steel sections resulting in cavitation damage when the vapor bubbles collapse upon entering regions of higher pressure.

Strong hydraulic vibrations created during transients might damage the water conductors as well as the measuring and control equipment. If resonance occurs, the entire power plant may be destroyed. Oscillations of the water masses between the reservoir and a surge tank may cause noise, concussions against the walls of the water conductors and tank, and compressible air to enter into the system. Control over the plant could be temporarily lost, with potentially disastrous consequences.

The design for control of transients should:

- Ensure the safety of the plant and personnel during any transient condition;
- Identify the most adverse transient conditions and take adequate measures to ensure that transient parameters remain within design values; and
- Establish reasonable design limits for speed rise and transient pressures.

Transients can create axial and radial forces acting on turbine runners, headcovers, bearing supports, and other structural components. Allowance must be made to accommodate the structural deflections resulting from transient conditions.

Protection

If properly controlled, positive and negative waterhammer pressures will be harmless. Many means exist for controlling the maximum and minimum pressures. The basic protective means may be to:

- Increase the size of the water conductors;
- Convert kinetic energy of motion into potential energy by providing surge tanks, reservoirs, or air cushion surge chambers;
- Slow down the angular acceleration of the unit by increasing the polar moment of inertia of the rotating parts—principally the generator;
- Regulate the opening and closing rate of valves or wicket gates;
- Add or subtract a quantity of water by providing regulator valves or synchronous bypass valves;
- Admit air through valves or aeration pipes;
- Limit the damage by providing calibrated "rupture" membranes (which burst at a predetermined overpressure);
- Change the natural frequency of the combined civil, electrical, and mechanical system;
- Select a specific speed for the turbine runner that will minimize the change in flow during excursions into overspeed; or
- Select a Pelton turbine instead of a Francis turbine.

The relative merits and shortcomings, as well as the costs, for providing alternative protection should be considered. The best solution may be a combination of alternatives. There are economic decisions to be made in optimizing the size of system components and the selection of materials. Also, economic decisions need to be made in determining the various combinations for controlling pressure and speed changes. For example, mass may be added to the generator and the timing of the governor stroke may also be changed. The designer should be aware that adding a surge tank, adding an air chamber, changing the penstock diameter, or rearranging the water conductors will change the natural frequency of the system. If the new natural frequency is closer to the frequency of pressure waves excited by the disturbance, the likelihood of resonance might be increased.

Table 12-1 gives the general characteristics of the various protection schemes. If properly controlled, positive and negative pressure surges will be harmless, and many means exist for controlling the maximum and minimum pressures. This table gives information about basic protective means. The relative merits and shortcomings, as well as the costs, for providing alternative protection should be considered. The best solution may be a combination of two or more methods.

Starting-Point, Steady-State Conditions

If all the operating parameters are constant with respect to time, the plant is in a steady state. Any change in operating condition (for example, rate-of-flow) will cause a transient that attenuates with time until another steady state is reached. While there are always hydraulic vibrations, in most practical cases the small, periodic disturbances may be omitted from the calculations of the transients. The operating conditions without accounting for the vibrations are then assumed to

TABLE 12-1: *General Characteristics of Waterhammer Protection Devices*

Device	When to Use	Protects Against High Pressure	Protects Against Low Pressure	Operating Reliability	Special Demands	Problems in Restarting	Frequency of Application	Cost
Surge Tank	High head with long tunnels/penstocks	Yes	Yes	Very good	None	None	Very often	Very high
Air Vessel	High head with long tunnels/penstocks	Yes	Yes	Good	Air compressor	Routine control	Rarely	Not very high
Increased Inertia GD^2 (WR^2)	All types	Yes	Yes	Excellent	None	None	Often	Not very high
Regulated Opening and Closure	All types	Yes	Yes	Good	High-pressure oil installation	Routine control	Very often	Low
One-Way Reservoir	Very long tunnel/penstock	No	Yes	Poor	Frequent maintenance and control	Routine control	Rarely	Not very high
Air Valves	Protection of penstocks	No	Yes	Poor	Frequent maintenance and control	Routine control	Very often	Low
Aeration Pipes	Protection tunnel/penstock	No	Yes	Good	None	None	Often	Low
Rupture Membrane	As a last measure of protection	Yes	No	Good	Arrange outlet for water discharge	Replace membrane	Rarely	Very low
Pressure Regulator Valves	All types	Yes	No	Very good	Air in large quantities if Howell-Bunger type valve	None	Often	Low

be the initial steady state.

The initial steady state is the basic reference for subsequent calculation of the transients. There will be a range of possible initial steady states for a given range of plant operating conditions (head and load) and turbine (or pump-turbine) performance characteristics. The designer should note that the reservoir levels may vary considerably—during floods, when there may be surcharge over the spillway gates, or at the end of an irrigation release season. The tailwater level may vary depending on operation of downstream reservoirs, control by downstream structures or obstacles in the channel, local inflows downstream backing water up to the tailrace, or variation of tidal water levels. Abnormal water levels, such as those that could occur after severe winter icing, after design-maximum floods, or during extreme high or low tides, should be considered.

Rigid Column Theory for Slow Changes in Flow

An initial evaluation of a hydroelectric plant may be made using rigid column theory. The basic parameter used is the water starting time, T_w, already defined in Chapter 4, Governors. This parameter describes the inertia of a moving column of water, and can be used to characterize the water column upstream of a hydraulic turbine when estimates of pressure rise must be made for turbine and penstock structural design purposes. The parameter has the physical meaning that if the water column is decelerated at a uniform rate, in a time equal to the water start time, there will be a pressure rise of 100 percent over the static pressure. The pressure rise will decrease proportionately as the closure time is extended. For example, for a closure time of $4 \times T_w$, the pressure rise is 25 percent of the head. In actual practice, a uniform gate movement does not give a uniform deceleration. This results in the pressure rise in the early part of the closure being slightly less than predicted, and slightly more during the final part of the closure.

Rapid Changes in Flow

Where a gate or valve closure occurs rapidly, compressibility effects must be taken into account. Flow changes at the turbine are propagated through the water column by pressure waves with a celerity or velocity of propagation of approximately 1,200 meters/second (4,000 feet/second). This velocity is reduced by the elasticity of the penstock wall and can be estimated using methods described in texts by Wylie and Streeter, Parmakian, or Chaudry.[1,4,6] The reduction may be very significant for a low-head penstock built with thin steel, wood stave, or fiberglass walls. If the turbine gate closure takes place in less time than is required for the wave to propagate to a free surface and back to the turbine, the closure is rapid, and rigid column theory cannot be used. Pressure rise is now controlled by the conversion of the kinetic energy in the moving water, which converts to pressure energy. For a uniform diameter conduit, the following relationship, expressed as Equation 12-1, is used to calculate the pressure rise.

EQUATION 12-1:

$$\Delta H = \frac{a\Delta V}{g}$$

where:
ΔH=increase in head;
a=celerity or velocity of propagation;
ΔV=change in flow velocity (initial flow velocity, in this case); and
g=acceleration of gravity.

Indeterminate Cases

Pressure transients can be estimated for simple conduits using simple or approximate methods where complete valve closure is occurring either very rapidly or very slowly. However, many engineering situations must deal with complex turbomachinery installed in complex conduits, with partial closures that cannot be described as either fast or slow. These situations are best analyzed using special-purpose computer programs. The people doing this work either should be experienced in similar calculations, or be prepared to spend sufficient time to properly understand and verify the results that are being obtained.

Computer Calculation Methods

There are basically two computational schemes in use for pressure transient calculations:

- Method of characteristics; and
- Method of implicit finite differences.

The first method is described by Wylie and Streeter [1] and by Chaudry.[6] The second method was originally described by Perkins et al.[8] Considerable effort has been expended to verify predictions based on these computational schemes, by comparison with recorded pressure transients at actual hydroelectric projects. The method of characteristics probably has been more widely used. However, the implicit finite difference method may use variable time steps and should be more efficient of computer time where slow surge effects must be calculated over extended time periods. For most engineering applications, either method should give acceptable results.

The engineer should analyze the transients for the range of initial steady-state combinations of head and flow (or discharge)—even though some of the steady states may not occur frequently. The analyses should include calculations of the pressure changes and speed changes (and surge tank or air chamber water level changes) after changes are made from the steady states.

There are more initial steady-state conditions to be analyzed for pumped-storage plants because the operating range in pump mode differs from the operating range in the turbine mode.

Special attention should be given to situations where more than one unit is fed by branches from a common penstock. In such cases, there will be reflections of pressure waves between the individual units as well as the primary transient pressure waves.

Influence of Control System

Role of a Governor

Transient conditions are, by definition, generated during any change of turbine flow. The nature and intensity of the transients depend on the overall characteristics of the hydropower scheme in question—including the type and characteristics of the turbine governor. A badly tuned governor may induce serious power and pressure oscillations in the system that are very annoying to the personnel and potentially dangerous for the plant.

Flow Regulation Devices

Wicket gates (in Francis, Kaplan, diagonal, or bulb turbines) or needle valves (in Pelton turbines) are the primary flow regulators. The devices are responsible for rapid closure, change of load, and rapid opening. If the flow regulation devices fail, upstream or downstream valves should take over by closing down to prevent the plant from eventual damage and loss of water. These valves may be turbine shut-off valves (located adjacent to the turbines), penstock valves (located at the top of the penstock), intake valves or gates (located near the upper reservoir), or draft tube gates. Analyses of the transient conditions must include how these valves operate in all circumstances—especially during rapid closure.

To allow for variability in closure timing, transient calculations should be made using the following assumptions:

- Compute maximum over-pressures with closure time scaled down by 10 percent; and
- Compute maximum overspeed with closure time scaled up by 10 percent.

The rapid closing and opening of the wicket gates are of fundamental importance in analyzing transient conditions.

Rapid closure. Rapid closure is an important operation that generates high stresses and strains in the system. If it is performed too quickly, the pressure rises will be excessive; if it is too slow, rotating parts will attain unacceptably high speeds. An optimum solution can only be found by detailed analysis. This analysis must take into account all the elements of the hydraulic system, characteristics of the turbines, flow-regulating equipment, and governors. Also, the relation between the servomotor stroke and the wicket gate openings should be included. In some cases, several analyses have to be made before obtaining a satisfactory solution. Decreasing the servomotor speed may attenuate or cushion high over-pressures, appearing toward the end of the closure time. This is described as a "two-speed" closure. Sometimes it is necessary to use a "three-speed" closure relationship.

The rapid closure analysis has to be performed at least twice:

• Once, in the design phase before the equipment manufacturers are known. This analysis has to be based on standard characteristics, experience, and good judgment;

• Secondly, after bidding when the contract has been awarded. The analysis is based on detailed and reliable data about the equipment and machine performance; and

• A third analysis may be required if the head or other site conditions during commissioning are different from the design values.

Usually the rapid closure of the wicket gates is initiated after a load rejection to prevent the turbine speed from increasing to the runaway value. As shown in Figure 12-2, this is achieved through rapid movement of the servomotor piston. Figure 12-2 shows the relative servomotor stroke, y, over the time for a typical rapid closure operation and also demonstrates the following parameters:

T_q=servomotor dead time(s);

T_f=minimum closure time from fully open position;

T_h=cushioning time(s); and

T_z=total closure time(s).

The servomotor piston moves with maximum speed until the transition position, y_h, is reached. Afterward, movement of the piston is retarded to attenuate pressure surges in the penstock. The value of the dead time, T_q, is usually 0.2 to 0.7 second for turbines and pump-turbines. The final governor values of T_f, T_h, and y_h are typically selected based on review of the resulting pressure and speed transients calculated from a range of trial governor valves. The transition position, y_h, is usually between 0.05 and 0.2, but there are numerous exceptions.

The servomotor stroke-time relationship shown in Figure 12-2 corresponds to rapid closure from full load. The same shape is also valid for quick closure from a partial load—only the starting point is moved from y=1 to $y=y_0$, as shown in Figure 12-3. The final position of the servomotor piston may not be zero, and, in such a case, the whole procedure is followed until the prescribed value of y is reached (y_{k1} or y_{k2}).

The servomotor speed depends, to a certain extent, on oil viscosity and temperature, the state of the mechanism, and the quality of maintenance. The engineer must consider all these external factors in the design phase. The transient calculation should:

• Compute maximum overpressure for the closure time, T_f, reduced by 10 percent; and

• Compute maximum overspeed for the closure time, T_f, increased by 10 percent.

The shape of the closure relationship shown in Figure 12-2 remains similar when the time scale is reduced or increased by 10 percent, respectively.

Slamming, caused by the uncontrolled rapid closure of the wicket gates, must be prevented if the governor loses control of the gate closure rate. Controlling

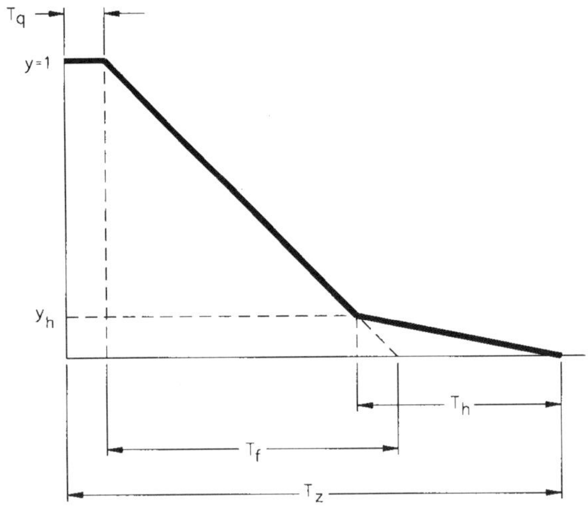

FIGURE 12-2: *Relative Servomotor Stroke versus Time for a Typical Rapid Closure from Full Load.* *Usually the rapid closure of the wicket gates is initiated after a load rejection to prevent the turbine speed from increasing to the runaway value. As shown in this figure this is achieved through rapid movement of the servomotor piston. In the figure, y=relative servomotor stroke; T_q=servomotor dead time(s); T_f=minimum closure time from fully open position; T_z=total closure time(s); and T_h=cushion time(s).*

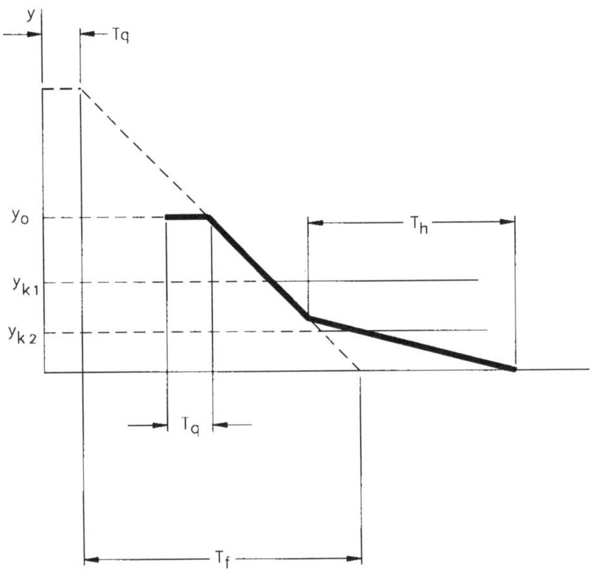

FIGURE 12-3: *Relative Servomotor Stroke versus Time for a Typical Rapid Closure Operation from Partial Load.* *The servomotor closing movement is similar to that of Figure 12-2, except that closure begins from a partial opening, y_o, and may end at a non-zero position such as y_{k1} or y_{k2}. In the figure, y=relative servomotor stroke; T_q=servomotor dead time(s); T_f=minimum closure time from fully open position; T_z=total closure time(s); and T_h=cushion time(s).*

the closure of upstream valves is very important. Uncontrolled slamming of these valves has caused many disasters. Closing speed of valves should be adjusted when the unit is not in operation and rechecked after any maintenance to components related to wicket gate timing. Hydraulic cylinders with fixed rate control orifices installed at the cylinder body should be employed to help prevent uncontrolled slamming if the control lines break.

Opening

If the opening of the wicket gates is performed too rapidly, serious vacuum conditions may be created, endangering the plant. Figure 12-4 shows a typical procedure that indicates the following parameters:

T_q=servomotor dead time(s); and

T_{f0}=minimum opening time(s) from closed (y=0) to full open (y=1).

The value of the dead time, T_q, is the same as for rapid closure—between 0.2 and 0.7 second. The value of T_{f0} has to be calculated, usually assuming that no vacuum is permitted that would damage the water conductors.

Special considerations. The rapid closing and opening rates for the wicket gates have to be carefully calculated. The rates may need to be adjusted based on measurements taken during commissioning. Any deviation from the accepted maximum servomotor speed might induce either excessively high over-pressures, vacuum conditions, or overspeed and, eventually, cause damage.

The piston's maximum displacement in the governor distributing valve is adjusted so that the maximum speed of the servomotor is equal to the value of rapid closure calculated during the design process. This timing adjustment is done when the unit is out of operation. Another approach is to use an independent flow control valve. Afterward, during normal operation, both the rapid closing and opening of the wicket gates should be checked. This should be repeated after every major maintenance or repair of components, which affect the gate timing.

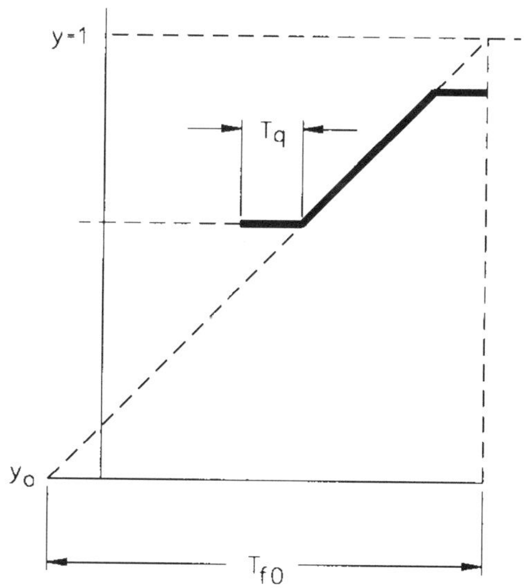

FIGURE 12-4: ***Relative Servomotor Stroke versus Time for a Typical Opening-Up of the Wicket Gates.*** *If the wicket gates are opened too rapidly, serious vacuum conditions may be created that endanger the plant. The servomotor dead time, T_q, is the same as for rapid closure—between 0.2 and 0.7 second. T_{fo} is the minimum opening time(s) from full closed to full open.*

DATA FOR TRANSIENT COMPUTATIONS

Turbine or Pump-Turbine

Transient, hydraulic oscillation and stability analyses depend on the turbine's performance characteristics. During the feasibility studies and the design phase, these characteristics may be derived from the available literature and data for plants of similar specific speed. As the design progresses, the actual performance characteristics should be obtained from hydraulic model tests.

The most common representation of the performance characteristics is based on unit quantities. In general, the performance data should cover the complete operating range of the turbine up to and beyond the zero efficiency curve. Special attention should be given to small values (including zero) of wicket gate opening.

Valves

Experimental data on head losses in valves, measured in steady-state conditions, have to be used for transient analyses. The head loss coefficient (ζ) is given by Equation 12-2.

EQUATION 12-2:

$$\zeta = \frac{1}{K^2} - 1$$

where:
K=flow coefficient for the valve that is usually presented graphically as a function of the valve opening, y.

The graph for a specific valve should be obtained from the manufacturer's data. The head loss, H_L, across the valve is given by Equation 12-3.

EQUATION 12-3:

$$H_L = \zeta \frac{v^2}{2g} = \zeta \frac{Q^2}{2gA^2}$$

where:
H_L=head loss, meters (feet);
Q=flow, cubic meters/second (cubic feet/second);
A=valve area, square meters (square feet); and
V=velocity, meters/second (feet/second).

Celerity

To calculate the celerity (speed of pressure waves) for the various sections of the hydraulic system, it is necessary to have data specific to the system.

For steel penstocks: diameter, thickness, modulus of elasticity, Poisson ratio, and support conditions.

For tunnels: dimensions of steel linings (as for steel penstocks), characteristic dimensions of the concrete, and the modulus of elasticity and Poisson ratio for steel, concrete, rock, or earth—whichever is appropriate for each tunnel.

The celerity may be from about 1,400 meters/second (4,700 feet/second) for a rigid conduit that is solidly encased to 1,000 meters/second (3,300 feet/second) for a relatively flexible, exposed steel penstock. The presence of free gases may reduce the celerity to less than 90 meters/second (300 feet/second) and will influence the magnitude of the waterhammer greatly.

Gas bubbles substantially decrease the value of the celerity, so it is important to know the quantity of air (or other gases) that are free or dissolved in the water. When the local pressure falls below saturated pressure, as in the case of the turbine draft tube, the dissolved air will be released reducing the celerity.

Hydraulic Losses

For any properly designed and constructed hydroelectric system, local head losses are not very important when analyzing hydraulic transients. Friction losses have minimal effect on the peak amplitudes but do affect the subsequent damping. Any formula for the friction head loss derived for steady-state computations may be used. The Darcy-Weisbach equation is the most popular formula. Handbooks on hydraulics have friction factor values for using in the friction head loss equations.

Surge Tanks

Various surge tank designs include cylindrical, differential, with or without lower and upper expansion galleries, or a gate shaft design. Any space that may be temporarily occupied by water during a transient operation should be regarded as a surge tank (i.e., aeration pipe, gate shaft, access shaft, overflow, or horizontal or sloping chamber). Any surge tank should be designed for overtopping the tank as well as for emptying due to transients. An alternative to open surface surge chambers is a closed chamber in which the water pushes against the air above.

To analyze transient phenomenon in systems using a surge tank, the following surge tank information is necessary:

- Shape of the surge tank (horizontal area as a function of elevation);
- Location of surge tank in the hydraulic system (there may be alternatives in the feasibility study);
- Overflow arrangements; and
- Connection between the surge tank and the hydraulic system.

Equation 12-4 gives the head loss, H_L, at the entrance to the surge tank.

EQUATION 12-4:

$$H_L = K_{st} Q_{st}^2$$

where:

$K_{st} = K_{in}$ for $Q_{st} > 0$; or K_{out} for $Q_{st} < 0$; and

Q_{st}=flow into surge tank (> 0); or flow out of surge tank (< 0).

Chapter 12

The value of K_{in} and K_{out} can be evaluated theoretically in simple cases, but it is best to determine the values using measurements on hydraulic models. In many situations, the flow throttling into or out of the surge tank is rather strong to keep the dimensions within acceptable limits. Special orifices, usually non-symmetrical, are designed to satisfy the limits, and evaluating the head loss coefficients, K_{in} and K_{out}, over the whole operating range on a hydraulic model is essential to establish their values. K_{in} and K_{out} are not constant but vary with the ratio of the flow into or out of the surge tank to the flow through the plant. If available or necessary, use these variable coefficients in the transient analysis.

During particularly violent surges, water may spill out of the surge tank. Simulate this effect in the analysis using the weir width and overflow constant. To avoid admitting large slugs of air into the water conduits, do not allow draining the surge tank under any condition.

Surge tanks do not necessarily attenuate hydraulic oscillations; in fact, in some cases, they even act as a resonator. For the analysis of hydraulic oscillation and stability of the whole system, it is necessary to have additional data about the surge tank to calculate the celerity in the surge tank (see the "Celerity" section in this chapter).

Pressure Regulator Valves

Pressure transients are frequently minimized by pressure regulator valves that open synchronously with the rapid closing of the wicket gates. This permits the flow in the water conductor to remain nearly constant while diverting part of the flow away from the turbine, thereby preventing excessive over-pressures. After the wicket gates are closed, then the pressure regulator valves are closed down slowly. Such an arrangement has a profound influence on the transient conditions and must be simulated carefully. The required valve flow versus stroke data (see the "Valves" section in this chapter) can be obtained from manufacturers.

Draft Tube Gate

If a draft tube gate capable of emergency closure against flow is incorporated in the plant and located close to the turbine, its presence can change the performance characteristics of the turbine. In this case, the turbine test model must incorporate the draft tube gate.

Branching

Usually, the head losses in the vicinity of branches are neglected in transient flow analyses since these branches are designed and constructed carefully. However, if it is considered that these losses might be significant, the head loss coefficients have to be determined through laboratory experiments for different flow ratios in the branches. However, reflections of waves into and out of the branches may significantly affect hydraulic transients experienced by the unit being analyzed. Whenever branches are present, the designer should analyze the effects of branches.

Governor

The governor and wicket gate data needed for normal transient analysis has been discussed previously in the Influence of Control System section. For stability analysis and hydraulic oscillation computation, further information is necessary including:

- Maximum servomotor stroke;
- Maximum inclination of runner blades (Kaplan, Deriaz);
- Time constant of the spool valve for the runner servomotor;
- f(y), relation between wicket gates opening and servomotor strokes; and
- $f(a_0)$, on-cam relation between wicket gates opening and runner blades opening.

Generator

The generator's influence on hydraulic transients is the magnitude of its polar moment of inertia, usually given as GD^2 (WR^2). While the total GD^2 (WR^2) of the rotating parts includes the contribution of the turbine runner and shaft, the generator's inertia is the much larger value. Large inertia values help governor stability and can reduce pressure transients following the rejection of electrical load. In some cases, greater than normal inertia can be purchased with the generator at nominal additional cost.

TRANSIENT CONDITIONS

Transient conditions can be classified into three hierarchies of increasing severity to the system:

- Normal operating conditions: These include start up, shutdown, increasing or decreasing load, or occasional loss of load. The design should permit no damage to the system during normal operating transients, and stresses on the structures should be within the normal allowable range. Note: Many small turbine-generator units and low-head units of even the greatest size have so little inherent inertia that they will reach runaway speed before flow can be reduced to diminish the shaft torque. That is a normal operating condition. However, in such cases, the maximum allowable stresses in the rotating parts during maximum runaway excursions (on-cam, in the case of Kaplan turbines) are usually specified as two-thirds of the nominal yield strength.
- Abnormal conditions: These may occur in any hydro plant. Under such conditions, unit loads on the materials may approach closer to the yield limits than is normally allowed, and
- Catastrophic cases: These are unexpected and highly improbable cases and do not fall under the category of transients (for example, an earthquake or a landslide rupturing a penstock). Elements of the system may be damaged, and there is no economical way to avoid the damage.

In situations where more than one unit feeds from the same water conductor, the normal and emergency conditions analyses should include cases where load rejection occurs almost simultaneously on all units.

Chapter 12

Normal Operating Conditions

The hydro plant should be designed to operate within normal allowable design stresses under normal operating conditions over the full range of operating heads. The normal operating conditions include:

- Normal steady-state conditions, either generating, pumping, or at synchronous speed with no load (speed no load);
- Starting up from standstill to synchronous speed in the generating mode, and then to accepting load;
- From generating to synchronous condensing and back to generating operation;
- Shutting down from generating under load;
- Shutting down from synchronous condensing;
- Starting up from standstill to synchronous speed in the pumping mode, and then accepting pumping load;
- Shutting down from pumping;
- From pumping to synchronous condensing and back to pumping;
- From synchronous condensing, pumping, and back to synchronous condensing;
- Shutting down from synchronous condensing, pumping direction;
- Changing from pumping with load to generating under load (a reversible unit is usually brought to standstill first, and then from standstill to generating under load);
- Loss of load while generating, followed by rapid closure of the wicket gates (load rejection); and
- Power failure during pumping, followed by rapid closure of the wicket gates.

Special attention must be given to the extremes: the maximum and minimum net heads, discharges, and power. Maximum changes of load, from speed no load to full load, should be analyzed based on the maximum speed of the servomotor piston. In addition, other changes (such as part-load changes and small variations in head, flow, and power) also should be analyzed. However, small changes are usually less dangerous to the system than the extreme cases. On turbines equipped with a pressure regulator, small load changes—especially at low wicket gate openings—can produce significant pressure changes. This is because backlash in the pressure regulator linkage may prevent a pressure regulator response to small wicket gate movements.

The most preliminary analyses and the engineer's experience are useful for deciding whether additional devices will be needed for diminishing the amplitudes of the pressure surges or preventing pressure surges. Those devices might be surge tank(s), pressure regulator valves, governor controls, or air chambers. If the system includes any of those devices, assume that the devices will function correctly when making the analyses for the normal cases. Reducing the rate at which valves or wicket gates close also can diminish or prevent pressure surges. Analyze the various cases to find which initial conditions, together with the various transient conditions, result in the most extreme high or low pressures or the most extreme high or low rotating speeds.

As general rules:

- Absolute pressure in the water conductors should not be less than one-half atmospheric pressure absolute at any time;
- Passive safety devices, such as surge tanks, should be designed for the most unfavorable working conditions;
- If there is no surge tank overflow spillway, the highest elevation of the surge tank should have a freeboard allowance above the highest transient water level; and
- The lowest elevation of the surge tank should be below the lowest transient water level at that station along the water conductor, above the junction with the water conductor, and above any orifices or structures controlling flow into and out from the surge tank.

Abnormal Conditions

Since a hydro plant is complex, it is possible that one or more of the components may fail at some time—either during a transient condition or during a steady state—and create a transient. Such events are abnormal cases. The allowable stresses under abnormal operating conditions may be higher than those under normal operating conditions. Abnormal cases can include:

- On machines with multiple speeds for closing the wicket gates, the rate changeover device fails to function;
- A pressure regulator valve may fail to operate properly;
- If the governor or gate controller fails, wicket gates may fail to close and the plant may have to be shut down by closing an intake or draft tube gate or a valve;
- Multiple units may reject the load when one of the units has blocked wicket gates;
- A draft tube gate used to prevent runaway in a low head plant may fail to operate;
- If the flow closure devices cannot alter flow sufficiently before the speed increases, a unit may accelerate to runaway speed; and
- Attempts to restart or shut down the turbine before pressure changes initiated from a previous transient have sufficiently damped down may initiate a second transient starting from higher or lower than normal pressure conditions.

On large units, the generator and the turbine are not normally designed to operate for a sustained period of time at the maximum runaway speed. This occurs when there is runaway under maximum head and, in the case of adjustable blade propeller turbines, with the pitch of the blades at a relatively low angle. The speeds at which protective closure devices should be activated must be determined after considering machinery inertia and time to reach maximum runaway, as well as the inertia of the water between the inlet and the turbine and the rate at which the protective device will cause the emergency closing. Some units, however, are specified with rotors and bearings designed to withstand maximum runaway speed. This is most common on small units and on low-head units.

Catastrophic Conditions

Unusual conditions may occur that are not caused by human error or machine failure resulting from imperfect design or maintenance. The pressure transients resulting from unusual conditions may be too unpredictable to be analyzed and used as criteria for designing the hydraulic system. However, the designer should analyze some of these cases to find out what may happen to the plant in such an event. If the damage cannot be prevented, it might be limited, for example, by installing rupture discs in the penstock.

If the local pressure falls below the vapor pressure of water, dissolved gases are released, and the water column may separate. It is important that water column separation and rejoining be avoided under any conditions. Analyses of the waterhammer effects in such situations are imprecise. Since the vapor is far more compressible than the water, pressure changes may cause rapid shocks that may damage, or even rupture, the water conductors. In addition, rejoining a water column beneath the turbine runner may lift the runner and coupled rotating parts and damage the machine. For plants with long tailrace tunnels, analyses should be made to find whether a surge tank downstream from the turbine is needed. Rejoining of a separated water column is like stopping a speeding freight train instantaneously.

Resonance caused by periodic oscillations of the wicket gates, known as "hunting," may occur if the governor is not adjusted properly. A case receiving attention in recent literature is the condition of resonant auto-oscillations of water masses that have leaked past a seal in a turbine inlet valve. This situation would occur only if penstock water pressure operates the seat on the valve body. Wicket gate opening speeds should not be so slow that adequate oil films do not develop on the bearings, but not so fast as to cause excessive thrust on the oil film on the thrust bearing.

By definition, periodic oscillations of wicket gates or auto-oscillations of leaking water are not hydraulic transient conditions. However, the designer should not overlook them and should design to avoid them.

Other periodic oscillations can exist at hydroelectric projects that, though not hydraulic transients, should be avoided by the designers. Among those are:

- Draft tube pulsations that may result from conditions external to the prime mover such as spillway discharge;
- Penstock lengths and diameters that happen to resonate at hydraulic and mechanical frequencies of the plant; and
- A host of natural frequencies throughout the mechanical, electrical, and civil features whose conjunction could cause resonant oscillations.

Presentation of the Results of Analyses of Transients

There are many ways of presenting the results of the analyses. Tables of numbers from the calculations are difficult to follow, if the analysis is extensive. Since most people think and understand in geometrical representations, graphical presentation of the conditions is usually desirable. An example of a graphical representation appears in Figure 12-1.

REFERENCES

[1]Wylie, B.E., and L.V. Streeter, *Fluid Transients,* Prentice Hall, Englewood Cliffs, New Jersey, 1993.

[2]Bergeron, L., *Du Coup de Belier en Hydraulique au Coup de Foudre en Electricite,* Dunod, Paris, France, 1950.

[3]Rich, G., *Hydraulic Transients,* McGraw-Hill, Inc., New York, New York, 1951.

[4]Parmakian, J., *Waterhammer Analysis,* Prentice-Hall, New York, New York, 1955.

[5]Pejovic, S., A.P. Boldy, and D. Obradovic, *Guidelines to Hydraulic Transient Analysis,* Gower Technical Press Limited, Aldershot, Hampshire GU11 3HR United Kingdom, 1987.

[6]Chaudry, H.M., *Applied Hydraulic Transients,* VNR, New York, New York, 1988.

[7]Zipparro, Vincent J., and Hans Hasen, Editors, *Davis' Handbook of Applied Hydraulics,* Fourth Edition, McGraw-Hill, Inc., New York, New York, 1993.

[8]Perkins, F.E., A.C. Tedrow, P.S. Eagleson, and A.T. Ippen, *Hydro-Power Plant Transients, Part II: Response to Load Rejection,* Report No. 71, MIT Hydrodynamics Laboratory, September 1964.

BIBLIOGRAPHY

Brown, J. Guthrie, Editor, *Hydroelectric Engineering in Practice,* Blackie, London, England, 1958.

Creager, W.P., and Joel D. Justin, *Hydroelectric Handbook,* 2nd Ed., John Wiley & Sons, New York, New York, 1950.

International Code for Testing of Speed Governing Systems for Hydraulic Turbines, IEC Publication 308, International Electrotechnical Commission, Geneva, Switzerland, 1970.

Jaeger, C., *Engineering Fluid Mechanics,* Blackie, London, England, 1956.

Jaeger, C., *Fluid Transients,* Blackie, London, England, 1977.

Kovalev, N.N., *Hydroturbines, Design and Construction, Israel Program for Scientific Translations,* translated from Russian, 1965.

Martin, C.S., "Stability of Pump Turbines During Transient Operation," *Proceedings of 5th International Conference on Pressure Surges,* Paper C3, Hanover, Germany, September 22-24, 1986, pages 61-71.

Mechanical Design of Hydro Plants, TVA Technical Report No. 24, Volume 3, Tennessee Valley Authority, U.S. Government Printing Office, Washington, 1960.

Mosonyi, Emil, *Water Power Development: High Head Power Plants,* Volumes 1 and 2, Third Edition, Akadémiai Kiadó, Budapest, Hungary, 1991.

Mosonyi, Emil, *Water Power Development: Low Head Power Plants,* Volume 1, Third Edition, Akadémiai Kiadó, Budapest, Hungary, 1987.

Nechleba, M., *Hydraulic Turbines: Their Design and Equipment,* Artia, Prague, 1957.

Pejovic, S., and A.P. Boldy, *Guidelines to Hydraulic Transient Analysis of Pumping Systems,* P&B Press, 1992.

Symposium on Waterhammer, American Society of Mechanical Engineers Hydraulic Division and American Society of Civil Engineers Power Division, 1933.

Welded Steel Penstocks, Engineering Monograph No. 3, U.S. Bureau of Reclamation, 1977.

Chapter 13
Inspection and Testing

Inspecting Hydropower Mechanical Systems

Hydropower project inspection programs are recommended to ensure that supplied equipment, materials, and construction processes satisfy design specifications. An inspection program must be developed early in the project's design and specification stages. This ensures that the manufacturing, delivery, and installation processes proceed according to plan.

The justifiable level of inspection depends on several factors such as the scope of project, the size and location of the hydropower plant, the size and number of mechanical components, use of local or foreign suppliers, the manufacturing and installation process, and the craft skill levels available. For any plant, defective work or products can have a substantial effect on the construction process and operating reliability. Mechanical components usually require long supply times, and this may require closer inspection levels to minimize the potential for unforeseen nonconforming events. Project schedules should include time for inspections. A complete inspection program should include reference to applicable industry and contractual requirements.

Inspection Scope

Inspection efforts should be established to assure that sufficient resources are focused on the appropriate parts of the hydropower project. Table 13-1 summarizes the type of hydropower equipment normally inspected, the location of inspection, and the typical responsibilities regarding an inspection.

Equipment Manufacturer's Shop Inspection

The purpose of inspecting equipment is to verify the equipment's compliance with contract specifications and approved drawings. Manufacturers should pro-

TABLE 13-1: *Hydropower Equipment To Inspect*
(Useful for Clarifying Responsibility and Equipment Inspection Needs)

	Manufacturer			Purchaser	
Equipment	Shop	Certified Report	Site Inspection	Shop	Site
Generator Assembly	Yes	Yes	Yes	Yes	Yes
Rotor	Yes	Yes	Yes		Yes
Stator	Yes	Yes			Yes
Shaft	Yes		Yes		Yes
Bearings	Yes				Yes
Turbine Assembly	Yes	Yes	Yes	Yes	Yes
Runner	Yes	Yes	Yes	Yes	Yes
Shaft	Yes	Yes	Yes		Yes
Bearings	Yes	Yes			Yes
Seals	Yes				Yes
Water Passages	Yes		Yes		Yes
Distributor Assembly	Yes	Yes	Yes		Yes
Shut-off Valves or Gates	Yes	Yes	Yes	Yes	Yes
Governor System	Yes	Yes	Yes	Yes	Yes
Lubrication System Components	Yes				Yes
Cooling System Components	Yes				Yes
Compressed Air System Components	Yes				Yes
Drainage System Components					Yes
HVAC System Components	Yes			Yes	Yes
Fire Protection System Components	Yes	Yes	Yes		Yes
Crane	Yes	Yes	Yes	Yes	Yes
Spillway Gates & Hoists	Yes	Yes	Yes	Yes	Yes

vide guaranties and warranties on their equipment. The manufacturer has the primary responsibility for inspecting and maintaining quality control during equipment fabrication.

An equipment failure will most seriously affect the hydropower project's owner. A consulting engineer, an installation contractor, and a construction manager frequently interact between the owner and the manufacturer. During equipment fabrication and installation, all of these people may be involved in various inspections to best represent their own particular responsibilities to the project.

Detailed dimensional checking is particularly important when the parts are not assembled at the shop. The effect of improperly fitting pieces is more serious when the correction is required at the power plant site. When practical and cost-effective, components or systems should be assembled and operationally tested in the manufacturer's shop. Shop assembly should be specified to the fullest extent practical, so that field assembly problems may be eliminated. All shop-assembled parts should be matchmarked in the factory to facilitate reassembly in the field.

Complete shop assembly may not be practical because of component size or shop and crane limitations. This makes dimensional inspections and any subassembly critically important for those components that cannot be assembled or tested in the shop.

Nondestructive examination (NDE) and nondestructive testing (NDT) of materials, welds, and fabricated parts often is appropriate for highly stressed areas and pressure components. It may be beneficial to test material samples from components in the manufacturing process. These examinations and tests are commonly performed in accordance with standards from the American Welding Society (AWS), American Society of Mechanical Engineers (ASME), and the American Society for Testing and Materials (ASTM). The standards may require special equipment and certified inspectors to perform the tests.

The recommended shop inspections are:

- Check the materials used to see that they are of the type and grades specified and free of rejectable defects;
- Check the fabrication details to ensure that the equipment is being built according to the contract documents and approved drawings;
- Check fit-up and assembly operations to see that various components are assembled properly and matchmarked to facilitate field assembly;
- Check the equipment features, accessories, and appurtenances to ensure that they match specifications;
- Ensure that all specified shop tests, including nondestructive tests, are properly executed using calibrated measuring devices and that the results conform with the specified criteria;
- Check all equipment for interfaces; and
- Inspect the areas of components that require painting or coating to verify that contract requirements are met and assure continued protection for handling and shipment. Care should be taken that no hazardous materials are used in protective coatings.

Field Inspections

Generally, the installation contractor(s) has the prime responsibility for inspecting the equipment at the construction site and the quality control of the installation. In addition, the field inspection should involve the owner, consulting engineer, construction manager, and manufacturers to ensure that their interests are upheld. Approved drawings and contract documents should determine the compliance of components. The recommended field inspections are:

- Inspect all materials received;
- Verify horizontal and vertical alignment of embedded parts and anchor bolt centerline dimensions and the location and alignment of stationary components such as soleplates;
- Verify all critical clearances for field-assembled rotating machinery;

• Examine piping material, size, and support for compliance. Verify the proper location (including ease of access) of piping components such as valves, gauges, filters, and instruments. In general, all piping shall be hydrostatically tested at 150 percent of the design working pressure (or as stated in the detailed test procedures) and approved before acceptance. Tests should be made before piping is encased in concrete or concealed. Tests for all piping shall conform to contract document.

Testing Hydropower Mechanical Systems

The testing process places powerhouse equipment in operation as an integrated system and demonstrates its ability to operate safely and satisfactorily in accordance with design criteria. All hydropower plants require test programs, and the complexity of the test program is in proportion to plant capacity.[1,2,3,4]

Equipment may have been designed in accordance with the highest standards and produced using the finest manufacturing practices, but this does not necessarily guarantee that equipment will operate properly when integrated in a system. Every hydropower project has unique design criteria. Unique characteristics of a particular installation can result in unknown and unexpected events during preliminary plant operation. For this reason, start-up testing should be a carefully planned, step-by-step procedure that provides adequate test data for a thorough analysis of all operating conditions. The analysis should indicate inconsistencies that could lead to detrimental results.

The testing program should be developed early in the project. Critical parts of the design should be reviewed to determine which items require monitoring during testing and to what extent. This is particularly important for new designs that lack proven operating records. For example, the design should provide pressure taps in the water passages, access holes for transducers, and visual ports for inspection. All major decisions concerning testing, particularly those affecting field construction, should be resolved by the time the powerhouse design is complete.

If start-up testing requirements are not covered fully in contract documents, discussions should be initiated with major equipment suppliers soon after receiving preliminary arrangement drawings. Early discussions allow suppliers to review proposed tests, assess their effect on equipment safety and warranty conditions, make suggestions for additional requirements and special tests, and arrange, where necessary, for supplier involvement in the start-up testing process.

A written description of test procedures should be produced before completing equipment installation. The description should contain, at a minimum, the step-by-step start-up testing procedure, the field organization required to implement the procedure, details of each test (including copies of standards and codes where applicable), lists of test instruments with information on their use, control procedures during tests, sample log sheets for recording data, test circuits and connections, and safety clearances. The importance of controlling the live circuits, including safety tagging procedures, cannot be overemphasized.

Where extensive automatic controls are provided, plant operating personnel should perform initial testing under manual control. Automatic controls should be energized in stages as the commissioning tests progress. The procedures must include instructions for arranging the controls for manual operation—including checklists for operators to follow. A complete description of testing equipment should be included, listing devices for each test and their location. Procedures for data acquisition should explain how to collect data from test instruments or plant metering.

Test instrumentation procurement, calibration, installation, and checkout must be included in construction schedules to ensure all test instruments are functional by the time the start-up program is ready to begin.

In summary, a properly planned and executed start-up testing program should consider:

- Personnel and equipment safety by increasing the possibility of discovering incipient failure conditions before serious damage occurs;
- Providing analytical data should problems occur;
- Verifying conformance with contractual requirements;
- Demonstrating equipment operation as a system, with limiting factors;
- Identifying potential operating constraints;
- Validating design features and provide data to validate plant capabilities;
- Training of power plant operating staff, particularly in procedures for abnormal conditions; and
- Establishing the baseline values for monitoring changes in unit capability or performance.

Field Organization

The size of the field organization required to conduct start-up testing depends largely on the complexity of the test. The start-up testing of a single 100-kW low-head hydroelectric generating unit, factory assembled and tested for installation in a remote area, may be a one-person operation with adequate instructions. Conversely, start-up testing for a 1,000-MW multi-unit high-head underground plant, where serious material damage and loss of life would result from equipment failure, requires a comprehensive start-up testing organization staffed with experienced specialist engineers supported by adequate test equipment. Establishing a start-up testing organization for projects between these extremes requires engineering judgment and close cooperation between those concerned. Engineers should write the start-up testing procedure and determine the field organization required to put the procedure into operation. Others may have specialized and important knowledge that must be taken into consideration. The test group should authorize each step in the procedure, ensure that test equipment is in proper working order, monitor each test, and ensure that test data is adequately analyzed and reviewed before proceeding to the next test in the procedures.

A method is required to prevent unauthorized operation of equipment while energized for start-up testing. Equipment in a condition that could cause serious

Chapter 13

damage if improperly operated should be locked, blocked, disconnected, or otherwise de-energized until the condition is made safe. Start-up procedures should include appropriate clearances, such as tagging or locking systems, to prevent unauthorized operation of equipment and protect personnel.

Test Instrumentation

Mostly, the type of equipment tested determines the extent of test instrumentation. For a small plant, start-up testing data recording requirements often include only manually logging the readings from operating instruments and measurements made with common mechanical and electrical tools such as steel scales, dial indicators, pressure gauges, continuity testers, and volt-ohm meters. As the size and complexity of equipment increases, the simple requirements may expand to include more sophisticated portable instruments for measuring pressures, speeds, vibration, and dynamic balance. Large installations, with a potential for serious consequences from equipment failure, require extensive start-up test instrumentation for sensing and recording essential data at all critical locations. Instrumentation in this last category can be classified into three principal groups: sensors, signal conditioners, and data recorders. Test instrumentation should have adequate range and frequency response with rigid mounting brackets.

Transducers

Transducers are devices that convert physical quantities into equivalent signals, commonly electrical, for measurement and recording purposes. Typical transducers required for testing are:

- Pressure transducers to monitor critical pressures in turbine water passages, cooling water systems, governor hydraulic control circuits, and compressed air operated equipment. Figure 13-1 shows some typical locations on a high-head pump-turbine;
- Sensors to measure shaft vibrations, displacement, and runout. Figure 13-2 shows typical locations on a vertical Francis unit;
- Linear variable differential transformers (LVDT) and linear potentiometers to monitor structural deflections and servomotors, wicket gates, and rotary-type turbine inlet valves. Figure 13-2 shows some locations of vibration and deflections sensors on a unit;
- Tachometer drives to measure the unit speed;
- Power transducers to measure generator output;
- Strain gauges measure material strain in penstocks, turbine headcovers, and generator support brackets;
- Temperature transducers to measure the temperature of bearings, cooling air and water, and structural parts;
- Electrical contacts to indicate events such as "start" and "stop" signals, open and closed positions of valves and other items, and certain stages in automatic control sequences;

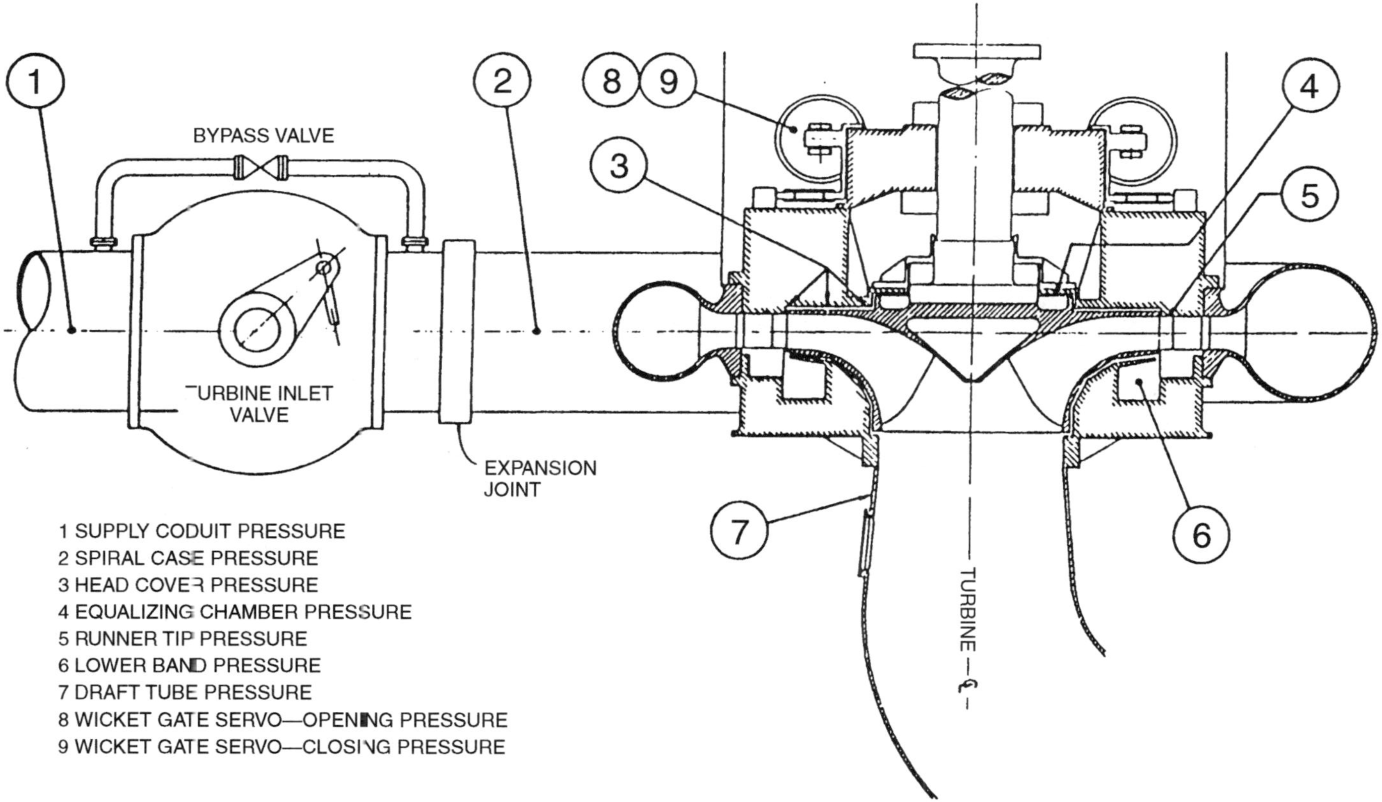

FIGURE 13-1: ***Pressure Tap Locations for Pressure Transducers.*** *This figure shows some typical pressure transducer locations on a high head pump-turbine. Pressure transducers monitor critical pressures in turbine water passages, cooling water systems, governor hydraulic control circuits, and compressed air operated equipment.*

(Adapted from "Commissioning Pumps and Pump-Turbines" by B. Eyden [12])

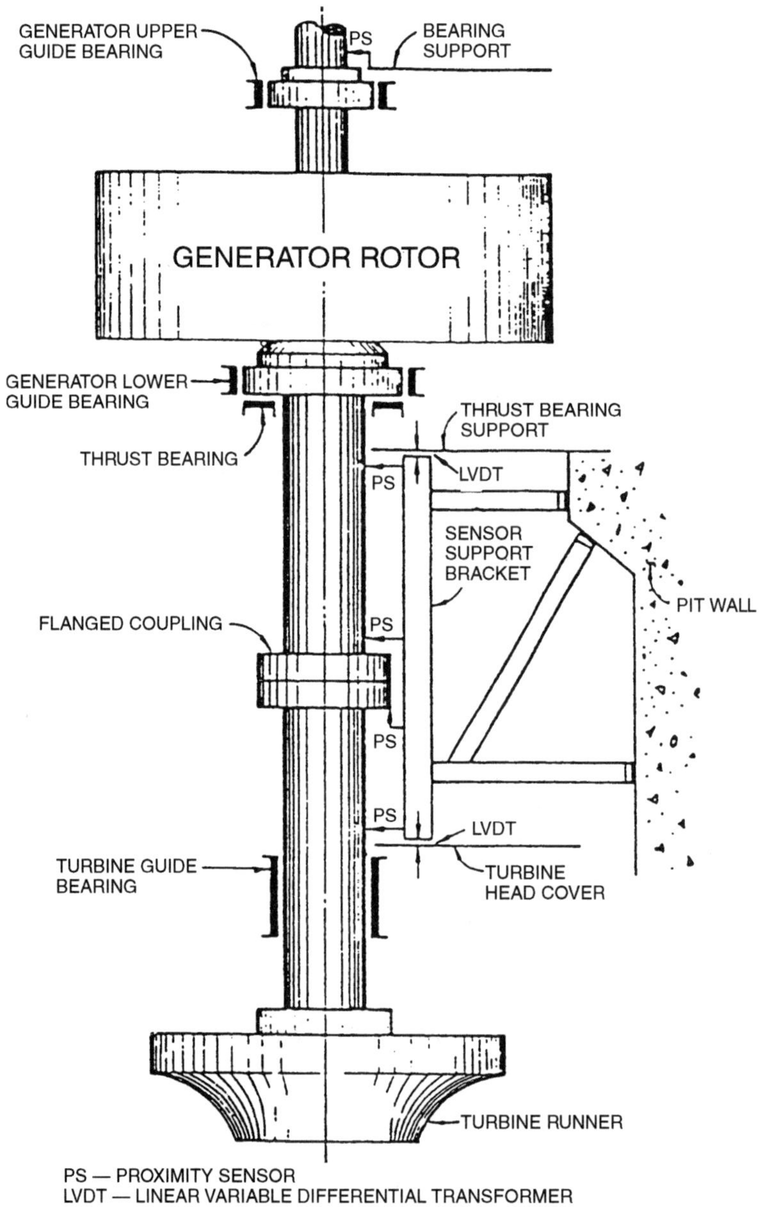

FIGURE 13-2: ***Locations of Vibration and Deflections Sensors on a Unit.*** *Proximity sensors monitor vibrations, displacement, and runout. Linear variable differential transformers (LVDT) and linear potentiometers monitor structural deflections. This figure shows the locations of vibration and deflection sensors on a vertical Francis unit.*

(Adapted from "Commissioning Pumps and Pump-Turbines" by B. Eyden [12])

• Accelerometers to monitor vibrations of structural members; and

• Flow transducers to monitor flows.

Powerhouse control panel meters and recorders should be reviewed to determine if manually logging their readings will be adequate during start-up testing. Tapping into appropriate control circuits may be required for measurements, but caution must be used to avoid safety problems or malfunctioning equipment.

Signal conditioners. Signal conditioners match transducer output characteristics to recording equipment. The choice of signal conditioners often depends as much on the purpose for which the test equipment will subsequently be used as on specific start-up testing requirements.

Data recording and devices. Test data recorders are available in a wide variety of types to meet many purposes. The choice is mostly a matter of selecting the lowest cost recording equipment that adequately meets test requirements. Consideration should be given to the purpose for which the recorders will be used after start-up testing.

Prerequisites to Equipment Start Up. Completing mechanical equipment and systems requires a detailed inspection and functional test of these systems. Usually the owner's or construction manager's personnel perform the pre-operational start-up inspection in cooperation with those who will be responsible for plant operation.

Start-up procedures for the main hydroelectric turbine-generator and supporting systems assume the following conditions:

• Shaft alignment has been verified;

• Plant structures have been tested as far as practical to ensure they will endure pressures and loads placed on them under operating conditions;

• Individual items of installed electrical and mechanical equipment have been appropriately tested to ensure, as far as practical, that they are in proper working order and meet performance requirements;

• Auxiliary systems are in operation and tested for providing electrical supplies, lubrication services, hydraulic operation, temperature control, and other essential services for each hydroelectric generating unit and its associated operating equipment;

• Turbine water passages are dry with headwater available on the upstream side of the intake gate or valve and tailwater available on the downstream side of the draft tube gates on a reaction turbine;

• Test instruments are installed, calibrated, and ready for operation; and

• Test personnel have been briefed on their jobs and safety procedures.

Initial Start-Up Procedures

Watering the hydraulic passages. The method for filling the water passages of a hydroelectric generating unit is one of the most critical operations in the start-up testing procedure, particularly for large units, high-head units, and long, complex waterways. Equipment failures and human error during the watering operation have resulted in numerous problems ranging from minor water leakages to disastrous project flooding. It is important, therefore, that watering be performed in a slow, methodical, and carefully controlled manner, with particular attention paid to detecting leakage.

The following precautions, as applicable, should be observed shortly before flooding the water passages:

- Inspect water passages, remove unnecessary materials, and sweep clean;
- Flush all piping systems connected to the water passages and ensure they are in proper working order;
- Check the operation of draft tube dewatering and powerhouse drainage systems and their associated pumps;
- Check the operation of all valves in, and attached to, the water passages and place them in appropriate positions ready for watering;
- On piping systems, verify that the piping, valves, and appurtenances comply with the contract requirements for material, size, support, and installation;
- Conduct pressure tests on the piping systems in accordance with ANSI/ASME B31.1.[5];
- After successfully completing the pressure tests, clean and flush the piping systems;
- Check the turbine shaft water seal for proper installation and adjustment;
- Check the operation and timing in the "dry" condition of flow control equipment such as turbine inlet valves, wicket gates on reaction turbines, nozzles and nozzle deflectors on impulse turbines, and pressure regulating valves. Leave all such items closed until required to be open in the watering procedure;
- Ensure that the air vents and vacuum breaker valves serving the water passages are in proper working order;
- Make a final check to ensure that the water passages are clear of personnel and unnecessary equipment, and then close all man-doors, manholes, and other access holes through which water could escape to the exterior;
- Make a final check to ensure that there are no loose bolts, tools, or equipment within the generator;
- Apply generator brakes, if provided;
- Place generator and turbine bearing lubrication systems in operation in case of inadvertent rotation; and
- Locate observers at points of potential water leakage and place appropriate test instruments in operation.

The process of flooding the water passages varies widely with the type of equipment. The general principle is to raise the water level gradually by admitting water, using a bypass valve or other method, until the passages are filled to maximum available tailwater level. Filling should be stopped at a suitable early point to test the operation of the draft tube dewatering system, where provided. This ensures that the draft tube dewatering system operates in an emergency condition. After removing the draft tube gates on a reaction turbine, filling may then continue from the headwater direction until the water passages are completely filled. Whatever method admits the water into the water passages, it should be under close control and capable of being quickly stopped and reversed if undesirable conditions are detected. Pressure, leakage, or foundation constraints may limit the rate at which the water passageway may be filled. The initial filling of a pumped-storage upper reservoir may need special provisions. For pump-turbines, initial operation may be as a pump with the penstocks partially filled to provide sufficient head for smooth pumping to fill the upper reservoir.

Mechanical run. The mechanical run is the first time that the hydroelectric generating unit rotating parts operate under the influence of water flow through the turbine. The mechanical run allows balancing of the rotating parts and tests the operation of unit mechanisms before applying generator electrical forces and higher hydraulic loads. In preparation, if the installation has intake gates, they should be raised enough to allow the water flow required to spin the unit to synchronous speed and then left in the partially raised position ready for emergency closure. Performing a bump test, described as follows, allows the unit to rotate. With high pressure oil lift pumps operating, open the gates enough for rotation. Make checks for unusual sounds and free rotation. While manually controlling the turbine flow control gates or valves, increase the turbine speed through several manufacturer-recommended speed levels (usually one-quarter, half, and three-quarters speed), then up to rated speed.

Shaft vibration is measured, and balance weights usually are added to the generator rotor to bring runouts within allowable values at synchronous speed. On units without pressure-lubricated thrust bearings, the unit may be required to go to half speed fairly quickly, after a brief pause at a few revolutions per minute to check for undesirable interferences with moving parts. Unit speed is held at synchronous speed until bearing temperatures have stabilized and are satisfactory. The unit is placed on automatic speed control at or near rated speed to check operation of the governor or other type of speed control equipment.

Important observations made during the mechanical run are the dynamic stability of the unit and the response of turbine water flow control devices.

Overspeed device test. This test checks the operation of overspeed devices. Shaft speed devices can be either electrical or mechanical and be configured for safety or normal control functions. In general, these devices shut down the unit if normal speed control equipment fails and the unit speed increases above a predetermined value. This test requires special care because it requires blocking, bypassing, or disconnecting normal rated speed-control devices in accordance with manufacturers' recommendations and conducting the test under manual control.

With bearings at normal operating temperatures, unit speed is increased smoothly and steadily until the first (lowest) overspeed protection device operates. The procedure is repeated for each overspeed device in ascending order of speed settings, with previous devices inactivated, until the test demonstrates the satisfactory operation of all overspeed devices.

Particular care should be given during overspeed testing to accurately record shaft runouts and the unit speed at which each overspeed device is activated. Care should be exercised, in setting overspeed devices, to avoid runaway speeds.

Excitation. The hydroelectric generating unit is started and accelerated to rated speed. Main field is applied to the synchronous generator and operating adjustments made to the excitation equipment. During this test, special attention should be given to the effect of the generator's magnetic field on the dynamic stability of the unit (for example, increased shaft runout and bearing temperatures or changes in shaft position).

Generator dryout. In epoxy-insulated windings, a dryout run sets the final cure of the insulating resins. In older, Class B windings, the dryout run heats the winding to expel any retained moisture. Armature dryout is preferable in a short-circuited condition.

Synchronizing and speed-no-load operation. The synchronizing test for a synchronous generator demonstrates that the turbine, governor, and synchronizing equipment can maintain proper control for matching unit speed with system frequency. With the unit at rated speed and the generator field applied, speed is manually matched to system frequency using a synchroscope or other frequency-matching device. The generator breaker is closed when a proper match of phase, frequency, and voltage is attained. The process is repeated using automatic frequency-matching equipment if such equipment is provided. An induction generator requires only maintaining operating speed before closing the generator breaker.

The unit is allowed to remain on the line at speed-no-load until all equipment adjustments are completed, and a stable operating condition is reached. All test instruments should be in operation for this test with data recorded after each equipment adjustment and during the time of generator breaker closure.

Power generation and load rejection. Load operation and rejection tests determine the hydroelectric generating unit's ability to accept and maintain electrical load in a satisfactory manner and to respond safely to a sudden loss of load. In preparation, any temporary obstructions that limit water flow through the turbine in previous tests, such as partially open intake gates and valves, are removed or opened to their widest positions.

The unit is started and placed on-line at speed-no-load. The bearings' temperatures are allowed to reach a stabilized and acceptable operating temperature. Then, load is applied slowly to the generator in a series of predetermined steps recommended by the equipment manufacturer. The unit is allowed to operate at the first load level until bearing temperatures have stabilized, and, then, by opening the generator breaker, the electrical load is rejected suddenly. Additional tests may be performed using the governor shutdown device. After test data for the run are analyzed and found satisfactory, the procedure is repeated at successively higher load levels until rated load is reached. Testing at one or more overload levels may be performed, if permitted. Gate timing, speed rise, and pressure rise are measured and compared to the predicted values at each load rejection.

All test instruments should be in operation for load operation and rejection. After each load rejection, particular attention should be given to the analysis of hydraulic and electrical transients to ensure allowable limits are not likely to be exceeded at the next load rejection. Similarly, shaft runouts should be monitored to prevent possible bearing overloads. Governor shutdown devices or the opening of the generator breaker initiate the load rejections. The former produces the maximum pressure rise and the latter produces the maximum speed rise. Normally, both tests are performed.

Careful observation of shaft runout is required during the heat run to determine possible changes due to generator temperature effects. Usually, a short circuit saturation test is performed following the dryout run.

Supplementary Tests

The test program described here contains the typical start-up tests applicable to most hydroelectric generating plants. The type of equipment at a specific plant and agreements with equipment manufacturers may require additional tests. These may be included as part of the start-up testing program or conducted separately at other times.

Runaway speed is a test similar to overspeed except that all speed restraints are removed, and the unit is allowed to reach its maximum steady-state runaway speed. Conduct of a runaway test is frequently not performed because of possible damage to the equipment. This test is not generally recommended.

Tests for spinning reserve and synchronous condensing are similar tests for units equipped for those operations. Where the runner is set below the tailwater, the tailwater is depressed below the runner by compressed air.

Multiple unit powerhouses with common turbine water supply and discharge conduits require load operation and rejection tests simultaneously on different combinations of units to determine complex hydraulic surge effects.

Turbine shut-off inlet valves and intake gates close under emergency conditions against maximum turbine discharge and should be tested.

A start-up test procedure sometimes includes an index test where connections for required instruments have been included in the turbine design.[6,7]

The start-up testing procedure normally does not include a turbine performance test to determine efficiency in accordance with national or international test codes.[6,8,9] A performance test, if required, is usually conducted subsequent to a "shakedown" period after the hydroelectric generating unit is placed in regular service or after the rated hydraulic range of plant conditions can be established.

Special needs, such as the effect of new and innovative equipment features, unusual operating conditions, or collecting data for future design purposes, may require other tests.

A pumped-storage unit requires testing in the pumping mode as well as the generating mode described previously. In addition, availability of a pony motor or other device for spinning the unit above rated speed with the pump-turbine dewatered may allow the mechanical run, overspeed, excitation, heat run, synchronizing, and any other tests to be performed without hydraulic effects in the runner. This allows easier separation of mechanical and hydraulic effects on dynamic balance of the rotating mass. Priming the penstock to allow the initial pumping operation start up may present additional complications to starting a pumped-storage unit. Additionally, the transfer from pumping to generating, transfer from generating to pumping, and stopping from the pumping mode must be tested, but are not covered in this guide.

Documentation

Results of all tests should be compiled into comprehensive start-up test reports. Parameters measured during start-up testing provide valuable baseline information for future operation and maintenance activities and for diagnostic reference if problems develop.

Turbine Performance and Efficiency Testing

Selecting the performance and efficiency test method should be included in the turbine's contract specification. Performance testing typically involves the measurement of head, power, and flow (in relative or absolute terms). Benefits include:

- Establishing a baseline performance against which later tests can be compared;
- Fine-tuning the operation (for example, gate opening/head relationship for the pump cycle of a pump-turbine or gate opening/blade tilt/head relationship for a Kaplan turbine); and
- Verifying guarantees.

Because the generator is used to determine the turbine output, the generator losses should be determined either by tests or by the manufacturer's calculations. Portable laboratory instruments and instrument transformers are required for these measurements. For a new unit, if the generator is small enough, losses may have been measured at the factory, and a field test for losses is not needed.

A field efficiency test requires the same data be measured as in model tests—that is, net head, output, speed, and water quantity (absolute or relative). However, a field efficiency test is performed in an industrial environment whereas the model test is performed in a laboratory. Even though test instrumentation with high accuracies are available, the environmental factor complicates field efficiency tests. High levels of test accuracy can be attained only when special measurement precautions are made. Even with special care, it is unlikely that field efficiency test accuracy will equal model test accuracy. To achieve the highest level of accuracy, equipment used in field efficiency should be fully calibrated against national standards both before and after the test. Then, after the systems have been configured at the test site, all measurement channels should be benchmark calibrated against known inputs to verify system performance.

There are two general categories of field efficiency tests. One is absolute efficiency with absolute flow measurement; the other is relative efficiency with relative flow measurement. The measurement of all other parameters are the same for both methods. In many cases, generator efficiency is not required for index testing. If it is necessary to establish turbine output, generator efficiency is needed. For an absolute efficiency test, an absolute flow measurement is obtained except when efficiency is measured by the thermodynamic method. Measurement of the large quantity of water passing through the turbine is the most difficult task of the field efficiency testing program and the one with the greatest measurement uncertainty, as seen in Table 13-2. Many methods have been used for flow measurement, and the requirements for achieving acceptable results from each are described in the codes listed in the reference and bibliography sections at the end of this chapter.

Flow Measurement

Current-meter flow measurement. A current meter registers a point velocity. By using a series of current meters on a fixed frame or a smaller number of meters on a movable frame, the velocity profile over the measuring section can be determined and converted to flow rate.

Current meters fall into two classes: those with a vertical axis and those with a

TABLE 13-2: *Uncertainty for Flow Measurement Methods*

Flow Measurement Method	Estimated Flow Rate Bias Error
Current Meter Method	
Open Channels	±1.5 percent
Closed Inlets and Conduits from 1.2 to 1.5 meters (4 to 5 feet) in diameter	±1.2 percent
Conduits of more than 1.5 meters (5 feet) in diameter	±1.0 percent
Pitot Meter Method	±1.5 percent
Pitot Tube Method	±1.5 percent
Salt Velocity Method	±1.0 percent
Pressure-Time Method	±1.0 percent
Ultrasonic Method (2 crossing planes with 4 paths each)	±1.0 percent
Venturi Meter Method	±1.5 percent
Weir Method	±1.5 to 2.0 percent
(If the bias error is 2 percent or greater, a Code test cannot be achieved.)	

Source: This list is from ASME PTC 18-1992, Performance Test Code Hydraulic Turbines.

Chapter 13

horizontal axis (axial flow). Usually the vertical axis current meters are used in streamflow. The axial type have been highly developed over the years for use in turbine flow measurement. The number of metering points is defined somewhat differently in the various test codes and depends upon the size of the water passage. On a very long unit where a large number of current meters would be needed if mounted on a fixed frame, the frame can be designed to traverse the water passage. This reduces the number of current meters required, but increases the time to measure one operating condition.

Considerable planning and preparation are needed to minimize field time for the frame and calibrations. The influence of intake structure on the flow requires special attention since local water velocity may not be parallel to the axis of the current meter. Also, debris easily damages individual meters, and they must be replaced.

Pitot tube flow measurement. The pitot tube is another point velocity method that, by traversing the water passage, can measure the velocity profile. This is classical laboratory technology that was adapted to the field in the initial efforts of prototype flow measurement.

A pitot tube measures velocity by taking the difference between the stagnation and piezometric pressures at a point in the streamflow. Many versions of pitot tubes exist. Although it is more adapted to model water passages, versions have been used in large diameter penstocks. Usually two pitot tubes that are 90 degrees apart traverse the penstock section. At least 18 measuring points are needed to define a velocity profile for each turbine load.

A uniform velocity distribution in the measuring section is important. This means that the pitot tube should be in a straight section of penstock 20 pipe diameters downstream from any bend or obstruction and five diameters upstream of similar disturbances. The difficulties of handling large pitots must be considered when planning tests on prototype-size turbines.

When using such devices, the differences between static and stagnation pressure levels could be small and require very careful measurement techniques.

Salt velocity flow measurement. The salt velocity method measures the transit time of a brine solution between two electrodes separated by a known distance in the penstock. It has been used successfully in both turbine and pump tests. The flow rate is computed using the measured transit time and the penstock volume calculated from field measurement.

One disadvantage in using this method is amount of hardware that must be securely anchored in the penstock. This method requires two electrode crosstrees and, perhaps, a turbulator in addition to the brine injection system with pop valves. Using a turbulator is required when the mixing distance for the brine solution is too short to ensure thorough mixing before reaching the downstream electrode. Also, be aware that the salt velocity method may not meet local environmental regulations.

Dye dilution flow measurement. Dye dilution methods have been in use since 1906. This method evaluates the change in concentration of the solution from the injection station to the measuring station. The method can be used in open canals or long penstocks. Equation 13-1 calculates the flow based on the change in concentration.

EQUATION 13-1:

$$Q = q\left[\frac{C_1 - C_2}{C_2 - C_0}\right]$$

where:
Q=flow rate, cubic meters/second (cubic feet/second);
q=flow rate of injected solution, cubic meters/second (cubic feet/second);
C_0=natural concentration existing in water, parts per million (ppm).
C_1=concentration of solution being injected, ppm;
C_2=concentration in resulting mixture at the measuring station, ppm; and

Different analyses methods have used sodium dichromate, rhodamine, fluorescein, sodium chloride, and lithium chloride as tracers. This method is particularly applicable to small hydro projects. Be aware that the dye dilution method may not meet local environmental regulations.

Pressure-time (Gibson) flow measurement. The method is basically an application of Newton's second law of motion, which states that "the total change of momentum is proportional to the impulse of the applied force." The momentum of a mass (M) of water in the penstock traveling at a velocity (v) is Mv. When the velocity of the mass is brought to zero by closing the wicket gate in (t) seconds, the change in momentum is MΔv.

Naturally, a force, F, is required to bring the flow to rest. Equation 13-2, from Newton's impulse-momentum law, gives the value of the force as the following.

EQUATION 13-2:

$$F = M\frac{dv}{dt}$$

where:
M=mass; and
dv/dt=rate of change of velocity.

The force involved is the pressure difference, Δp, over a length, L, of conduit multiplied by the cross sectional area, A, of the conduit. Since velocity is approaching zero, the flow rate can be computed by integrating Equation 13-2. This gives Equation 13-3.

EQUATION 13-3:

$$Q = A \int_0^t \frac{\Delta p}{\rho L} \, dt$$

where:
A=cross sectional area;
t=the time required to bring the flow to rest;
ρ=the water density;
Q=the flow rate being measured; and
Δp=pressure difference; and
L=distance along the conduit between pressure measurements.

The Gibson method involves measuring the pressure difference between two sets of pressure taps located in the penstock as a function of time and evaluating the simple integral in Equation 13-3. The actual equation contains more terms than shown in Equation 13-3.

One major advantage to the pressure-time method for measuring flow is that no apparatus needs to be installed inside the penstock to interfere with the flow. One major disadvantage on old units is that the penstock/spiral case is subjected to a series of pressure waves.

Ultrasonic meters flow measurement. This is the most recently developed method and is increasing in use. Most acoustic methods measure the transient time of ultrasonic pulses from one transducer to another across a penstock. The acoustic path is not diametrically across the penstock but at a prescribed angle. The transit time decreases for the acoustic path that travels with the flow and increases for the path that travels against the flow. For accurate measurement, acoustic meters require long, straight conduits and flows that do not have entrained gases or sediment. Single path meters provide less accuracy than multi-path meters.

The most accurate meters are multi-path meters with eight or more acoustic paths across the penstock. The system can be permanently mounted or temporarily mounted to serve more than one conduit or powerhouse.

Since a digital processor or computer system calculates the flow values, the acoustic methods hold considerable interest for on-line monitoring of flow and efficiency. This method may be used in place of calibrated Winter-Kennedy taps, frequently used to monitor flow.

Venturi meter flow measurement. The differential pressure across a specially built Venturi tube translates into flow rate by a calibrated rating curve. Smaller Venturi meters can be calibrated against weight tanks. It can be very difficult to calibrate very large Venturi meters; therefore, some codes allow Venturi coefficients to be assigned to carefully constructed meters.

A Venturi meter should be preceded by eight inlet diameters of straight pipe before and three diameters after the meter. Venturi meters cause a head loss in the penstock that is undesirable in a permanent turbine installation. For temporary testing in small units, the Venturi meter could be used, but manufacturing and calibrating the meter are expensive.

Weir flow measurement. By measuring the height of water level above a weir, the flow rate can be calculated from a variety of formulae. Although this is one of the oldest flow-measuring techniques, there is renewed interest in it for small hydro units. Usually, the weir is placed in the tailrace and is a permanent structure.

Volumetric flow measurement. This method measures the change in volume of water in the reservoir over a measured period of time. To do this requires a somewhat regular form of reservoir in which the volume can be measured accurately. The inflow or outflow, except for the test unit, must be very small, and the reservoir water level cannot be influenced by wind. Some captive pump-storage reservoirs are suitable for volumetric tests. Photogrammetry (the process of using aerial photography to make accurate measurements of reservoirs' volume) can evaluate large, irregular-shaped reservoirs.

This method is not applicable except where the reservoir is captive, or self contained, since any inflow from an upstream river and evaporation from the surface cannot be precisely known. Many hours of fixed-gate operation are required to achieve a single test point.

Thermodynamic method. The principle of this method is the conservation of energy with all turbine losses assumed to be dissipated as heat in the water flow. By measuring the temperature differential of the water upstream and downstream of the hydro machinery with precision thermometry, the efficiency can be calculated directly. The thermodynamic method differs from all the previously mentioned methods because absolute flow rate is not measured. The losses are measured and flow is calculated from the efficiency measurement. In other methods, the flow is measured and the efficiency is calculated. Temperature differential is measured by platinum resistance gauges, thermistors, or quartz crystals to an accuracy of 0.001 degree Celsius (0.002 degree Fahrenheit) or better. The accuracy of this method decreases with head and generally is applied in plants with heads above 100 meters (325 feet).

This method is subject to problems not encountered by other acceptance test methods. Sometimes it is necessary to test at night to avoid thermal problems caused by sunshine on exposed penstock. To maintain the draft tube temperature, cooling water passing must be temporarily routed to another location instead of into the draft tube. A draft tube traverse is very difficult to perform.

Relative flow/relative efficiency measurement—the index method. The test methods described previously are "absolute" methods because they involve either a direct or indirect measurement of the quantity of water. A second, and considerably easier, testing method is the index test in which the flow is not measured, either directly or indirectly, but is related to a difference in measured pressures, which is known to vary as the square of the "flow."

The method is defined, with examples, in papers published by the American Society of Civil Engineers and the American Society of Mechanical Engineers.[7,10] The principle is based on flow (Q) being proportional to approximately the square root of the difference in pressure between two piezometric taps or, as shown in Equation 13-4.

EQUATION 13-4:

$$Q = K\Delta p^n$$

where:

Q=flow;
K=a field determined constant;
Δp=differential pressure; and
$0.48 \leq n \leq 0.52$

The codes give the locations for the piezometric taps.[6,11]

The turbine outputs are obtained by measuring the generator outputs and adding applicable losses from the generator manufacturer's data. For small head differences, the turbine outputs are then corrected to a common head by similitude relations described in Chapter 3, Hydraulic Turbines. Each turbine output is then divided by the square root of the measured differential pressure to produce a relative index or efficiency of that test point. Plotting kW (corrected to a common head) versus $kW/\Delta p^n$ forms an approximation to a performance curve. This method is useful for determining the shape of the performance curve, confirming a baseline for future diagnostic comparison, and verifying or revising the following:

- Kaplan blade-to-gate control cam;
- Gate opening/head relationship of the pump cycle of a pump-turbine; and
- Best gate generating efficiency.

Model tests at manufacturers' and independent laboratories have consistently demonstrated efficiency accuracies of approximately ±0.25 percent. Field efficiency test methods cannot claim test accuracies of this caliber, as seen in Table 13-2. Therefore, some owners and equipment suppliers prefer to rely on the laboratory test and the manufacturer's step-up method to establish field performance and, then, confirm the shape of the efficiency curve by conducting a field index test.

In many cases, a relative efficiency test can be done first to help in a decision regarding the need for an absolute test.

Assuming the maximum relative efficiency, $kW/\Delta p^n$ (corrected to a common head) of this curve, is equal to the predicted maximum efficiency, a value of flow is found for that point, leading to calculating a value for K. Thereafter, all other flow values can be determined for each test point by holding K constant.

Table 13-3 summarizes the turbine field efficiency test methods discussed previously and application to generation and pumping.

TABLE 13-3: ***Summary of Field Test Methods***

Method	Comments	Generating	Pumping
Current Meters	Can be used on almost any plant. However, it requires many test measurement points.	X	X
Pitot Tube	Usually used on small units only. Not practical for large units. Requires a modest run of straight pipe.	X	X
Salt Velocity	Requires straight run of penstock. Requires placing equipment into the water passage. Possible environmental concern.	X	X
Dye Dilution	Requires a long penstock or draft tube tunnel. Possible environmental concern.	X	X
Pressure-Time	Requires two sets of taps in a long straight run of penstock.	X	
Ultrasonic	Requires a run of straight pipe.	X	X
Weir	Usually used on small units. Many times, the weir is in the tailrace to provide the proper tailwater level on the unit. Large errors.	X	
Volumetric	Requires captive pond. Pump-turbine is the only hydroelectric equipment with this condition.	X	X
Thermodynamic	Dependent upon plant head.	X	X
Index Test	Applicable to any unit.	X	X

Test Codes

Test codes, while not mandatory, can provide mutual agreement of rules for conduct between owner and manufacturer for making guarantees, testing, and interpreting the test results.[6,11] Commercial matters such as performance guarantees should be resolved during the initial contract discussions rather than during the field testing.

If periodic tests are run to monitor the performance or condition of the turbine, methods should be used that are comparable to the baseline test.

REFERENCES

[1] *Guide for Commissioning, Operation and Maintenance of Hydraulic Turbines,* First Edition, IEC Publication 545, International Electrotechnical Commission, Geneva, Switzerland, 1976.

[2] *Installation of Vertical Hydraulic-Turbine-Driven Generators and Reversible Generator-Motors for Pumped-Storage Projects,* NEMA Publication No. MG 5.2-1972, National Electrical Manufacturers' Association, Washington, D.C., 1972.

[3] *American Standard General Requirements for Synchronous Machines,* C50.10-1977, American Standards Association, New York, New York, 1977.

[4] *Test Procedures for Synchronous Machines,* IEEE No. 115, Institute of Electrical and Electronics Engineers, New York, New York, March 1965.

[5] *Code for Pressure Piping-Power Piping,* ANSI/ASME B31.1, American Society of Mechanical Engineers, New York, 1977.

[6] *Performance Test Code Hydraulic Turbines,* Power Test Code PTC18-1992, American Society of Mechanical Engineers, New York, 1992.

[7] Voaden, G.H., "Index Testing of Hydraulic Turbines," *Transactions of the ASME,* July 1951, pages 481-497.

[8] *Guide for commissioning, operation and maintenance of hydraulic turbines,* IEC Publication 805, International Electrotechnical Commission, Geneva, Switzerland, 1978.

[9] *Pumping Mode of Pump-Turbines,* ASME Performance Test Code, ANSI/ASME PTC 18.1-1978, American Society of Mechanical Engineers, New York, New York, 1978.

[10] Winter I.A., "Improved Type of Flow Meter for Hydraulic Turbines," *ASCE Proceedings,* Volume 59, No. 4, Part 1, April 1933.

[11] *International code for the field acceptance tests to determine the hydraulic performance of hydraulic turbines, storage pumps and pump-turbines,* IEC Publication No. 41, International Electrotechnical Commission, Geneva, Switzerland, 1991.

[12] Eyden, B., "Commissioning Pumps and Pump-Turbines," *International Water Power & Dam Construction,* August 1984.

Bibliography

Allen, Charles M., PhD, and E.A. Taylor, "The Salt Velocity Method of Water Management," *ASME Transactions,* Volume 45, American Society of Mechanical Engineers, New York, New York, 1923, pages 285-341.

ASME Boiler and Pressure Vessel Code, Section VIII, Revision 1, American Society of Mechanical Engineers, New York, New York, 1995.

CEA Guide to Erection and Alignment for Hydropower, Canadian Electrical Association.

Determination of Prototype Performance from Model Tests of Hydraulic Machines with Scale Effects, IEC Publication 995, International Electrotechnical Commission, Geneva, Switzerland, 1991.

Gibson, Norman R., "The Gibson Method and Apparatus for Measuring the Flow in Water in Closed Conduits," *ASME Transactions,* American Society of Mechanical Engineers, New York, New York, December 1923, pages 343-392.

International Code for Field Acceptance Test of Storage Pumps, IEC Publication 198-1966, International Electrotechnical Commission, Geneva, Switzerland 1966.

International Code for Model Acceptance Tests of Hydraulic Turbines with Supplement 193A and Amendment No. 1, IEC Publication 193, International Electrotechnical Commission, Geneva, Switzerland, 1972.

Johnson, G.D., and C.P. Kittredge, *Evaluation of Measurement Uncertainties in Performance Testing of Hydraulic Turbines and Pump-Turbines,* ASME Paper No. 73-WA/PTC-1, American Society of Mechanical Engineers, New York, New York, 1973.

Speed-Governing Systems, ASME PTC 29-1965, American Society of Mechanical Engineers, New York, 1965.

Speed-Governing Systems, IEC Publication 308-1970, International Electrotechnical Commission, Geneva, Switzerland, 1970.

Symposium on Laboratory Testing of Hydraulic Turbine Models in Relation to Field Performance, 1957, ASME Paper No. 57-A-124, *ASME Transactions,* Volume 80, American Society of Mechanical Engineers, New York, New York, pages 1,525-1,543, translated 1958.

Thermodynamic Method for Measuring Efficiency of Hydraulic Turbines, Storage Pumps and Other Pump-Turbines, IEC Publication No. 607, International Electrotechnical Commission, Geneva, Switzerland, current version.

Yamabe, and Yamaguchi, "Index Method for Pumping Operation of Reversible Pump-Turbine," *ASME Transactions,* Paper No. 68-WA/FE-5, Volume 91, American Society of Mechanical Engineers, 1968, pages 103-107.

Checklist for Chapter 13

1. Are all of the design criteria and operating conditions specified?
2. What codes were specified to cover the equipment or material?
4. What performance guarantees, standards, and methods of verification are specified?
5. What are the warranty requirements, including starting time and duration period?
6. Has the vendor been notified that manufacturing site inspections will be required? What is the vendor's manufacturing schedule?
7. Have vendor-certified inspection reports been required?
8. Has the schedule of construction manager fabrication inspections been specified?
9. Is there a requirement for "on-demand" construction manager inspections?
10. Has the clear right to reject defective work been specified?
11. Have erection or installation manuals been required?
12. Have operation and maintenance manuals been required?
13. Has the services of an erection engineer or supervisor been required?
14. Have the provisions for field correction of defective parts been specified?
15. Has a method for coordinating vendors been developed?
16. Has a delivery schedule including milestones been developed or required?
17. Have fabrication hold points been identified to facilitate inspection?
18. Have requirements for group inspections been specified?
19. Has coordination and advance notice of inspection requirements been specified?

Chapter 14
Maintenance and Operation Considerations

Hydro plants need formal operation and maintenance programs to protect the safety of personnel, equipment, and the economic investment of the owners. The design and specification documents should reflect the programs' considerations. These considerations should be applied throughout the plant design and when implementing the maintenance and operation programs. Both programs should be flexible enough to accommodate changes as the knowledge of the plant's response increases and the staff becomes more capable.

Typical objectives of maintenance and operations programs include reduced costs, high reliability, availability, low incidence of forced outages, and the optimized use of the facility. Toward this end, the design engineer should collaborate with the plant operator to determine the necessary program structure. Certain aspects of the plant's design may require special recognition by the operations or maintenance programs. A site-specific design may create the need for unique or unusual operations and maintenance response.

Plant design criteria and parameters should be incorporated into the maintenance and operations procedures. And, the plant's design should accommodate the intended operations and maintenance programs. The knowledge of the specifics of design, intended operations, and mode of maintenance must be coordinated.

Newer installations incorporate many new technologies and materials that may not be familiar to existing operation or maintenance staff. New equipment (or modifications being supplied to existing operations staff) should accommodate the safe clearance procedures and lock-out methods used within that organization. Operators should be suitably trained prior to initiating start-up and testing operations. That training should incorporate all applicable standards for staff, public safety, and operating requirements.

Maintenance Considerations

Some components of all units will be expected to last the full time between scheduled major overhauls—normally 20 to 25 years. Some elements may require periodic maintenance to achieve full effective performance. Other elements have instruments to identify possibly damaging operation or changing conditions. Where chronic or repetitive problems occur, an effort should be made to identify the underlying problems and provide corrective treatment to eliminate or minimize subsequent problems. This effort is often a design modification and should be discussed with the designer or manufacturer before implementing, if possible. Typically, all plants have three modes of maintenance to some degree: breakdown maintenance, preventive maintenance, and predictive maintenance.

Breakdown Maintenance

Breakdown maintenance is a response to a component failure. Those programs emphasizing this mode of maintenance also may extend the period of operation between internal inspections well beyond the time the designer may have intended. Warranty periods usually dictate an early inspection requirement. Beyond the early inspection(s), the procedure for breakdown maintenance must be modified if potentially damaging events are to be avoided. Some components are accepted routinely as being "permanent." These components include headcover or bearing fasteners, operating linkage for gates or blades, hydraulic rams, and main shafts. However, any major dismantling should include inspecting a sampling of the many components that normally have no limiting life.

Preventive Maintenance

Preventive maintenance involves scheduled maintenance at repetitive intervals, usually identified by the calendar. If the maintenance staff appropriately records information that details conditions found during inspection and repair, they may be able to extend the intervals between repairs. The designer should identify any of those items requiring an outage for inspection or repair that would not normally be able to withstand such extended operation.

The typical periodic damage requiring internal inspection in hydraulic turbines involves surfaces damaged by cavitation, corrosion or electrochemical reaction, and erosion or cracking. Access to the runner's intake and discharge should be provided on all units large enough to allow such access. Smaller units should incorporate devices for handling and dismantling that are appropriate to the size of the installation. Other systems such as the oil, governor, exciter, brushes, or auxiliary systems may require more frequent periodic attention to provide reliable operation.

The equipment manufacturer will usually provide schedules for inspections and maintenance that are developed typically from "conservative" scenarios. Consequently, the maintenance plan should be able to accommodate changing these maintenance intervals. Proper analysis of early conditions is essential. Often, the warranty requirements provide for an inspection within the first year. This inspection should include the manufacturer's representatives as well as design and maintenance personnel.

Early internal inspections should identify the allowable period of operation between repairs. Normally, subsequent inspections would occur at intervals of less than the normal repair interval to avoid unexpected problems. For those units with designs that create difficulty in installing inspection platforms in the draft tube, a brief inspection from the draft tube door or through the wicket gates may have to suffice.

Predictive Maintenance

Predictive maintenance is the process of providing indications of maintenance requirements. Normally, various sensors are installed in a unit. Often resistance thermal devices (RTDs) and gas bulbs can identify operating temperatures in bearings, cooling systems, and auxiliary motors. These sensors are part of a predictive maintenance system. Historical data developed from these sensors (such as the thermal indicators) will provide information for developing a maintenance schedule for a particular component. Analyzing the data from these sensors, then, will allow scheduling outages in an appropriate time.

Condition monitoring is an important addition that should be implemented at a new or existing plant to provide a more comprehensive predictive program. Proximity probes and accelerometers often are installed to identify unusual vibration. The signals from these devices may be displayed continuously or periodically. Usually, permanent records are obtained periodically. Other devices, (for example, partial discharge, radio frequency, infrared, acoustical, and optical positioning devices) may be used in the unit at appropriate locations. The combined information obtained from analyzing the various outputs can establish the need for an impending outage.

With appropriate lead time, many functions normally scheduled for a subsequent outage may be performed during this newly identified outage requirement. When combined with the previous operating history, data from these sensors also can determine the likely risk from extending operations due to delaying an upcoming scheduled outage. Flow devices may be interfaced with the unit's watt-hour meter and head measurement to provide direct reading, on-line performance measurements. After a baseline is established, this data can be a guide to enhanced operations as well as deteriorating conditions in the unit.

Ultimately, the optimum maintenance program will minimize the components subject to breakdown maintenance and incorporate the total history and current status from predictive elements into workable, scheduled events. This should lead to increased reliability, increased intervals between outages, and decreased outage duration.

Design Considerations

The design of certain major electrical and mechanical components should be based upon their life cycle value. Many turbine components will have a service life exceeding 50 years. Some plant systems may have a very limited life. A balance between initial cost, design life, efficiency, and maintenance requirements needs to be agreed upon early in the planning of a project.

Using Stainless Steel

Modern design should incorporate materials appropriate to the installation. For turbine runners and major plant components, the material of choice should incorporate weldability and repairability on site without needing extensive heat treatment following repair. Many hydro locations are remote, and servicing components externally may require lengthy transport.

Cavitation repair welding of runners must be performed with due consideration to the properties of the base and weld metals. Care must be exercised to avoid distortion of the blades or inducing cracking in the base material due to excessive heat input or material incompatibility. Special precautions must be taken when welding near highly stressed areas of runners. Additional precautions are required during handling and finishing to prevent surface contamination of the stainless alloys from carbon steel tools, rigging equipment, and dust. This contamination can result in surface and crevice corrosion.

Oil Maintenance

Minimizing the different types of oil for bearings, governor systems, hydraulic systems, and gear boxes throughout the plant reduces the possibility of erroneous usage, and decreases replacement and storage problems. Identifying the most severe applications (typically, the governor or gear box) will indicate the inspection and testing interval required for the oil system(s). The inspection and test interval should be modified as the application history is developed. The plant should have suitable oil sampling taps and filtering equipment. Instructions for sampling and filtering should be provided for incorporating into the operation and maintenance procedures.

Annual tests for viscosity, acidity, and water content are recommended. The frequency of oil filtering can be based on test results. Often, bearing oil can be used indefinitely. For most turbine oils, there will be no oil breakdown if oil temperature does not exceed 60 degrees Celsius.

Maintaining Components

Intake Structures

Some intakes may include elaborate mechanisms to draw water from a specific level to control gas entrainment or temperatures in the downstream flows. Some intakes may also have gas injection systems to provide nitrogen stripping or adequate oxygenation downstream. Gas injection systems or other environmentally-oriented additions may affect plant maintenance or operation. Entrainment of many small bubbles may decrease the potential for cavitation in the turbine, but may increase corrosion.

In cold climates, canal-type or river take-off intakes may be subject to frazil or anchor ice problems. Any very cold climate entrance channel must be deep enough to withstand an anchor ice buildup without causing structure flooding or overtopping. If sufficient pondage cannot be incorporated to stop these problems with a surface ice cover, a method of treatment should be identified. A system to flush the trashracks with air or warmer water could be used underwater. Using non-metallic materials in the rack may minimize ice buildup and make removal easier. Operational modification(s) also may be required, such as periodic shut-downs until an ice cover forms.

If construction is completed as reservoir waters rise, construction debris may accumulate both inside and outside of the structure. In this case, an underwater inspection should be performed prior to unit start up. The inspection should begin outside the racks from the surface to the bottom. If the inspection verifies that there is no major or dangerous debris inside the racks, that may be the extent of inspection. Depending upon the hazards and depth involved, tethered cameras or divers may be a part of the inspection program.

Periodic re-coating of metallic structures may be necessary. Coatings for the racks and water control mechanisms that are alternately submerged and exposed should be tough, reflective, and suitably thick to prevent corrosion damage. Some trashrack designs have a corrosion allowance rather than a coating to avoid reliability or re-coating problems within the design life of that element.

Head Gates

The head gate is often the last line of defense in controlling the turbine waterway. If the gate is considered an emergency closure device for a potentially ruptured waterway, it should be given a periodic test to simulate emergency closure and to confirm that capability. Simulated emergency closure usually is initiated remotely with full output on the unit(s) it supplies. Verifying the functionality should be scheduled early in the start-up phase of operations. Periodic intervals for retesting should be established to verify complete closure and control function.

If a long tunnel is located downstream of a head gate, a modified procedure, primarily to test the remote controls, may be used. Closure is initiated and stopped without full shut-off to avoid rock falls or collapsing the tunnel's liners. Cable-suspended gates or gravity gates may require periodic maintenance to avoid incomplete closure. Inadequate gate weight during high flow conditions, debris accumulation, or frozen wheel bearings may cause incomplete closure. The emergency closure should not be used to provide closure for a subsequent maintenance outage. The leakage past the gate from a simulated emergency closure is normally much greater than the leakage obtained from a balanced head closure.

Critical areas for head gates are the seats and seals, hoists, dogging devices, grout packing behind the seats, wheels, seal actuators, seal actuator lines, bypass or filling valves, gate operators, and emergency power sources. Inspecting the structure downstream of the intake gate should be performed periodically, usually every five to ten years. An immediate inspection of the gate seat area should occur if concrete or metallic materials are found in the spiral case or drainage valve grate. Separate gate bypass lines often suffer severe corrosion from long exposure to non-flowing water. Those valves and lines should be checked whenever upstream stoplogs or bulkheads are installed. Spare non-metallic seal material may break down rapidly in the ozone-rich atmosphere often found around power facilities. If such spares are stored, they should be kept in cool, dark storage areas away from electrical machinery.

The penstock or head gate vent is a critical part of the installation from a maintenance perspective. The adit or grill of the vent should be in such a position that it cannot be covered. It should not face in a direction or be in a location where water erupting from it would hazard personnel or equipment. The vent line should enter the top of the penstock immediately downstream of the head gate. No temporary covers should be applied to the vent openings when the waterway may be operated.

Turbine Shut-Off Valve

The turbine may have either a spherical or butterfly valve. This valve may be used each time the unit is shut down. It can provide isolation for multiple units on a single penstock or primary lower waterway control in case of an emergency. Seats, seals, operating links, bearings or bushings, power source, and hydraulic lines are the primary maintenance concerns. The valve should have an initial test to verify closure with flow through the unit. Its function should be verified periodically through test or normal frequent operation.

Turbine. If suitably sized for internal maintenance work, the reaction turbine should have adequate access to both intake and discharge sides of the runner. Inspections should be performed at intervals that will allow adequate control of any damage that may occur. The first inspection should be within 12 months or 8,000 operating hours of initial operation. The turbine should have a draft tube access platform and its method of installation should be detailed. All access doors should have a valve or petcock near the bottom of the door to test the water level behind the door.

The spiral case inspection should include observation of the stay vane area, wicket gates, and adjacent facing (wear) plates. The source of any metallic or concrete debris found in the water case should be determined. The upstream edges of any interruptions or guides for water flow should be examined for impact damage from hard particles. Hard particle damage may evidence itself as chips or dents in the leading edge (from large solids), or a high degree of polish or wear (from fine abrasive material). The leading edge of the turbine blades should be examined for damage from hard particles. The area downstream of the leading edge of each runner blade and the band between the blades should be examined for cavitation damage. Should damage be found in these areas, the cause should be determined. If forebay, tailrace, or power output conditions were outside of design parameters, the damage may not be from inappropriate design, installation, or manufacture.

If leading edge damage is found in Francis units, the cause should be identified and corrected, if possible. On new installations, leading edge damage often occurs because of a reduced wicket gate position for the prevailing operation. The manufacturer or designer should be consulted if such problems occur. If that operation is expected to continue, the leading edge entrance angle may be modified to accommodate that wicket gate angle. [1]

Variations in cavitation damage from blade to blade may be from poor quality control during the runner's manufacture. If the damage is uniform throughout all of the blades, it is likely a design or operational problem. It is important to control the amount of material loss at an early stage before contours are lost to the point that correction in place becomes extremely difficult. One way to determine whether material loss occurs is by using templates. A prudent choice is to specify a supply of templates for both the leading and trailing edge portions of the blades. These templates should have accurate locating features to allow proper positioning on the installed wheel. Field personnel may choose to create their own templates before any damage occurs to the wheel.

Wicket gates and runners should be visually inspected for possible cracking. From a maintenance standpoint, coatings should be relatively thin (less than 0.25 millimeter or 0.01 inch) and brittle, rather than tough. Tough coatings may conceal cracks and hard particle impacts that would be indicators of possible problems. Facing plates should be inspected for possible galling owing to contact with the wicket gates.

Fabricated runners should be examined closely in the area of all weldments, the upper fillet near the trailing edge, and the areas adjacent to all overlays. Fabricated headcovers should be given close attention at all welded joints.

In the past there have been problems with seal or wear rings that were installed on Francis runners with a shrink process to fix them to the wheel. The shrink installation may lose its clamping tension or fasteners, or it may be subject to cracking. Excessive cavitation repair welding can reduce the runner periphery sufficiently to loosen removable ring installations. Many runners made entirely of stainless steel will have integral rings designed into the finished runner.

The clearances between the rotating and stationary wear rings should be measured and compared to initial clearances. The choice of materials for replacing these stationary and rotating components should be carefully selected to minimize the potential for galling if contact occurs.

Kaplan units may experience trunnion wear that may allow blade contact. Material choices for blade tips and throat rings should include combinations that are non-galling such as those used for Francis unit ring installations. Restoring efficiency can be obtained by closing blade tip and throat ring clearances to the original values. Certain non-galling combinations of aluminum bronze and stainless steel may provide good bearing qualities should contact occur. These materials offer both cavitation resistance and anti-galling capabilities as an overlay on one or both surfaces. These material combinations are suggested to avoid serious damage should contact occur. The use of such galling resistant materials does not diminish the need for a properly aligned, fitted, and balanced unit.

The optimal galling and cavitation resistance can be added to a different base alloy on any type of unit. With certain stainless steel material, machining, coincidentally, will "cold work" the surface to higher strength and add or enhance the surface's compressive stress retention. Over the years, stress relaxation could occur in the rotating rings. This might cause loosening or reduce the original clearances to an unacceptable amount.

Unusual events resulting in excessive vibration, severe contact between rotating and stationary parts, or shock loading owing to improper operation require extensive investigations. Under those conditions, inspection should include dismantling, as necessary, to confirm the integrity of all stressed components within the unit.

Carbon steel reaction turbine runners, with or without a stainless steel overlay, should have corrosion protection. This corrosion control should be restored periodically to minimize loss of the carbon steel from corrosion or electrolytic action at the edge of the overlays.

Shafts

Turbine, generator, and intermediate shafts have a history of good performance with few problems. Several instances of very rapid stops due to electrical or mechanical locks have been known to occur with little shaft damage other than misaligned match marks from top to bottom. The shafts and their couplings should be able to withstand all normal activities. Should extreme shock or vibration be experienced, shaft couplings and fasteners should be closely examined for offset, cracks, and misalignment. Any change from original conditions should be well documented and retained for reference.

Shaft seals or water-lubricated bearings supplied from an external high pressure source require additional consideration. Normally, the primary filtered water supply should not be chlorinated. The facing material for the non-metallic seal is usually an austenitic stainless steel that may be damaged by chlorinated water. Also, associated piping is likely to be austenitic and could fail if exposed to chlorinated water.

Packing glands should be adjusted to allow a small amount of leakage above the headcover. Water supplies to the lantern ring should not be compromised by sediments or corrosion within the piping. This piping should be flushed periodically to assure good flow.

Cranes

Requirements from the Occupational Safety and Health Administration (OSHA) and the American National Standards Institute (ANSI) govern inspection requirements for cranes. One major consideration for crane installations is a detailed inspection of the hooks. This may entail visual, magnetic particle, and dye checks of the hooks. No paint should be on the hooks. Hoist brakes should be carefully checked in accordance with ASME/ANSI B30.2.[2] All cranes should have load cells to identify cable loading. This helps to avoid destructive lifts due to binding of the device being lifted.

Generator

While the generator is normally considered an electrical element, it contains many mechanical elements. Mechanical generator components include the bearings and lubrication systems, bearing support, rotor spider and rim, stator frame, soleplates and anchors, air housing, ventilation and surface air coolers, brakes, and jacks.

Vertical shaft generators normally incorporate a thrust bearing as a major mechanical component. Usually, two guide bearings are provided if the generator rotor is supported from above (suspended type) and one guide bearing if the rotor is supported from below (umbrella type). Horizontal units may have a thrust bearing if the turbine has an axial force.

The thrust bearing must perform for long periods with little maintenance. Many common elements exist whether the thrust bearing design has shoes supported by multiple springs, multiple plates, or a pivoted jack shaft. The fluid used for bearing operation must be adequately cooled, remain free of contaminants, and provide high lubricity throughout the range of bearing operating temperatures. Lubricating fluids should provide good electrical resistance, since stray currents traveling through the shoe babbitt can ruin a bearing. Graphite-filled greases should be avoided for installation or manual rotation.

Most vertical generators will be equipped with a pressure lubrication pump of 2,000 kPa (300 psi) or less to provide an oil film of 0.04 to 0.08 millimeter (0.0015 to 0.003 inch) between the shoes and the thrust runner for low speed and start-up operation. Some very large designs use pressures up to 21,000 kPa (3,000 psi) and provide direct water cooled shoes to prevent surface contact from thermal warping with a 0.013-millimeter (0.0005-inch) oil film. Highly sophisticated systems tend to create their own needs that may be beyond the capacity of existing staff.

Thrust bearing failures usually occur very rapidly. After a bearing failure, all bearing components, including oil lines and pumps, must be cleaned very thoroughly to remove any traces of the failure. Often, replacement shoes and runners are available as spares to help shorten the downtime of the unit. When reinstalling the new components, check the total height of the shoes and runner plate combination to ensure the proper positioning of the rotating elements within the stator and turbine case.

The stator coils must be supported and restrained to resist the high magnetic forces created during generation. Some generator installations may require adjusting the support mechanisms after early operation to restore the stator shape. Initially, inspect and check the stator structure components and fasteners after the first three, six, and twelve months, then follow this inspection sequence using the recommended intervals for the machine.

Visual capabilities within the stator and rotor portions of the generator often are quite limited. To enhance the ability to identify problems, greater light reflection and contrast with problem areas is required. When specifying a stator color, use a pastel color (preferably light green or blue) to show up problem deposits better. The deposits typically found are white (resin, mica, or fiberglass powder), rust red (fretting), and black (burnt or carbonized insulation materials). The light colors also will readily identify dust and oil accumulations.

Cooling the stator and rotor is critical to maintaining the rated and uniform operating temperatures corresponding to the thermal capability of construction and insulation bonding materials. Often, circulated ambient plant air cools small older generators. Newer generators usually have an enclosure or air housing and air-to-water heat exchangers (or air coolers), which use recirculated air through the stator and rotor. Circulation through the generator often is accomplished with blades mounted to the rotor. These blades and static baffles are near the air gap and magnetic field and should be checked periodically for joint or fastener integrity. Air coolers and associated piping should be checked for corrosion and sediment deposits both internally and externally. When several tubes have been plugged due to leakage, the coolers should be dismantled and refurbished. Units cooled with ambient plant air should have adequate filtration to minimize dirt accumulation.

Generators equipped with gas-discharge fire suppression systems should have adequate pressure relief. Check relief doors for freedom of movement and periodically lubricate the doors. Releasing gas for fire suppression may cause loss of generator bearing oil.

Bearing vent systems should exit to traps or containers away from the housing or exciter brush area to minimize oil mist accumulations in those areas. The cause of excessive accumulations of oil or dust should be determined and steps taken to reduce or eliminate this potential fire hazard.

Generator guide bearings should have sufficient reservoirs and sealing capability to avoid shutdown due to loss of small amounts of oil. Normal oil sampling should be performed during operation without causing appreciable oil level loss.

Hydro generators are equipped with a braking system to decelerate a rotating unit and resist rotation from wicket gate leakage. A common design uses brake shoes mounted on the lower bracket that operate against a rotor-mounted brake ring. Brake systems cause dust and create very high local temperatures. The dust can mix with oil droplets from the bearings to form adherent deposits that are quite flammable. Usually, unit braking occurs with a single continuous brake application until the unit is stopped. Brake application usually occurs at about 40 percent full speed upon shutdown. Intermittent brake application tends to increase the time required to stop the unit, but may be necessary if excessive heating occurs.

The jacking system may be used to raise the machine rotating parts a sufficient distance to permit adjusting or removing thrust bearing components and to relieve pressure on the thrust bearing during unit outages. The maximum lift of the rotor by the jacking systems should not allow rotating turbine and/or generator components to contact non-rotating components. The seal problem is nominally a turbine consideration, but requires coordination with the generator design. Blocks for supporting the rotor system should be mounted in such a way that excessive force is not needed and personnel should not encounter hazards to their fingers to move the blocks into place. The blocks should not be in a location that would allow them to contact any portion of the rotating system during operation if their mountings should fail.

Governor

Piping systems should be free of any reverse slopes that could cause air entrapment. Systems that depend upon many mechanical linkages must be routinely inspected to prevent lost motion, friction, and binding. Reducing the number of mechanical components is often a primary justification for upgrading the unit to an electronic control system. Timing of the pump cycling provides an indication of losses within the system and pump condition. Periodic tests of deadband and deadtime should be made to assure accurate control.

Draft Tube

Draft tube portions designed for removal of the runner should be examined during early operation to identify unusual flexing at some power outputs. Operations modifications or additions of air or flow directing vanes may be necessary to eliminate or minimize excessive draft tube vibration.

Draft Tube Gates

One, two, or three gates, bulkheads, or sets of stoplogs may be needed to close off the draft tube. These closures are usually cable-suspended gravity devices intended for balanced pressure closure. Some low head installations may be operated hydraulically to provide full flow closure capabilities. Problems with sealing due to debris in the bottom seal area are common. Tailrace floors should be a concrete monolith or consist of rock large enough to be stable with reverse flows up to 4 meters per second (13 feet per second). Equalizing the pressure across the gates for withdrawal may be done with bypass lines, valves within the gate, or from the high pressure side of the turbine.

Inspection and Maintenance Requirements

This section provides the recommendations for an operating hydroelectric power plant inspection and maintenance plan. It identifies those aspects that a fully functional preventive maintenance program would normally address. The periodicity of these functions will vary with the site requirements. The flexibility of the maintenance program should allow for changes based upon historical information developed during inspection, maintenance, and operation. Improved methods and technologies often will tend to increase the time between outages and decrease the outage length.

Inspection: Procedures that can be accomplished while the equipment is running or during a short duration outage. The magnitude of work is routine in nature and should include operational data, vibration measurements, and fluid sampling.

Minor Preventive Maintenance: Work that requires an equipment outage of short duration. The degree of work is such that adjustments or component replacement or repair is justified to reduce the probability of costly maintenance or repair. This activity includes test and inspection requirements.

Major Preventive Maintenance or Overhaul Maintenance: Work that needs an equipment outage of substantial duration. Significant work or modification is required to improve equipment reliability. This activity includes minor preventive maintenance and inspection requirements. This is work that entails major dismantling, and, possibly, upgrading, uprating, or replacing major components. This may extend for several (six to nine) months. Overhaul maintenance usually is performed at the end of the predicted life for a major component, such as the runner or generator stator. Minimum operation periods prior to overhauling conventional turbine and generator assemblies is usually 20 to 25 years, and for reversible pump-turbines about ten to 15 years.

Typical Maintenance Intervals

Periodic tests and inspections should be performed and recorded. For initial operations, this is usually at one year intervals. As the equipment stabilizes and trends can be determined, this could extend to two years or longer. As the equipment ages and approaches a major overhaul, it may be appropriate to shorten the interval for testing and inspecting. If on-line machine condition monitoring equipment is installed, the trend information should be evaluated to modify the frequency of permanent retention inspections and tests. Tests that would normally be kept for the lifetime of a given component include DC Hipot or ramp test of the stator, oil acidity and contamination, and runner bucket modification performance.

Each installation will be subject to many variables affecting the maintenance program. Turbine runner damage or stator replacement are often the indicators for major outages. Internal inspection is often a time-consuming process and may evolve into minor runner repair. It is prudent to make repairs and modifications prior to significant material and contour loss. This periodic internal repair may start at one year intervals and be extended to four or more years after the unit and its application have stabilized.

Typically, major outages or overhauls are timed to occur just prior to the predicted end of major equipment life. On-line monitoring or frequent periodic monitoring is critical to identify the need and the timing for major work. Historical data from similar units in the same or similar installations often helps to identify approximate life spans. When the generator rotor is removed, the turbine is exposed for greater ease of dismantling; consequently, often a turbine overhaul or upgrade is scheduled simultaneously.

Some components and their normal maintenance and inspection intervals

Spillway Gates: Check the gate equipment at least annually or after major usage. During the ensuing interval prior to next use, rehabilitate as necessary.

Turbine Intake Gates: Annually inspect the actuator and controls; test the remote capability. Visually inspect exposed ways and guides during low reservoir levels or use an underwater inspection if above water conditions or debris in the turbine indicate a need. Perform major work, as needed, during scheduled unit outages. Usually, stoplog installations are required to perform major work on the gate. Hoist work may be performed with the gate on installed "dogs" (supports), on temporary supports, or in the closed position.

Turbine Shut-off Valves: Check piping, motors, controls, and any associated energy storage systems annually. Perform an internal inspection whenever the waterway is empty and secured.

Turbines: Perform an internal inspection periodically (at intervals indicated by repair records) or after any unusual internal noises, wicket gate shear pin breakage, or known debris intrusion. Have a maintenance interval appropriate to minimal repair or correction needs. Peaking operation tends to accelerate damage. If the plant's operation has emphasized peaking, a shorter interval between repairs may be necessary.

Governors: Check oil quality, pumps, linkage or electronic sensing, ballheads, feedback linkage, and valve vibration devices annually. Perform a five-year check of response time, fully dismantling the hydraulic amplifier and replacing components such as ballhead bearings, valves, speed sensing devices, cylinders, pumps, or impellers as necessary. Record all replacements.

Cooling Water Systems: Functionality is verified with each start or daily operation. Annual inspection and maintenance of joints, valves, strainers, and pumps will assure good performance. Dismantle and inspect pump(s) when a major turbine outage occurs and repair as needed.

Compressed Air Systems: The station's air and brake air are usually part of the daily check for proper functioning. A quarterly inspection of the oil levels and draining of the accumulators should be adequate for most installations. Major maintenance of the compressors depends on the type involved and the severity of service, usually at intervals of one to three years (at 1,500 hours per year or less) is considered adequate. Construction activity within a plant often increases air consumption and causes dusty conditions. This creates a need for more frequent maintenance of the compressed air system(s).

Lubricating Oil: The governor's or speed increaser's oil is usually the most highly stressed oil in the system. Filtering and testing intervals adequate for that oil will be adequate for all other systems. Usually, this is coincident with minor turbine maintenance. If the governor oil system is unusually dirty, provide an additional in-line filter or parallel loop with an improved filter.

Sewage Treatment Systems: Plants are supplied with a variety of sewage systems. Inspecting the exposed piping, metering, gas injection, or other associated transport systems should be done on scheduled intervals. Maintenance is usually on a demand basis. Usage and plant efficiency determine frequency of sludge removal requirements.

Heating, Ventilating, and Air Conditioning (HVAC): Annual checks of component function are normally sufficient. Thermal deviations from normal can require sporadic component maintenance or replacement.

Cranes, Fire Protection, and Elevators: These must be inspected and maintained according to state, federal, and industry standards.

Maintenance Requirements

Generators

Vertical shaft generators may be suspended or umbrella-type and embody many mechanical elements. As indicated earlier, usually two guide bearings are provided if the generator rotor is suspended from above the rotor and only one guide bearing if the rotor is supported from below. Horizontal units may incorporate a thrust bearing if the turbine has an axial reaction force.

Inspection

1. Prior to shutdown, sample the oil from each individual oil pot. Obtain permanent records of partial discharge, radio frequency, optical gap, vibration, and any other readings the machine may be equipped to provide.
2. After shutdown, make visual exams of the brush and brake areas for excessive dust or dust and oil accumulations. Clean and record if these conditions are present.
3. Visually examine the upper and lower end turns and into the air gap to the extent possible. Look for reddish brown (fretting) and white or black deposits. Do not smell, taste, or disturb any of these potentially toxic deposits.
4. Examine brakes and brake plates for excess wear and any indication of excessive heating, warping, or "chatter" (low-frequency vibration).

5. Examine all oil and water systems for external leaks.
6. Examine spider arms for cracking at the shaft connection. If fasteners are used, check for lost heads or indications of movement.
7. For units with gas-discharge fire suppression systems, check for open pressure relief vents or easy-opening vent doors.

Minor Preventive Maintenance. This consists primarily of responding to any leaks or dust accumulations found during the inspection. Thrust bearing resistance tests should be made. General cleanup is advised.

Major Preventive Maintenance or Overhaul. In addition to inspections and minor preventive maintenance, this may include unit alignment or balance checks and adjustment, guide bearing maintenance and adjustment, and electrical tests of the generator. Dismantling procedures for the generator should follow the manufacturer's instructions.

1. If the rotor (or rotor spider) is to be removed "as found," alignment and plumb readings should be taken prior to dismantling.
2. Move the rotor to the generator erection base and thoroughly examine the pole pieces and stator.
3. Check conventional stator construction for loose coils and wedges.
4. Examine thrust bearing shoes for damage and scrape any minor damage found.
5. Examine the runner plate to shaft support plate connection for any sign of relative movement (fretting).
6. Examine guide shoes for excessive wear at one elevation or direction.
7. Adjust thrust bearing and guide shoes if a tilt is found. The thrust bearing support may require shimming to bring the unit back to plumb. To accommodate the new center, two-piece guide bearings may require a rotation and redowel of the turbine headcover.

The usual reason for a major dismantling of the generator is to replace all or a portion of the stator coils.

Compressors

Most hydroelectric power plants have several compressed air systems. Usually, electric motors power the compressors. The following inspections and maintenance should be performed as a minimum.

Inspection

1. Drain condensate from traps and accumulators.
2. Check operating condition of drive (belts).
3. Note unusual noises or vibrations.
4. Check the unloaders relative to the start-up process.
5. Check the oil level and condition.
6. Clean the strainers or air filters.
7. Record the compressor's running hours.

Minor Preventive Maintenance

1. Note pressure and time compressor from cut-in to cut-out. "Cut-in" is start time. Pressure should be noted at that time. Record run time to "cut-out" (stop). At that time, pressure should be recorded.
2. Check any alarms or previous trouble reports.
3. Clean the condenser or heat exchanger.
4. Check and, if necessary, adjust the pulley system(s).
5. Check the piping for leaks.
6. Check the over-pressure safety valve.

Major Preventive Maintenance

1. Conduct a detailed inspection of the air compressor, lubrication system, and condenser for system leaks.
2. Inspect piston rings, cylinder bore, bearings, valves and springs, rotors, and seals. Replace, if necessary.
3. Test and calibrate gauges, alarms, and relay systems.

Governors

Electronic governors have caused the methods used in governor inspection and maintenance to change. The speed-sensing and signal-processing devices are quite different from the sensitive, high maintenance mechanical components they have replaced. Many older units still use rotating ballheads for speed sensing, and those elements still require frequent maintenance.

Inspection

1. Observe for oil leaks and vibration during normal operation.
2. Clean or replace strainers.
3. Observe for proper functioning during unit synchronizing.
4. Check the oil motor vibrator (dither) for proper function.

Minor Preventive Maintenance

1. Check and calibrate the pressure control and protection devices on the system.
2. Check the condition of drive motors, and repair if required.
3. Sample oil, make a visual and olfactory examination, and have a spectral analysis if any unusual conditions appear.

Major Preventive Maintenance

1. Dismantle valves, linkages, and pumps; inspect for unusual conditions or excessive wear.
2. Replace with new or rebuilt components where needed.
3. Be sure to drop accumulator pressure to ambient prior to or immediately after disabling governor pump.
4. Empty and clean the governor sump and accumulator tank. The accumulator may need state- or ASME-certified reinspection or a repeat of the hydrostatic test.

5. Perform an internal inspection for accumulations in servomotors.
6. Inspect piping and the linkage at servos, restoring linkage and joints, and realign where necessary.
7. Check all components for proper fit and operation.
8. Perform post-overhaul operational tests. Adjust for optimal operation and acceptable pressure rise in the turbine case.
9. Check the piping for positive slope; modify as necessary to eliminate possible gas pockets.
10. Upgrade to an electronic governor if the age or condition justifies it.

Turbine

Inspection. The inspection should determine the current condition and extent of required maintenance. Planning and scheduling for subsequent maintenance will be based upon the information provided from these inspections.

1. Record detailed information for each blade, hub, band location, throat ring, wicket gates, and spiral or water case. This should include wicket gate timing and stroke and correlation with the governor scale. The timing usually covers a span from 25 to 75 percent of servo travel to avoid slamming gates and encountering the cushion. The total time is then considered twice that interval.
2. Place permanent numbers on each blade or bucket outside of a normal cavitation or wear area to identify repetitive problem areas.
3. Take seal or wear ring or blade tip clearances. For those units with access ports only, an attempt to stop the wheel in the same relative orientation each time adds to the value of the readings.

Inspection time requirements usually depend upon the access time required (such as draft tube platform installation, dewatering, and gate or stoplog installations). Smaller units may be accessed in a single day; larger units may take a week or more.

Minor Preventive Maintenance. The runner repair and modification required usually determines the extent of effort required during this outage. The time required to perform minor maintenance could vary from less than one week to more than three weeks.

1. After access is provided, some dry-out time may be required prior to initiating the actual work.
2. Usually, work is limited to repairing the wheel and throat ring (draft tube) and checking the underwater wicket gate bushing lubrication.
3. Excessive welding and thermal stressing of the blades can lead to cracking; therefore, it is better to repair minor damage more often than to allow large areas to become involved in cavitation damage.
4. Crack repair should be performed carefully with a program that will minimize the occurrence of subsequent cracking.

Major Preventive Maintenance or Overhaul. In addition to Inspections and Minor Preventive Maintenance, the shaft packing may be replaced, guide bearing may be scraped or adjusted, the wicket gate lubrication system upgraded, and the servo and gate linkage bearings replaced. This would add one to two weeks to the Minor Maintenance period.

When a major dismantling is scheduled or forced, it presents the opportunity to inspect, repair, and upgrade all components not normally accessible. Five to eight months could be required to overhaul medium to large units. The following activities should be considered during major unit overhaul:

1. Replace those replaceable or renewable elements of seal or wear rings with appropriate materials and machine them to restore minimal clearance where needed.
2. Check couplings, bearing journals, packing sleeves, and servomotors.
3. For upgrades, consider a wicket gate replacement to open up the waterway and a runner replacement to uprate and enhance operational capability.
4. If existing wicket gates are to be reused, consider rehabilitating them to include a layer of cavitation and gall-resistant material on the ends of the vanes; replace the opposing facing plates with material with a low-galling potential; clad the mating surfaces of the vanes with similar low galling material; line bore the wicket gate bushing installation bores; fit greaseless, reduced clearance bushings to all shaft journals and install; clad or install stainless sleeves on the gate journals if they are carbon steel or are worn significantly; and install replaceable seals on the underwater bushings if water conditions permit.
5. Check and repair air admission systems.
6. Repaint the spiral case, gates, and runner (unless it is an all stainless steel runner).

Computer-Based Maintenance Programs

The primary advantage of computer-based maintenance programs is to provide long-term, readily available storage and the retrievability of data and procedures. A primary reason for having such a program is easy access to operational and maintenance history and to procedures for specific treatment by the crews performing the work. The rapid advances in machinery technology applied to the personal computer (PC) systems and their programs make the PC the best choice for local systems. Generic programs may require considerable customizing to provide specific formatting or priorities. Local modification to any great extent tends to make those programs less flexible as new editions develop.

Installing any programs will require a significant period of time to fully input the existing history, procedures, inventory, crew dispatch information, trouble report response, and other data. When fully loaded, pertinent sections of the program should be accessible to all levels of staff from journeyman to superintendents and to financial and personnel managers with suitable security provisions. This access should include data retrieval and input. The data put into the PC must be reviewed at least once by a technically knowledgeable person prior to permanent retention. Often, a second review is required before that input becomes a permanent record.

Operations Considerations

The transmission and distribution system and the other types of plants connected in the system will determine how much the units will be used. Large, flexible systems may require only that plants be base loaded to the extent that water is available. However, small, isolated systems will require operation over the full range of loads to accommodate local power needs. All systems are likely to grow with time and require more energy from the plant. The ultimate limiting factor will be the available water. Operation at less than optimum operating conditions and loading will tend to increase the need for maintenance.

Isolated systems typically require synchronous machines with "black start" capability. Isolated systems often have an increase in power needs that is quite dramatic. Initial power demand could, therefore, be provided with a single unit in a multiple-unit plant. The economic decision for adding a second unit would be governed by the rate of power usage growth. Initial installation of a single unit large enough to provide satisfactory water control and ultimate demand would subject that unit to excessive low power output at off-peak and during the early years of operation. The planner should consider different size units to optimize delivery of low loads as well as peak requirements.

Installations on large existing transmission systems can be designed to supply either base or peaking output. Unless the system is extremely hydro-rich, peaking may be the likely mode. Turbines operating in peaking modes will require more maintenance and should be designed for more extensive useful operating ranges. Draft tube surging should be kept to a narrow range or none at all; efficiency should be high over a broad range; and reliability should be emphasized. Reversible pump-turbine installations will likely require twice the maintenance attention of conventional turbines because of stress reversals and the many stops and starts inherent in the operating mode. Operations staff as well as designers must recognize the interaction of operating modes and maintenance demands. Surging and vibration may require moving through certain ranges rapidly to avoid detrimental operation.

Aids for operation should be proportional to the size and complexity of the installation. Large, diverse installations should be instrumented fully to identify on-line operating conditions and component status. Appropriate alarm and shutdown responses should be installed. Normal plant operation is often from a remote site, usually near the load center rather than the generation site. Usually, on-site staff must be trained both as operators and maintenance staff for those situations where communications or control circuits are disrupted.

Small plants on large systems may be allowed simply to separate from the system, shut down, and wait for attention by a crew from a remote site to restore operations. It is usually critical to restore small plants to operation on isolated systems, and, therefore, they should be staffed to provide the necessary attention as quickly as possible.

When developing operating procedures, focus on the information needed by the plant operators. For each procedure, operators need to know exactly what to do to accomplish a desired result. The procedures, therefore, must provide clear, step-by-step instructions complete with flow diagrams, specific equipment identifiers and locations, and descriptions of other plant or system responses.

Operating procedures will first be used during the start-up process for the various components and systems. During this phase, they serve the dual purpose of explaining how the plant was designed to operate as well as providing a training medium for the operating staff. Later, operating procedures will provide training for new operators and a refresher training for experienced operators.

The procedures should be reviewed whenever an operating problem occurs. This review will evaluate if the procedure is correct and appropriate and if it was followed. For this reason, procedures should be in a format that is easily revised according to the needs of the owner or operator. A typical procedure will have the following components:

1. Title
2. Number or electronic file name
3. Revision date
4. Approval signatures
5. General description
6. Step-by-step procedure
7. Flow diagrams and schematics

The general description should facilitate the operator's understanding of a specific component or system within the larger plant or total system. Flow diagrams or schematics will assist in this understanding. It is helpful to indicate the time required to execute a procedure or major portions of that procedure. The step-by-step procedure should explain the actions required to achieve a given result. Where appropriate, the system response time to a given action should be noted. Components 1 through 4 control the document and assure the correct revision is in use.

Maintenance procedures should be made available to the operators to assist in understanding the effect of some operating modes upon the units.

References

[1]Spicher, Thomas, *Hydro Wheels: A Guide to Maintaining & Improving Hydro Units,* Second Edition, *Hydro Review* Magazine, HCI Publications, Kansas City, Missouri, 1994.

[2]ASME/ANSI B30.2, *Overhead and Gantry Cranes (Top Running Bridge, Single or Multiple Girder, Top Running Trolley Hoist),* American National Standards Institute, 1990.

BIBLIOGRAPHY

Behal and Melilli, editors, *Stainless Steel Castings, American Society for Testing and Materials ASTM STP 756, Metallurgical Characteristics of a Large Hydraulic Runner Casting of Type 13Cr-Ni Stainless Steel; Weld Procedure, Filler Metal, and Post-Weld Heat Treatment: Their Effect on the Hardness and Quality of Welds in CA6NM Alloy,* Compiled from Papers Presented at a American Society for Testing and Materials Conference in Bal Harbour, Florida, November 12-13, 1980.

Cavitation Pitting Mitigation in Hydraulic Turbines, Volumes 1 and 2, EPRI Report AP-4719, Electric Power Research Institute, Palo Alto, California, 1986.

Hydraulic Turbine Cavitation Pitting Damage, CEA Report RP000 G 136, Prepared by Crippen Associates, Canadian Electrical Association, 1981.

Hydropower Reliability Study, EPRI report EM-3435, Electric Power Research Institute, Palo Alto, California, 1984.

Hydroelectric Operations & Maintenance, Proceedings from Workshop, 1982-1983, Electric Power Research Institute, Palo Alto, California.

Hydrogenerator Design Manual, U.S. Bureau of Reclamation, Denver, Colorado, April, 1992.

March, P., and S. Vigander, *Some TVA Experiences with Flow-Induced Vibrations,* Tennessee Valley Authority.

Richman, R.H., and W.P. McNaughton, *Compilation of Cavitation-Erosion Behavior with Mechanical Properties of Metals,* EPRI Report RP 2866-1, Prepared by Daedalus Associates, Electric Power Research Institute, Palo Alto, California, 1988.

Water Operation and Maintenance Bulletin, No. 160, U.S. Bureau of Reclamation, 1992.

Wheeler Units 9-11 Vibration Studies: A Summary of the Vibration Problems, TVA Report No. 3-509, Tennessee Valley Authority.

In addition to the foregoing documents, the following types of references can be useful:

- Plant-specific operational limitations and regulations issued by the Federal Energy Regulatory Commission;
- Federal and state regulatory agency directives; and
- Manufacturers' recommended operations and maintenance manuals.

Appendix

Appendix
Standards Organizations

This appendix lists the addresses and other pertinent information about organizations that publish the technical standards cited in this guide. This information is provided to assist you in acquiring standards that may be of interest. Most of these organizations require prepayment of publication orders by check, money order, or credit card. With payment by credit card, most accept orders by telephone.

Organization	Form of Payment/Notes
ANSI American National Standards Institute 141 West 42nd Street New York, NY 10036 Phone: (212) 642-4900	Check Money Order American Express MasterCard Visa International orders must include Federal Express account number. Members order through existing account.
ASCE American Society of Civil Engineers United Engineering Center 345 East 47th Street New York, NY 10017-2398 Phone: (800) 548-2723 Fax: (212) 705-7300	Check Money Order American Express MasterCard Visa ASCE members include ID number with order.
ASHRAE, Inc. American Society of Heating, Refrigerating and Air-Conditioning Engineers Customer Service Department 1791 Tullie Circle N.E. Atlanta, GA 30329 Phone: (800) 527-4723 (404) 636-8400 Fax: (404) 321-5478	Check Money Order American Express MasterCard Visa Diners Club Discounts available to members

ASME American Society of Mechanical Engineers Codes and Standards Department 345 East 47th Street M/S 7A New York, NY 10017 Phone: (212) 705-8570	Check Money Order American Express MasterCard Visa Diners Club Discover Euro Card
AWWA American Water Works Association Customer Service Department 6666 West Quincy Avenue Denver, CO 80235 Phone: (800) 926-7337 Fax: (303) 347-0804	Check Money Order American Express MasterCard Visa Diners Club Accompany mail orders of less than $35 with check or money order Billing available for larger orders
CMAA Crane Manufacturers Association of America c/o Material Handling Institute 8720 Red Oak Boulevard, Suite 201 Charlotte, NC 28217 Phone: (704) 522-8644	Check Money Order MasterCard Visa Company purchase orders accepted for orders larger than $100
EPRI Electric Power Research Institute Distribution Center EPRI Distribution Center 207 Coggins Drive Pleasant Hill, CA 94523 Phone: (510) 934-4212 Fax: (510) 944-0510	Sale or distribution of materials is restricted; to order, contact EPRI.

Hydraulic Institute

Department 854-0107W
P.O. Box 94020
Paratine, IL 60094-4020

Phone: (708) 364-6206

Check
Money Order
MasterCard
Visa

Purchase orders from members accepted

IEC

International Electrotechnical Commission
IEC Central Office
3, rue de Varembe
P.O. Box 131
1211 Geneva 20, Switzerland

Phone: (41) 22-9190111
Fax: (41) 22-9190300

Checks on Swiss banks
Eurocheques
American Express
MasterCard
Visa

Materials may be ordered from IEC national committee offices in 41 countries. In the U.S. and Canada, use the American National Standards Institute

---

IEEE

Institute of Electrical and Electronics Engineers
Customer Service Department
P.O. Box 1331
Piscataway, NJ 08855

Phone: (800) 678-4333
(212) 705-7900
Fax: (908) 981-0345

Check
Money Order
MasterCard
American Express
Visa

Mail or fax company purchase order (subject to approval)

NFPA

National Fire Protection Association
Customer Service Department
P.O. Box 9146
Quincy, MA 02269

Phone: (800) 344-3555 (U.S.)
(800) 668-2955 (Canada)

Check
Money Order
MasterCard
American Express
Visa

Index

Index